Ford Festiva & Aspire Automotive Repair Manual

by Jeff Kibler and John H Haynes

Member of the Guild of Motoring Writers

Models covered:
All Ford Festiva and Aspire models
1988 through 1997

(9L5 - 36030)

Haynes Publishing Group
Sparkford Nr Yeovil
Somerset BA22 7JJ England

Haynes North America, Inc
861 Lawrence Drive
Newbury Park
California 91320 USA

Acknowledgements
Wiring diagrams provided exclusively for Haynes North America, Inc. by Valley Forge Technical Communications. Technical writers who contributed to this project include Mike Stubblefield, Rob Maddox and Jay Storer.

© **Haynes North America, Inc. 1998**
With permission from J.H. Haynes & Co. Ltd.

A book in the Haynes Automotive Repair Manual Series

Printed in the U.S.A.

All rights reserved. No part of this book may be reproduced or transmitted in any form or by any means, electronic or mechanical, including photocopying, recording or by any information storage or retrieval system, without permission in writing from the copyright holder.

ISBN 1 56392 287 8

Library of Congress Catalog Card Number 98-84255

While every attempt is made to ensure that the information in this manual is correct, no liability can be accepted by the authors or publishers for loss, damage or injury caused by any errors in, or omissions from, the information given.

"Ford" and the Ford logo are registered trademarks of Ford Motor Company. Ford Motor Company is not a sponsor or affiliate of Haynes Publishing Group or Haynes North America, Inc. and is not a contributor to the content of this manual.

Contents

Introductory pages

About this manual	0-5
Introduction	0-5
Vehicle identification numbers	0-6
Buying parts	0-7
Maintenance techniques, tools and working facilities	0-7
Booster battery (jump) starting	0-16
Jacking and towing	0-17
Automotive chemicals and lubricants	0-18
Conversion factors	0-19
Safety first!	0-20
Troubleshooting	0-21

Chapter 1
Tune-up and routine maintenance — 1-1

Chapter 2   Part A
Engine — 2A-1

Chapter 2   Part B
General engine overhaul procedures — 2B-1

Chapter 3
Cooling, heating and air conditioning systems — 3-1

Chapter 4
Fuel and exhaust systems — 4-1

Chapter 5
Engine electrical systems — 5-1

Chapter 6
Emissions and engine control systems — 6-1

Chapter 7   Part A
Manual transaxle — 7A-1

Chapter 7   Part B
Automatic transaxle — 7B-1

Chapter 8
Clutch and driveaxles — 8-1

Chapter 9
Brakes — 9-1

Chapter 10
Suspension and steering systems — 10-1

Chapter 11
Body — 11-1

Chapter 12
Chassis electrical system — 12-1

Wiring diagrams — 12-19

Index — IND-1

Haynes mechanic, author and photographer with 1997 Aspire

About this manual

Its purpose

The purpose of this manual is to help you get the best value from your vehicle. It can do so in several ways. It can help you decide what work must be done, even if you choose to have it done by a dealer service department or a repair shop; it provides information and procedures for routine maintenance and servicing; and it offers diagnostic and repair procedures to follow when trouble occurs.

We hope you use the manual to tackle the work yourself. For many simpler jobs, doing it yourself may be quicker than arranging an appointment to get the vehicle into a shop and making the trips to leave it and pick it up. More importantly, a lot of money can be saved by avoiding the expense the shop must pass on to you to cover its labor and overhead costs. An added benefit is the sense of satisfaction and accomplishment that you feel after doing the job yourself.

Using the manual

The manual is divided into Chapters. Each Chapter is divided into numbered Sections, which are headed in bold type between horizontal lines. Each Section consists of consecutively numbered paragraphs.

At the beginning of each numbered Section you will be referred to any illustrations which apply to the procedures in that Section. The reference numbers used in illustration captions pinpoint the pertinent Section and the Step within that Section. That is, illustration 3.2 means the illustration refers to Section 3 and Step (or paragraph) 2 within that Section.

Procedures, once described in the text, are not normally repeated. When it's necessary to refer to another Chapter, the reference will be given as Chapter and Section number. Cross references given without use of the word "Chapter" apply to Sections and/or paragraphs in the same Chapter. For example, "see Section 8" means in the same Chapter.

References to the left or right side of the vehicle assume you are sitting in the driver's seat, facing forward.

Even though we have prepared this manual with extreme care, neither the publisher nor the author can accept responsibility for any errors in, or omissions from, the information given.

NOTE

A **Note** provides information necessary to properly complete a procedure or information which will make the procedure easier to understand.

CAUTION

A **Caution** provides a special procedure or special steps which must be taken while completing the procedure where the Caution is found. Not heeding a Caution can result in damage to the assembly being worked on.

WARNING

A **Warning** provides a special procedure or special steps which must be taken while completing the procedure where the Warning is found. Not heeding a Warning can result in personal injury.

Introduction

Festiva and Aspire models are available in two and four-door hatchback body styles.

Early four-cylinder engines are equipped with a carburetor, while later four-cylinder engines are equipped with port fuel injection.

The transversely mounted engine drives the front wheels through either a five-speed manual or a four-speed automatic transaxle via independent driveaxles.

Independent suspension, featuring coil spring/strut damper units, is used on the front wheels, while a semi independent suspension using coil spring/strut dampers and a trailing arm is used at the rear. The rack and pinion steering unit is mounted behind the engine with power-assist available as an option.

The brakes are disc at the front and drums at the rear, with power assist standard. An Anti-lock Brake System (ABS) became available on later models.

Vehicle identification numbers

Modifications are a continuing and unpublicized process in vehicle manufacturing. Since spare parts manuals and lists are compiled on a numerical basis, the individual vehicle numbers are essential to correctly identify the component required.

Vehicle Identification Number (VIN)

This very important identification number is located on a plate attached to the dashboard inside the windshield on the driver's side of the vehicle (see illustration). The VIN also appears on the Vehicle Certificate of Title and Registration. It contains information such as where and when the vehicle was manufactured, the model year and the body style.

Chassis number

The chassis number is a repetition of the VIN number stamped on the firewall in the engine compartment (see illustration).

Vehicle Safety Certification label

The Vehicle Safety Certification label is attached to the driver's side door end or post (see illustration). The label contains the name of the manufacturer, the month and year of production, the Gross Vehicle Weight Rating (GVWR), the Gross Axle Weight Rating (GAWR) and the certification statement.

Engine identification number

The engine identification number can be found stamped on a pad to the right of the distributor on the cylinder block (see illustration).

Transaxle identification number

The transaxle identification information can be found stamped on a pad located on the transaxle bellhousing (see illustration).

Vehicle Emissions Control Information (VECI) label

The emissions control information label is found under the hood, normally on the bottom side of the hood. This label contains information on the emissions control equipment installed on the vehicle, as well as tune-up specifications (see Chapter 6).

The Vehicle Identification Number (VIN) is stamped into a metal plate fastened to the dashboard on the driver's side - it is visible through the windshield

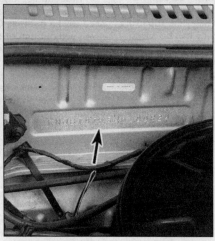

The chassis number is stamped into the firewall

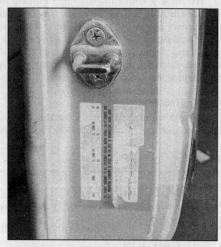

The Vehicle Safety Certification label is affixed to the drivers door pillar

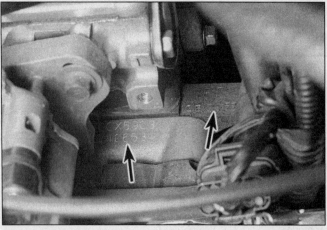

The engine identification number is stamped on a pad at the rear of the engine block

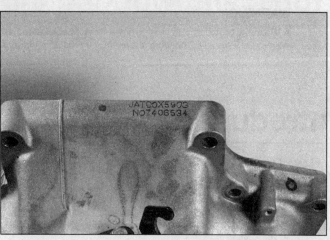

The transaxle identification number is stamped on the top of the bellhousing

Buying parts

Replacement parts are available from many sources, which generally fall into one of two categories - authorized dealer parts departments and independent retail auto parts stores. Our advice concerning these parts is as follows:

Retail auto parts stores: Good auto parts stores will stock frequently needed components which wear out relatively fast, such as clutch components, exhaust systems, brake parts, tune-up parts, etc. These stores often supply new or reconditioned parts on an exchange basis, which can save a considerable amount of money. Discount auto parts stores are often very good places to buy materials and parts needed for general vehicle maintenance such as oil, grease, filters, spark plugs, belts, touch-up paint, bulbs, etc. They also usually sell tools and general accessories, have convenient hours, charge lower prices and can often be found not far from home.

Authorized dealer parts department: This is the best source for parts which are unique to the vehicle and not generally available elsewhere (such as major engine parts, transmission parts, trim pieces, etc.).

Warranty information: If the vehicle is still covered under warranty, be sure that any replacement parts purchased - regardless of the source - do not invalidate the warranty!

To be sure of obtaining the correct parts, have engine and chassis numbers available and, if possible, take the old parts along for positive identification.

Maintenance techniques, tools and working facilities

Maintenance techniques

There are a number of techniques involved in maintenance and repair that will be referred to throughout this manual. Application of these techniques will enable the home mechanic to be more efficient, better organized and capable of performing the various tasks properly, which will ensure that the repair job is thorough and complete.

Fasteners

Fasteners are nuts, bolts, studs and screws used to hold two or more parts together. There are a few things to keep in mind when working with fasteners. Almost all of them use a locking device of some type, either a lockwasher, locknut, locking tab or thread adhesive. All threaded fasteners should be clean and straight, with undamaged threads and undamaged corners on the hex head where the wrench fits. Develop the habit of replacing all damaged nuts and bolts with new ones. Special locknuts with nylon or fiber inserts can only be used once. If they are removed, they lose their locking ability and must be replaced with new ones.

Rusted nuts and bolts should be treated with a penetrating fluid to ease removal and prevent breakage. Some mechanics use turpentine in a spout-type oil can, which works quite well. After applying the rust penetrant, let it work for a few minutes before trying to loosen the nut or bolt. Badly rusted fasteners may have to be chiseled or sawed off or removed with a special nut breaker, available at tool stores.

If a bolt or stud breaks off in an assembly, it can be drilled and removed with a special tool commonly available for this purpose. Most automotive machine shops can perform this task, as well as other repair procedures, such as the repair of threaded holes that have been stripped out.

Flat washers and lockwashers, when removed from an assembly, should always be replaced exactly as removed. Replace any damaged washers with new ones. Never use a lockwasher on any soft metal surface (such as aluminum), thin sheet metal or plastic.

Fastener sizes

For a number of reasons, automobile manufacturers are making wider and wider use of metric fasteners. Therefore, it is important to be able to tell the difference between standard (sometimes called U.S. or SAE) and metric hardware, since they cannot be interchanged.

All bolts, whether standard or metric, are sized according to diameter, thread pitch and length. For example, a standard 1/2 - 13 x 1 bolt is 1/2 inch in diameter, has 13 threads per inch and is 1 inch long. An M12 - 1.75 x 25 metric bolt is 12 mm in diameter, has a thread pitch of 1.75 mm (the distance between threads) and is 25 mm long. The two bolts are nearly identical, and easily confused, but they are not interchangeable.

In addition to the differences in diameter, thread pitch and length, metric and standard bolts can also be distinguished by examining the bolt heads. To begin with, the distance across the flats on a standard bolt head is measured in inches, while the same dimension on a metric bolt is sized in millimeters (the same is true for nuts). As a result, a standard wrench should not be used on a metric bolt and a metric wrench should not be used on a standard bolt. Also, most standard bolts have slashes radiating out from the center of the head to denote the grade or strength of the bolt, which is an indication of the amount of torque that can be applied to it. The greater the number of slashes, the greater the strength of the bolt. Grades 0 through 5 are commonly used on automobiles. Metric bolts have a property class (grade) number, rather than a slash, molded into their heads to indicate bolt strength. In this case, the higher the number, the stronger the bolt. Property class numbers 8.8, 9.8 and 10.9 are commonly used on automobiles.

Strength markings can also be used to distinguish standard hex nuts from metric hex nuts. Many standard nuts have dots stamped into one side, while metric nuts are marked with a number. The greater the number of dots, or the higher the number, the greater the strength of the nut.

Metric studs are also marked on their ends according to property class (grade). Larger studs are numbered (the same as metric bolts), while smaller studs carry a geometric code to denote grade.

Bolt strength marking (standard/SAE/USS; bottom - metric)

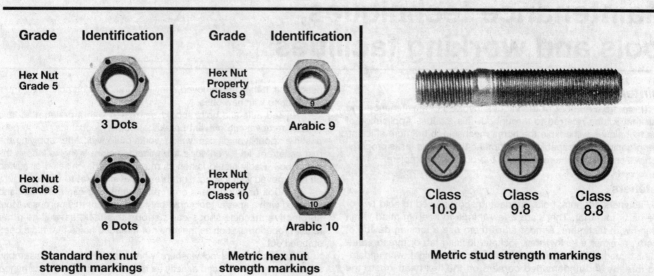

Standard hex nut strength markings

Metric hex nut strength markings

Metric stud strength markings

Maintenance techniques, tools and working facilities

It should be noted that many fasteners, especially Grades 0 through 2, have no distinguishing marks on them. When such is the case, the only way to determine whether it is standard or metric is to measure the thread pitch or compare it to a known fastener of the same size.

Standard fasteners are often referred to as SAE, as opposed to metric. However, it should be noted that SAE technically refers to a non-metric fine thread fastener only. Coarse thread non-metric fasteners are referred to as USS sizes.

Since fasteners of the same size (both standard and metric) may have different strength ratings, be sure to reinstall any bolts, studs or nuts removed from your vehicle in their original locations. Also, when replacing a fastener with a new one, make sure that the new one has a strength rating equal to or greater than the original.

Tightening sequences and procedures

Most threaded fasteners should be tightened to a specific torque value (torque is the twisting force applied to a threaded component such as a nut or bolt). Overtightening the fastener can weaken it and cause it to break, while undertightening can cause it to eventually come loose. Bolts, screws and studs, depending on the material they are made of and their thread diameters, have specific torque values, many of which are noted in the Specifications at the beginning of each Chapter. Be sure to follow the torque recommendations closely. For fasteners not assigned a specific torque, a general torque value chart is presented here as a guide. These torque values are for dry (unlubricated) fasteners threaded into steel or cast iron (not aluminum). As was previously mentioned, the size and grade of a fastener determine the amount of torque that can safely be applied to it. The figures listed

Metric thread sizes	Ft-lbs	Nm
M-6	6 to 9	9 to 12
M-8	14 to 21	19 to 28
M-10	28 to 40	38 to 54
M-12	50 to 71	68 to 96
M-14	80 to 140	109 to 154
Pipe thread sizes		
1/8	5 to 8	7 to 10
1/4	12 to 18	17 to 24
3/8	22 to 33	30 to 44
1/2	25 to 35	34 to 47
U.S. thread sizes		
1/4 - 20	6 to 9	9 to 12
5/16 - 18	12 to 18	17 to 24
5/16 - 24	14 to 20	19 to 27
3/8 - 16	22 to 32	30 to 43
3/8 - 24	27 to 38	37 to 51
7/16 - 14	40 to 55	55 to 74
7/16 - 20	40 to 60	55 to 81
1/2 - 13	55 to 80	75 to 108

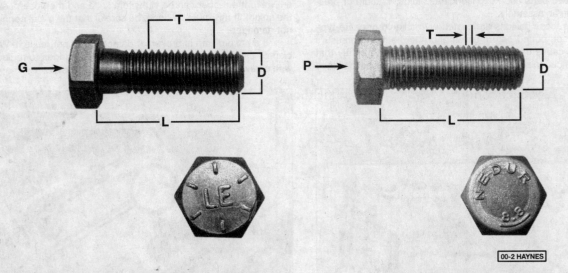

Standard (SAE and USS) bolt dimensions/grade marks
- G   Grade marks (bolt strength)
- L   Length (in inches)
- T   Thread pitch (number of threads per inch)
- D   Nominal diameter (in inches)

Metric bolt dimensions/grade marks
- P   Property class (bolt strength)
- L   Length (in millimeters)
- T   Thread pitch (distance between threads in millimeters)
- D   Diameter

here are approximate for Grade 2 and Grade 3 fasteners. Higher grades can tolerate higher torque values.

Fasteners laid out in a pattern, such as cylinder head bolts, oil pan bolts, differential cover bolts, etc., must be loosened or tightened in sequence to avoid warping the component. This sequence will normally be shown in the appropriate Chapter. If a specific pattern is not given, the following procedures can be used to prevent warping.

Initially, the bolts or nuts should be assembled finger-tight only. Next, they should be tightened one full turn each, in a criss-cross or diagonal pattern. After each one has been tightened one full turn, return to the first one and tighten them all one-half turn, following the same pattern. Finally, tighten each of them one-quarter turn at a time until each fastener has been tightened to the proper torque. To loosen and remove the fasteners, the procedure would be reversed.

Component disassembly

Component disassembly should be done with care and purpose to help ensure that the parts go back together properly. Always keep track of the sequence in which parts are removed. Make note of special characteristics or marks on parts that can be installed more than one way, such as a grooved thrust washer on a shaft. It is a good idea to lay the disassembled parts out on a clean surface in the order that they were removed. It may also be helpful to make sketches or take instant photos of components before removal.

When removing fasteners from a component, keep track of their locations. Sometimes threading a bolt back in a part, or putting the washers and nut back on a stud, can prevent mix-ups later. If nuts and bolts cannot be returned to their original locations, they should be kept in a compartmented box or a series of small boxes. A cupcake or muffin tin is ideal for this purpose, since each cavity can hold the bolts and nuts from a particular area (i.e. oil pan bolts, valve cover bolts, engine mount bolts, etc.). A pan of this type is especially helpful when working on assemblies with very small parts, such as the carburetor, alternator, valve train or interior dash and trim pieces. The cavities can be marked with paint or tape to identify the contents.

Whenever wiring looms, harnesses or connectors are separated, it is a good idea to identify the two halves with numbered pieces of masking tape so they can be easily reconnected.

Gasket sealing surfaces

Throughout any vehicle, gaskets are used to seal the mating surfaces between two parts and keep lubricants, fluids, vacuum or pressure contained in an assembly.

Many times these gaskets are coated with a liquid or paste-type gasket sealing compound before assembly. Age, heat and pressure can sometimes cause the two parts to stick together so tightly that they are very difficult to separate. Often, the assembly can be loosened by striking it with a soft-face hammer near the mating surfaces. A regular hammer can be used if a block of wood is placed between the hammer and the part. Do not hammer on cast parts or parts that could be easily damaged. With any particularly stubborn part, always recheck to make sure that every fastener has been removed.

Avoid using a screwdriver or bar to pry apart an assembly, as they can easily mar the gasket sealing surfaces of the parts, which must remain smooth. If prying is absolutely necessary, use an old broom handle, but keep in mind that extra clean up will be necessary if the wood splinters.

After the parts are separated, the old gasket must be carefully scraped off and the gasket surfaces cleaned. Stubborn gasket material can be soaked with rust penetrant or treated with a special chemical to soften it so it can be easily scraped off. A scraper can be fashioned from a piece of copper tubing by flattening and sharpening one end. Copper is recommended because it is usually softer than the surfaces to be scraped, which reduces the chance of gouging the part. Some gaskets can be removed with a wire brush, but regardless of the method used, the mating surfaces must be left clean and smooth. If for some reason the gasket surface is gouged, then a gasket sealer thick enough to fill scratches will have to be used during reassembly of the components. For most applications, a non-drying (or semi-drying) gasket sealer should be used.

Hose removal tips

Warning: *If the vehicle is equipped with air conditioning, do not disconnect any of the A/C hoses without first having the system depressurized by a dealer service department or a service station.*

Hose removal precautions closely parallel gasket removal precautions. Avoid scratching or gouging the surface that the hose mates against or the connection may leak. This is especially true for radiator hoses. Because of various chemical reactions, the rubber in hoses can bond itself to the metal spigot that the hose fits over. To remove a hose, first loosen the hose clamps that secure it to the spigot. Then, with slip-joint pliers, grab the hose at the clamp and rotate it around the spigot. Work it back and forth until it is completely free, then pull it off. Silicone or other lubricants will ease removal if they can be applied between the hose and the outside of the spigot. Apply the same lubricant to the inside of the hose and the outside of the spigot to simplify installation.

As a last resort (and if the hose is to be replaced with a new one anyway), the rubber can be slit with a knife and the hose peeled from the spigot. If this must be done, be careful that the metal connection is not damaged.

If a hose clamp is broken or damaged, do not reuse it. Wire-type clamps usually weaken with age, so it is a good idea to replace them with screw-type clamps whenever a hose is removed.

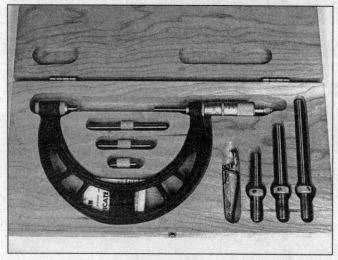

Micrometer set

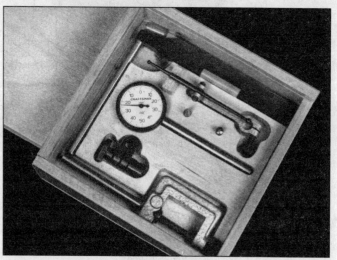

Dial indicator set

Maintenance techniques, tools and working facilities

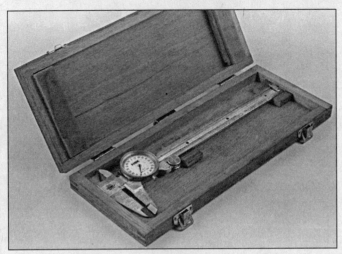

Dial caliper

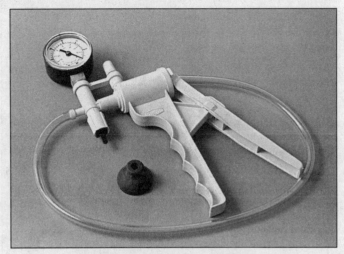

Hand-operated vacuum pump

Tools

A selection of good tools is a basic requirement for anyone who plans to maintain and repair his or her own vehicle. For the owner who has few tools, the initial investment might seem high, but when compared to the spiraling costs of professional auto maintenance and repair, it is a wise one.

To help the owner decide which tools are needed to perform the tasks detailed in this manual, the following tool lists are offered: *Maintenance and minor repair, Repair/overhaul* and *Special*.

The newcomer to practical mechanics should start off with the *maintenance and minor repair* tool kit, which is adequate for the simpler jobs performed on a vehicle. Then, as confidence and experience grow, the owner can tackle more difficult tasks, buying additional tools as they are needed. Eventually the basic kit will be expanded into the *repair and overhaul* tool set. Over a period of time, the experienced do-it-yourselfer will assemble a tool set complete enough for most repair and overhaul procedures and will add tools from the special category when it is felt that the expense is justified by the frequency of use.

Maintenance and minor repair tool kit

The tools in this list should be considered the minimum required for performance of routine maintenance, servicing and minor repair work. We recommend the purchase of combination wrenches (box-end and open-end combined in one wrench). While more expensive than open end wrenches, they offer the advantages of both types of wrench.

Combination wrench set (1/4-inch to 1 inch or 6 mm to 19 mm)
Adjustable wrench, 8 inch
Spark plug wrench with rubber insert
Spark plug gap adjusting tool
Feeler gauge set
Brake bleeder wrench
Standard screwdriver (5/16-inch x 6 inch)
Phillips screwdriver (No. 2 x 6 inch)
Combination pliers - 6 inch
Hacksaw and assortment of blades
Tire pressure gauge
Grease gun
Oil can
Fine emery cloth
Wire brush
Battery post and cable cleaning tool
Oil filter wrench
Funnel (medium size)
Safety goggles
Jackstands (2)
Drain pan

Note: *If basic tune-ups are going to be part of routine maintenance, it will be necessary to purchase a good quality stroboscopic timing light*

Timing light

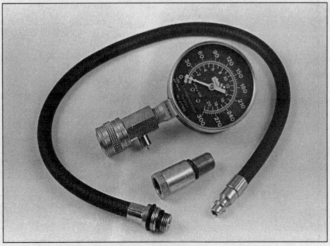

Compression gauge with spark plug hole adapter

0-12 Maintenance techniques, tools and working facilities

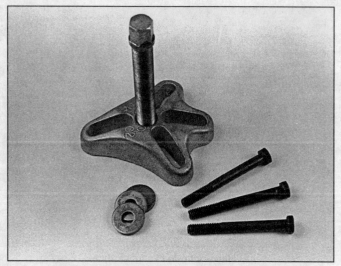
Damper/steering wheel puller

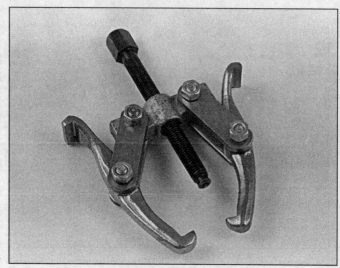
General purpose puller

Hydraulic lifter removal tool

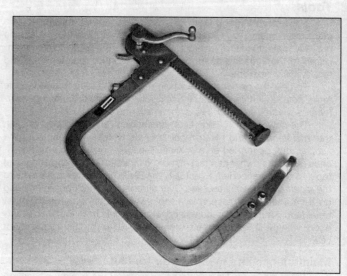
Valve spring compressor

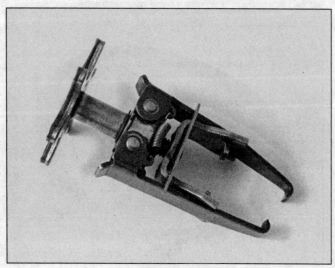
Valve spring compressor

Ridge reamer

Maintenance techniques, tools and working facilities

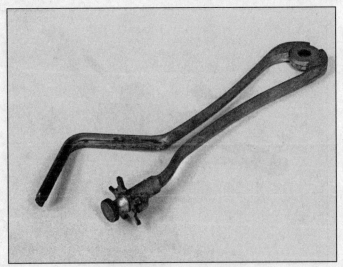
Piston ring groove cleaning tool

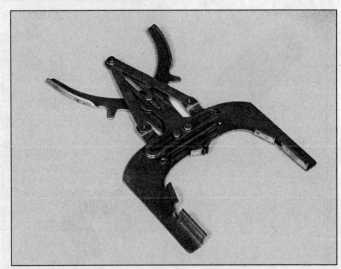
Ring removal/installation tool

Ring compressor

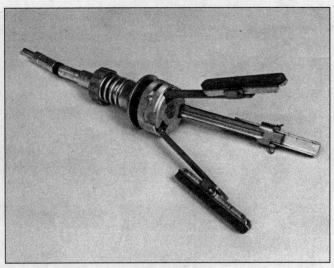
Cylinder hone

and combination tachometer/dwell meter. Although they are included in the list of special tools, it is mentioned here because they are absolutely necessary for tuning most vehicles properly.

Repair and overhaul tool set

These tools are essential for anyone who plans to perform major repairs and are in addition to those in the maintenance and minor repair tool kit. Included is a comprehensive set of sockets which, though expensive, are invaluable because of their versatility, especially when various extensions and drives are available. We recommend the 1/2-inch drive over the 3/8-inch drive. Although the larger drive is bulky and more expensive, it has the capacity of accepting a very wide range of large sockets. Ideally, however, the mechanic should have a 3/8-inch drive set and a 1/2-inch drive set.

Socket set(s)
Reversible ratchet
Extension - 10 inch
Universal joint
Torque wrench (same size drive as sockets)
Ball peen hammer - 8 ounce
Soft-face hammer (plastic/rubber)
Standard screwdriver (1/4-inch x 6 inch)
Standard screwdriver (stubby - 5/16-inch)
Phillips screwdriver (No. 3 x 8 inch)
Phillips screwdriver (stubby - No. 2)
Pliers - vise grip
Pliers - lineman's
Pliers - needle nose
Pliers - snap-ring (internal and external)
Cold chisel - 1/2-inch
Scribe
Scraper (made from flattened copper tubing)
Centerpunch
Pin punches (1/16, 1/8, 3/16-inch)
Steel rule/straightedge - 12 inch
Allen wrench set (1/8 to 3/8-inch or 4 mm to 10 mm)
A selection of files
Wire brush (large)
Jackstands (second set)
Jack (scissor or hydraulic type)

Note: *Another tool which is often useful is an electric drill with a chuck capacity of 3/8-inch and a set of good quality drill bits.*

Special tools

The tools in this list include those which are not used regularly, are expensive to buy, or which need to be used in accordance with

Maintenance techniques, tools and working facilities

Brake hold-down spring tool

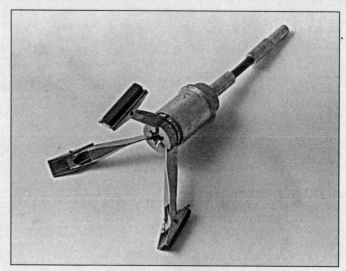

Brake cylinder hone

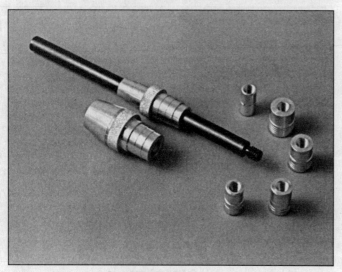

Clutch plate alignment tool

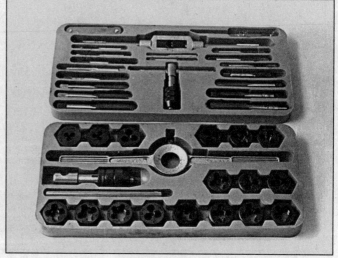

Tap and die set

their manufacturer's instructions. Unless these tools will be used frequently, it is not very economical to purchase many of them. A consideration would be to split the cost and use between yourself and a friend or friends. In addition, most of these tools can be obtained from a tool rental shop on a temporary basis.

This list primarily contains only those tools and instruments widely available to the public, and not those special tools produced by the vehicle manufacturer for distribution to dealer service departments. Occasionally, references to the manufacturer's special tools are included in the text of this manual. Generally, an alternative method of doing the job without the special tool is offered. However, sometimes there is no alternative to their use. Where this is the case, and the tool cannot be purchased or borrowed, the work should be turned over to the dealer service department or an automotive repair shop.

Valve spring compressor
Piston ring groove cleaning tool
Piston ring compressor
Piston ring installation tool
Cylinder compression gauge
Cylinder ridge reamer
Cylinder surfacing hone
Cylinder bore gauge
Micrometers and/or dial calipers
Hydraulic lifter removal tool
Balljoint separator
Universal-type puller
Impact screwdriver
Dial indicator set
Stroboscopic timing light (inductive pick-up)
Hand operated vacuum/pressure pump
Tachometer/dwell meter
Universal electrical multimeter
Cable hoist
Brake spring removal and installation tools
Floor jack

Buying tools

For the do-it-yourselfer who is just starting to get involved in vehicle maintenance and repair, there are a number of options available when purchasing tools. If maintenance and minor repair is the extent of the work to be done, the purchase of individual tools is satisfactory. If, on the other hand, extensive work is planned, it would be a good idea to purchase a modest tool set from one of the large retail chain stores. A set can usually be bought at a substantial savings over the individual tool prices, and they often come with a tool box. As additional tools are needed, add-on sets, individual tools and a larger tool box can be pur-

Maintenance techniques, tools and working facilities

chased to expand the tool selection. Building a tool set gradually allows the cost of the tools to be spread over a longer period of time and gives the mechanic the freedom to choose only those tools that will actually be used.

Tool stores will often be the only source of some of the special tools that are needed, but regardless of where tools are bought, try to avoid cheap ones, especially when buying screwdrivers and sockets, because they won't last very long. The expense involved in replacing cheap tools will eventually be greater than the initial cost of quality tools.

Care and maintenance of tools

Good tools are expensive, so it makes sense to treat them with respect. Keep them clean and in usable condition and store them properly when not in use. Always wipe off any dirt, grease or metal chips before putting them away. Never leave tools lying around in the work area. Upon completion of a job, always check closely under the hood for tools that may have been left there so they won't get lost during a test drive.

Some tools, such as screwdrivers, pliers, wrenches and sockets, can be hung on a panel mounted on the garage or workshop wall, while others should be kept in a tool box or tray. Measuring instruments, gauges, meters, etc. must be carefully stored where they cannot be damaged by weather or impact from other tools.

When tools are used with care and stored properly, they will last a very long time. Even with the best of care, though, tools will wear out if used frequently. When a tool is damaged or worn out, replace it. Subsequent jobs will be safer and more enjoyable if you do.

How to repair damaged threads

Sometimes, the internal threads of a nut or bolt hole can become stripped, usually from overtightening. Stripping threads is an all-too-common occurrence, especially when working with aluminum parts, because aluminum is so soft that it easily strips out.

Usually, external or internal threads are only partially stripped. After they've been cleaned up with a tap or die, they'll still work. Sometimes, however, threads are badly damaged. When this happens, you've got three choices:

1) Drill and tap the hole to the next suitable oversize and install a larger diameter bolt, screw or stud.
2) Drill and tap the hole to accept a threaded plug, then drill and tap the plug to the original screw size. You can also buy a plug already threaded to the original size. Then you simply drill a hole to the specified size, then run the threaded plug into the hole with a bolt and jam nut. Once the plug is fully seated, remove the jam nut and bolt.
3) The third method uses a patented thread repair kit like Heli-Coil or Slimsert. These easy-to-use kits are designed to repair damaged threads in straight-through holes and blind holes. Both are available as kits which can handle a variety of sizes and thread patterns. Drill the hole, then tap it with the special included tap. Install the Heli-Coil and the hole is back to its original diameter and thread pitch.

Regardless of which method you use, be sure to proceed calmly and carefully. A little impatience or carelessness during one of these relatively simple procedures can ruin your whole day's work and cost you a bundle if you wreck an expensive part.

Working facilities

Not to be overlooked when discussing tools is the workshop. If anything more than routine maintenance is to be carried out, some sort of suitable work area is essential.

It is understood, and appreciated, that many home mechanics do not have a good workshop or garage available, and end up removing an engine or doing major repairs outside. It is recommended, however, that the overhaul or repair be completed under the cover of a roof.

A clean, flat workbench or table of comfortable working height is an absolute necessity. The workbench should be equipped with a vise that has a jaw opening of at least four inches.

As mentioned previously, some clean, dry storage space is also required for tools, as well as the lubricants, fluids, cleaning solvents, etc. which soon become necessary.

Sometimes waste oil and fluids, drained from the engine or cooling system during normal maintenance or repairs, present a disposal problem. To avoid pouring them on the ground or into a sewage system, pour the used fluids into large containers, seal them with caps and take them to an authorized disposal site or recycling center. Plastic jugs, such as old antifreeze containers, are ideal for this purpose.

Always keep a supply of old newspapers and clean rags available. Old towels are excellent for mopping up spills. Many mechanics use rolls of paper towels for most work because they are readily available and disposable. To help keep the area under the vehicle clean, a large cardboard box can be cut open and flattened to protect the garage or shop floor.

Whenever working over a painted surface, such as when leaning over a fender to service something under the hood, always cover it with an old blanket or bedspread to protect the finish. Vinyl covered pads, made especially for this purpose, are available at auto parts stores.

Booster battery (jump) starting

Observe these precautions when using a booster battery to start a vehicle:

a) *Before connecting the booster battery, make sure the ignition switch is in the Off position.*
b) *Turn off the lights, heater and other electrical loads.*
c) *Your eyes should be shielded. Safety goggles are a good idea.*
d) *Make sure the booster battery is the same voltage as the dead one in the vehicle.*
e) *The two vehicles MUST NOT TOUCH each other!*
f) *Make sure the transaxle is in Neutral (manual) or Park (automatic).*
g) *If the booster battery is not a maintenance-free type, remove the vent caps and lay a cloth over the vent holes.*

Connect the red jumper cable to the positive (+) terminals of each battery **(see illustration)**.

Connect one end of the black jumper cable to the negative (-) terminal of the booster battery. The other end of this cable should be connected to a good ground on the vehicle to be started, such as a bolt or bracket on the body.

Start the engine using the booster battery, then, with the engine running at idle speed, disconnect the jumper cables in the reverse order of connection.

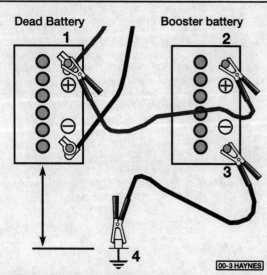

Make the booster battery cable connections in the numerical order shown (note that the negative cable of the booster battery is NOT attached to the negative terminal of the dead battery)

Jacking and towing

Warning: *The jack supplied with the vehicle should only be used for changing a tire or placing jackstands under the frame. Never work under the vehicle or start the engine while this jack is being used as the only means of support.*

The vehicle should be on level ground. Place the shift lever in Park, if you have an automatic, or Reverse if you have a manual transaxle. Block the wheel diagonally opposite the wheel being changed. Set the parking brake.

Remove the spare tire and jack from stowage. Remove the wheel cover and trim ring (if so equipped) with the tapered end of the lug nut wrench by inserting and twisting the handle and then prying against the back of the wheel cover. Loosen the wheel lug nuts about 1/4-to-1/2 turn each.

Place the scissors-type jack under the side of the vehicle and adjust the jack height until it fits in the notch in the vertical rocker panel flange nearest the wheel to be changed. There is a front and rear jacking point on each side of the vehicle **(see illustration)**.

Turn the jack handle clockwise until the tire clears the ground. Remove the lug nuts and pull the wheel off. Replace it with the spare.

Install the lug nuts with the beveled edges facing in. Tighten them snugly. Don't attempt to tighten them completely until the vehicle is lowered or it could slip off the jack. Turn the jack handle counterclockwise to lower the vehicle. Remove the jack and tighten the lug nuts in a diagonal pattern.

Install the cover (and trim ring, if used) and be sure it's snapped into place all the way around.

Stow the tire, jack and wrench. Unblock the wheels.

Towing

As a general rule, the vehicle should be towed from the front with the front (drive) wheels off the ground. If the vehicle must be towed from the rear, place the front wheels on a towing dolly. **Caution:** *Never tow a front wheel drive vehicle from the rear with the front wheels on the ground.*

Equipment specifically designed for towing should be used. It should be attached to the main structural members of the vehicle, not the bumpers or brackets. Do not use the tie-down hook loops at the front or the rear of the vehicle for towing. These hooks loops are designed for securing the vehicle during transport, if used for towing, damage to the front or rear bumper may occur.

The ignition key must be in the ACC position, since the steering lock mechanism isn't strong enough to hold the front wheels straight while towing. Pace the shift lever in neutral and release the parking brake.

Safety is a major consideration when towing and all applicable state and local laws must be obeyed. A safety chain system must be used at all times.

The jack fits under the rocker panel (there are two jacking points on each side of the vehicle, indicated by a notch in the rocker panel flange)

Automotive chemicals and lubricants

A number of automotive chemicals and lubricants are available for use during vehicle maintenance and repair. They include a wide variety of products ranging from cleaning solvents and degreasers to lubricants and protective sprays for rubber, plastic and vinyl.

Cleaners

Carburetor cleaner and choke cleaner is a strong solvent for gum, varnish and carbon. Most carburetor cleaners leave a dry-type lubricant film which will not harden or gum up. Because of this film it is not recommended for use on electrical components.

Brake system cleaner is used to remove brake dust, grease and brake fluid from the brake system, where clean surfaces are absolutely necessary. It leaves no residue and often eliminates brake squeal caused by contaminants.

Electrical cleaner removes oxidation, corrosion and carbon deposits from electrical contacts, restoring full current flow. It can also be used to clean spark plugs, carburetor jets, voltage regulators and other parts where an oil-free surface is desired.

Demoisturants remove water and moisture from electrical components such as alternators, voltage regulators, electrical connectors and fuse blocks. They are non-conductive and non-corrosive.

Degreasers are heavy-duty solvents used to remove grease from the outside of the engine and from chassis components. They can be sprayed or brushed on and, depending on the type, are rinsed off either with water or solvent.

Lubricants

Motor oil is the lubricant formulated for use in engines. It normally contains a wide variety of additives to prevent corrosion and reduce foaming and wear. Motor oil comes in various weights (viscosity ratings) from 0 to 50. The recommended weight of the oil depends on the season, temperature and the demands on the engine. Light oil is used in cold climates and under light load conditions. Heavy oil is used in hot climates and where high loads are encountered. Multi-viscosity oils are designed to have characteristics of both light and heavy oils and are available in a number of weights from 5W-20 to 20W-50.

Gear oil is designed to be used in differentials, manual transmissions and other areas where high-temperature lubrication is required.

Chassis and wheel bearing grease is a heavy grease used where increased loads and friction are encountered, such as for wheel bearings, balljoints, tie-rod ends and universal joints.

High-temperature wheel bearing grease is designed to withstand the extreme temperatures encountered by wheel bearings in disc brake equipped vehicles. It usually contains molybdenum disulfide (moly), which is a dry-type lubricant.

White grease is a heavy grease for metal-to-metal applications where water is a problem. White grease stays soft under both low and high temperatures (usually from -100 to +190-degrees F), and will not wash off or dilute in the presence of water.

Assembly lube is a special extreme pressure lubricant, usually containing moly, used to lubricate high-load parts (such as main and rod bearings and cam lobes) for initial start-up of a new engine. The assembly lube lubricates the parts without being squeezed out or washed away until the engine oiling system begins to function.

Silicone lubricants are used to protect rubber, plastic, vinyl and nylon parts.

Graphite lubricants are used where oils cannot be used due to contamination problems, such as in locks. The dry graphite will lubricate metal parts while remaining uncontaminated by dirt, water, oil or acids. It is electrically conductive and will not foul electrical contacts in locks such as the ignition switch.

Moly penetrants loosen and lubricate frozen, rusted and corroded fasteners and prevent future rusting or freezing.

Heat-sink grease is a special electrically non-conductive grease that is used for mounting electronic ignition modules where it is essential that heat is transferred away from the module.

Sealants

RTV sealant is one of the most widely used gasket compounds. Made from silicone, RTV is air curing, it seals, bonds, waterproofs, fills surface irregularities, remains flexible, doesn't shrink, is relatively easy to remove, and is used as a supplementary sealer with almost all low and medium temperature gaskets.

Anaerobic sealant is much like RTV in that it can be used either to seal gaskets or to form gaskets by itself. It remains flexible, is solvent resistant and fills surface imperfections. The difference between an anaerobic sealant and an RTV-type sealant is in the curing. RTV cures when exposed to air, while an anaerobic sealant cures only in the absence of air. This means that an anaerobic sealant cures only after the assembly of parts, sealing them together.

Thread and pipe sealant is used for sealing hydraulic and pneumatic fittings and vacuum lines. It is usually made from a Teflon compound, and comes in a spray, a paint-on liquid and as a wrap-around tape.

Chemicals

Anti-seize compound prevents seizing, galling, cold welding, rust and corrosion in fasteners. High-temperature ant-seize, usually made with copper and graphite lubricants, is used for exhaust system and exhaust manifold bolts.

Anaerobic locking compounds are used to keep fasteners from vibrating or working loose and cure only after installation, in the absence of air. Medium strength locking compound is used for small nuts, bolts and screws that may be removed later. High-strength locking compound is for large nuts, bolts and studs which aren't removed on a regular basis.

Oil additives range from viscosity index improvers to chemical treatments that claim to reduce internal engine friction. It should be noted that most oil manufacturers caution against using additives with their oils.

Gas additives perform several functions, depending on their chemical makeup. They usually contain solvents that help dissolve gum and varnish that build up on carburetor, fuel injection and intake parts. They also serve to break down carbon deposits that form on the inside surfaces of the combustion chambers. Some additives contain upper cylinder lubricants for valves and piston rings, and others contain chemicals to remove condensation from the gas tank.

Miscellaneous

Brake fluid is specially formulated hydraulic fluid that can withstand the heat and pressure encountered in brake systems. Care must be taken so this fluid does not come in contact with painted surfaces or plastics. An opened container should always be resealed to prevent contamination by water or dirt.

Weatherstrip adhesive is used to bond weatherstripping around doors, windows and trunk lids. It is sometimes used to attach trim pieces.

Undercoating is a petroleum-based, tar-like substance that is designed to protect metal surfaces on the underside of the vehicle from corrosion. It also acts as a sound-deadening agent by insulating the bottom of the vehicle.

Waxes and polishes are used to help protect painted and plated surfaces from the weather. Different types of paint may require the use of different types of wax and polish. Some polishes utilize a chemical or abrasive cleaner to help remove the top layer of oxidized (dull) paint on older vehicles. In recent years many non-wax polishes that contain a wide variety of chemicals such as polymers and silicones have been introduced. These non-wax polishes are usually easier to apply and last longer than conventional waxes and polishes.

Conversion factors

Length (distance)
Inches (in)	X 25.4	= Millimeters (mm)	X	0.0394	= Inches (in)
Feet (ft)	X 0.305	= Meters (m)	X	3.281	= Feet (ft)
Miles	X 1.609	= Kilometers (km)	X	0.621	= Miles

Volume (capacity)
Cubic inches (cu in; in^3)	X 16.387	= Cubic centimeters (cc; cm^3)	X	0.061	= Cubic inches (cu in; in^3)
Imperial pints (Imp pt)	X 0.568	= Liters (l)	X	1.76	= Imperial pints (Imp pt)
Imperial quarts (Imp qt)	X 1.137	= Liters (l)	X	0.88	= Imperial quarts (Imp qt)
Imperial quarts (Imp qt)	X 1.201	= US quarts (US qt)	X	0.833	= Imperial quarts (Imp qt)
US quarts (US qt)	X 0.946	= Liters (l)	X	1.057	= US quarts (US qt)
Imperial gallons (Imp gal)	X 4.546	= Liters (l)	X	0.22	= Imperial gallons (Imp gal)
Imperial gallons (Imp gal)	X 1.201	= US gallons (US gal)	X	0.833	= Imperial gallons (Imp gal)
US gallons (US gal)	X 3.785	= Liters (l)	X	0.264	= US gallons (US gal)

Mass (weight)
Ounces (oz)	X 28.35	= Grams (g)	X	0.035	= Ounces (oz)
Pounds (lb)	X 0.454	= Kilograms (kg)	X	2.205	= Pounds (lb)

Force
Ounces-force (ozf; oz)	X 0.278	= Newtons (N)	X	3.6	= Ounces-force (ozf; oz)
Pounds-force (lbf; lb)	X 4.448	= Newtons (N)	X	0.225	= Pounds-force (lbf; lb)
Newtons (N)	X 0.1	= Kilograms-force (kgf; kg)	X	9.81	= Newtons (N)

Pressure
Pounds-force per square inch (psi; lbf/in^2; lb/in^2)	X 0.070	= Kilograms-force per square centimeter (kgf/cm^2; kg/cm^2)	X	14.223	= Pounds-force per square inch (psi; lbf/in^2; lb/in^2)
Pounds-force per square inch (psi; lbf/in^2; lb/in^2)	X 0.068	= Atmospheres (atm)	X	14.696	= Pounds-force per square inch (psi; lbf/in^2; lb/in^2)
Pounds-force per square inch (psi; lbf/in^2; lb/in^2)	X 0.069	= Bars	X	14.5	= Pounds-force per square inch (psi; lbf/in^2; lb/in^2)
Pounds-force per square inch (psi; lbf/in^2; lb/in^2)	X 6.895	= Kilopascals (kPa)	X	0.145	= Pounds-force per square inch (psi; lbf/in^2; lb/in^2)
Kilopascals (kPa)	X 0.01	= Kilograms-force per square centimeter (kgf/cm^2; kg/cm^2)	X	98.1	= Kilopascals (kPa)

Torque (moment of force)
Pounds-force inches (lbf in; lb in)	X 1.152	= Kilograms-force centimeter (kgf cm; kg cm)	X	0.868	= Pounds-force inches (lbf in; lb in)
Pounds-force inches (lbf in; lb in)	X 0.113	= Newton meters (Nm)	X	8.85	= Pounds-force inches (lbf in; lb in)
Pounds-force inches (lbf in; lb in)	X 0.083	= Pounds-force feet (lbf ft; lb ft)	X	12	= Pounds-force inches (lbf in; lb in)
Pounds-force feet (lbf ft; lb ft)	X 0.138	= Kilograms-force meters (kgf m; kg m)	X	7.233	= Pounds-force feet (lbf ft; lb ft)
Pounds-force feet (lbf ft; lb ft)	X 1.356	= Newton meters (Nm)	X	0.738	= Pounds-force feet (lbf ft; lb ft)
Newton meters (Nm)	X 0.102	= Kilograms-force meters (kgf m; kg m)	X	9.804	= Newton meters (Nm)

Vacuum
Inches mercury (in. Hg)	X 3.377	= Kilopascals (kPa)	X	0.2961	= Inches mercury
Inches mercury (in. Hg)	X 25.4	= Millimeters mercury (mm Hg)	X	0.0394	= Inches mercury

Power
Horsepower (hp)	X 745.7	= Watts (W)	X	0.0013	= Horsepower (hp)

Velocity (speed)
Miles per hour (miles/hr; mph)	X 1.609	= Kilometers per hour (km/hr; kph)	X	0.621	= Miles per hour (miles/hr; mph)

*Fuel consumption**
Miles per gallon, Imperial (mpg)	X 0.354	= Kilometers per liter (km/l)	X	2.825	= Miles per gallon, Imperial (mpg)
Miles per gallon, US (mpg)	X 0.425	= Kilometers per liter (km/l)	X	2.352	= Miles per gallon, US (mpg)

Temperature
Degrees Fahrenheit = (°C x 1.8) + 32   Degrees Celsius (Degrees Centigrade; °C) = (°F - 32) x 0.56

*It is common practice to convert from miles per gallon (mpg) to liters/100 kilometers (l/100km), where mpg (Imperial) x l/100 km = 282 and mpg (US) x l/100 km = 235

Safety first!

Regardless of how enthusiastic you may be about getting on with the job at hand, take the time to ensure that your safety is not jeopardized. A moment's lack of attention can result in an accident, as can failure to observe certain simple safety precautions. The possibility of an accident will always exist, and the following points should not be considered a comprehensive list of all dangers. Rather, they are intended to make you aware of the risks and to encourage a safety conscious approach to all work you carry out on your vehicle.

Essential DOs and DON'Ts

DON'T rely on a jack when working under the vehicle. Always use approved jackstands to support the weight of the vehicle and place them under the recommended lift or support points.

DON'T attempt to loosen extremely tight fasteners (i.e. wheel lug nuts) while the vehicle is on a jack - it may fall.

DON'T start the engine without first making sure that the transmission is in Neutral (or Park where applicable) and the parking brake is set.

DON'T remove the radiator cap from a hot cooling system - let it cool or cover it with a cloth and release the pressure gradually.

DON'T attempt to drain the engine oil until you are sure it has cooled to the point that it will not burn you.

DON'T touch any part of the engine or exhaust system until it has cooled sufficiently to avoid burns.

DON'T siphon toxic liquids such as gasoline, antifreeze and brake fluid by mouth, or allow them to remain on your skin.

DON'T inhale brake lining dust - it is potentially hazardous (see *Asbestos* below).

DON'T allow spilled oil or grease to remain on the floor - wipe it up before someone slips on it.

DON'T use loose fitting wrenches or other tools which may slip and cause injury.

DON'T push on wrenches when loosening or tightening nuts or bolts. Always try to pull the wrench toward you. If the situation calls for pushing the wrench away, push with an open hand to avoid scraped knuckles if the wrench should slip.

DON'T attempt to lift a heavy component alone - get someone to help you.

DON'T rush or take unsafe shortcuts to finish a job.

DON'T allow children or animals in or around the vehicle while you are working on it.

DO wear eye protection when using power tools such as a drill, sander, bench grinder, etc. and when working under a vehicle.

DO keep loose clothing and long hair well out of the way of moving parts.

DO make sure that any hoist used has a safe working load rating adequate for the job.

DO get someone to check on you periodically when working alone on a vehicle.

DO carry out work in a logical sequence and make sure that everything is correctly assembled and tightened.

DO keep chemicals and fluids tightly capped and out of the reach of children and pets.

DO remember that your vehicle's safety affects that of yourself and others. If in doubt on any point, get professional advice.

Asbestos

Certain friction, insulating, sealing, and other products - such as brake linings, brake bands, clutch linings, torque converters, gaskets, etc. - may contain asbestos. Extreme care must be taken to avoid inhalation of dust from such products, since it is hazardous to health. If in doubt, assume that they do contain asbestos.

Fire

Remember at all times that gasoline is highly flammable. Never smoke or have any kind of open flame around when working on a vehicle. But the risk does not end there. A spark caused by an electrical short circuit, by two metal surfaces contacting each other, or even by static electricity built up in your body under certain conditions, can ignite gasoline vapors, which in a confined space are highly explosive. Do not, under any circumstances, use gasoline for cleaning parts. Use an approved safety solvent.

Always disconnect the battery ground (-) cable at the battery before working on any part of the fuel system or electrical system. Never risk spilling fuel on a hot engine or exhaust component. It is strongly recommended that a fire extinguisher suitable for use on fuel and electrical fires be kept handy in the garage or workshop at all times. Never try to extinguish a fuel or electrical fire with water.

Fumes

Certain fumes are highly toxic and can quickly cause unconsciousness and even death if inhaled to any extent. Gasoline vapor falls into this category, as do the vapors from some cleaning solvents. Any draining or pouring of such volatile fluids should be done in a well ventilated area.

When using cleaning fluids and solvents, read the instructions on the container carefully. Never use materials from unmarked containers.

Never run the engine in an enclosed space, such as a garage. Exhaust fumes contain carbon monoxide, which is extremely poisonous. If you need to run the engine, always do so in the open air, or at least have the rear of the vehicle outside the work area.

If you are fortunate enough to have the use of an inspection pit, never drain or pour gasoline and never run the engine while the vehicle is over the pit. The fumes, being heavier than air, will concentrate in the pit with possibly lethal results.

The battery

Never create a spark or allow a bare light bulb near a battery. They normally give off a certain amount of hydrogen gas, which is highly explosive.

Always disconnect the battery ground (-) cable at the battery before working on the fuel or electrical systems.

If possible, loosen the filler caps or cover when charging the battery from an external source (this does not apply to sealed or maintenance-free batteries). Do not charge at an excessive rate or the battery may burst.

Take care when adding water to a non maintenance-free battery and when carrying a battery. The electrolyte, even when diluted, is very corrosive and should not be allowed to contact clothing or skin.

Always wear eye protection when cleaning the battery to prevent the caustic deposits from entering your eyes.

Household current

When using an electric power tool, inspection light, etc., which operates on household current, always make sure that the tool is correctly connected to its plug and that, where necessary, it is properly grounded. Do not use such items in damp conditions and, again, do not create a spark or apply excessive heat in the vicinity of fuel or fuel vapor.

Secondary ignition system voltage

A severe electric shock can result from touching certain parts of the ignition system (such as the spark plug wires) when the engine is running or being cranked, particularly if components are damp or the insulation is defective. In the case of an electronic ignition system, the secondary system voltage is much higher and could prove fatal.

Troubleshooting

Contents

Symptom	Section
Engine	
Engine backfires	15
Engine diesels (continues to run) after switching off	18
Engine hard to start when cold	3
Engine hard to start when hot	4
Engine lacks power	14
Engine lopes while idling or idles erratically	8
Engine misses at idle speed	9
Engine misses throughout driving speed range	10
Engine rotates but will not start	2
Engine runs with oil pressure light on	17
Engine stalls	13
Engine starts but stops immediately	6
Engine stumbles on acceleration	11
Engine surges while holding accelerator steady	12
Engine will not rotate when attempting to start	1
Oil puddle under engine	7
Pinging or knocking engine sounds during acceleration or uphill	16
Starter motor noisy or excessively rough in engagement	5
Engine electrical system	
Alternator light fails to come on when key is turned on	21
Alternator light fails to go out	20
Battery will not hold a charge	19
Fuel system	
Excessive fuel consumption	22
Fuel leakage and/or fuel odor	23
Cooling system	
Coolant loss	28
External coolant leakage	26
Internal coolant leakage	27
Overcooling	25
Overheating	24
Poor coolant circulation	29
Clutch	
Clutch pedal stays on floor	36
Clutch slips (engine speed increases with no increase in vehicle speed)	32
Grabbing (chattering) as clutch is engaged	33
High pedal effort	37
Noise in clutch area	35
Pedal travels to floor - no pressure or very little resistance	30
Transaxle rattling (clicking)	34
Unable to select gears	31
Manual transaxle	
Clicking noise in turns	41
Clunk on acceleration or deceleration	40
Hard to shift	48

Symptom	Section
Knocking noise at low speeds	38
Leaks lubricant	47
Noise most pronounced when turning	39
Noisy in all gears	45
Noisy in neutral with engine running	43
Noisy in one particular gear	44
Slips out of gear	46
Vibration	42
Automatic transaxle	
Engine will start in gears other than Park or Neutral	53
Fluid leakage	49
General shift mechanism problems	51
Transaxle fluid brown or has burned smell	50
Transaxle slips, shifts roughly, is noisy or has no drive in forward or reverse gears	54
Transaxle will not downshift with accelerator pedal pressed to the floor	52
Driveaxles	
Clicking noise in turns	55
Shudder or vibration during acceleration	56
Vibration at highway speeds	57
Brakes	
Brake pedal feels spongy when depressed	65
Brake pedal travels to the floor with little resistance	66
Brake roughness or chatter (pedal pulsates)	60
Dragging brakes	63
Excessive brake pedal travel	62
Excessive pedal effort required to stop vehicle	61
Grabbing or uneven braking action	64
Noise (high-pitched squeal when the brakes are applied)	59
Parking brake does not hold	67
Vehicle pulls to one side during braking	58
Suspension and steering systems	
Abnormal noise at the front end	74
Abnormal or excessive tire wear	69
Cupped tires	79
Erratic steering when braking	76
Excessive pitching and/or rolling around corners or during braking	77
Excessive play or looseness in steering system	83
Excessive tire wear on inside edge	81
Excessive tire wear on outside edge	80
Hard steering	72
Poor returnability of steering to center	73
Rattling or clicking noise in steering gear	84
Shimmy, shake or vibration	71
Suspension bottoms	78
Tire tread worn in one place	82
Vehicle pulls to one side	68
Wander or poor steering stability	75
Wheel makes a thumping noise	70

Troubleshooting

Engine

1 Engine will not rotate when attempting to start

1. Battery terminal connections loose or corroded (Chapter 1).
2. Battery discharged or faulty (Chapter 1).
3. Automatic transaxle not completely engaged in Park (Chapter 7) or clutch pedal not completely depressed (Chapter 8).
4. Broken, loose or disconnected wiring in the starting circuit (Chapters 5 and 12).
5. Starter motor pinion jammed in flywheel ring gear (Chapter 5).
6. Starter solenoid faulty (Chapter 5).
7. Starter motor faulty (Chapter 5).
8. Ignition switch faulty (Chapter 12).
9. Starter pinion or flywheel teeth worn or broken (Chapter 5).
10. Defective fusible link (see Chapter 12).

2 Engine rotates but will not start

1. Fuel tank empty.
2. Battery discharged (engine rotates slowly) (Chapter 5).
3. Battery terminal connections loose or corroded (Chapter 1).
4. Leaking fuel injector(s), faulty fuel pump, pressure regulator, etc. (Chapter 4).
5. Broken or stripped timing belt (Chapter 2).
6. Ignition components damp or damaged (Chapter 5).
7. Worn, faulty or incorrectly gapped spark plugs (Chapter 1).
8. Broken, loose or disconnected wiring in the starting circuit (Chapter 5).
9. Broken, loose or disconnected wires at the ignition coils or faulty coils (Chapter 5).
10. Defective crankshaft sensor or ECM (see Chapter 6).

3 Engine hard to start when cold

1. Battery discharged or low (Chapter 1).
2. Malfunctioning fuel system (Chapter 4).
3. Faulty coolant temperature sensor or intake air temperature sensor (Chapter 6).
4. Fuel injector(s) leaking (Chapter 4).
5. Faulty ignition system (Chapter 5).
6. Choke inoperative (Chapter 4).

4 Engine hard to start when hot

1. Air filter clogged (Chapter 1).
2. Fuel not reaching the fuel injection system (Chapter 4).
3. Corroded battery connections, especially ground (Chapter 1).
4. Faulty coolant temperature sensor or intake air temperature sensor (Chapter 6).

5 Starter motor noisy or excessively rough in engagement

1. Pinion or flywheel gear teeth worn or broken (Chapter 5).
2. Starter motor mounting bolts loose or missing (Chapter 5).

6 Engine starts but stops immediately

1. Loose or faulty electrical connections at ignition coil (Chapter 5).
2. Insufficient fuel reaching the fuel injector(s) (Chapters 4).
3. Vacuum leak at the gasket between the intake manifold/plenum and throttle body (Chapter 4).
4. Fault in the engine control system (Chapter 6).
5. Intake air leaks, broken vacuum lines (see Chapter 4)

7 Oil puddle under engine

1. Oil pan gasket and/or oil pan drain bolt washer leaking (Chapter 2).
2. Oil pressure sending unit leaking (Chapter 2).
3. Valve covers leaking (Chapter 2).
4. Engine oil seals leaking (Chapter 2).

8 Engine lopes while idling or idles erratically

1. Vacuum leakage (Chapters 2 and 4).
2. Leaking EGR valve (Chapter 6).
3. Air filter clogged (Chapter 1).
4. Fuel pump not delivering sufficient fuel to the fuel injection system (Chapter 4).
5. Leaking head gasket (Chapter 2).
6. Timing belt and/or pulleys worn (Chapter 2).
7. Camshaft lobes worn (Chapter 2).

9 Engine misses at idle speed

1. Spark plugs worn or not gapped properly (Chapter 1).
2. Faulty spark plug wires (Chapter 1).
3. Vacuum leaks (Chapters 2 and 4).
4. Faulty ignition coil (Chapter 5).
5. Uneven or low compression (Chapter 2).
6. Faulty fuel injector(s) (Chapter 4).

10 Engine misses throughout driving speed range

1. Fuel filter clogged and/or impurities in the fuel system (Chapter 1).
2. Low fuel output at the fuel injector(s) (Chapter 4).
3. Faulty or incorrectly gapped spark plugs (Chapter 1).
4. Leaking spark plug wires (Chapters 1 or 5).
5. Faulty emission system components (Chapter 6).
6. Low or uneven cylinder compression pressures (Chapter 2).
7. Burned valves (Chapter 2).
8. Weak or faulty ignition system (Chapter 5).
9. Vacuum leak in fuel injection system, throttle body, intake manifold or vacuum hoses (Chapter 4).

11 Engine stumbles on acceleration

1. Spark plugs fouled (Chapter 1).
2. Problem with fuel injection system (Chapter 4).
3. Fuel filter clogged (Chapters 1 and 4).
4. Fault in the engine control system (Chapter 6).
5. Intake manifold air leak (Chapters 2 and 4).
6. EGR system malfunction (Chapter 6).

12 Engine surges while holding accelerator steady

1. Intake air leak (Chapter 4).
2. Fuel pump or fuel pressure regulator faulty (Chapter 4).
3. Problem with fuel injection system (Chapter 4).
4. Problem with the emissions control system (Chapter 6).

Troubleshooting 0-23

13 Engine stalls

1. Idle speed incorrect (Chapter 1).
2. Fuel filter clogged and/or water and impurities in the fuel system (Chapters 1 and 4).
3. Ignition components damp or damaged (Chapter 5).
4. Faulty emissions system components (Chapter 6).
5. Faulty or incorrectly gapped spark plugs (Chapter 1).
6. Faulty spark plug wires (Chapter 1).
7. Vacuum leak in the fuel injection system, intake manifold or vacuum hoses (Chapters 2 and 4).

14 Engine lacks power

1. Worn camshaft lobes (Chapter 2).
2. Burned valves or incorrect valve timing (Chapter 2).
3. Faulty spark plug wires or faulty coil (Chapters 1 and 5).
4. Faulty or incorrectly gapped spark plugs (Chapter 1).
5. Problem with the fuel injection system (Chapter 4).
6. Plugged air filter (Chapter 1).
7. Brakes binding (Chapter 9).
8. Automatic transaxle fluid level incorrect (Chapter 1).
9. Clutch slipping (Chapter 8).
10. Fuel filter clogged and/or impurities in the fuel system (Chapters 1 and 4).
11. Emission control system not functioning properly (Chapter 6).
12. Low or uneven cylinder compression pressures (Chapter 2).
13. Restricted exhaust system (Chapters 4).

15 Engine backfires

1. Emission control system not functioning properly (Chapter 6).
2. Faulty spark plug wires or coil(s) (Chapter 5).
3. Problem with the fuel injection system (Chapter 4).
4. Vacuum leak at fuel injector(s), intake manifold or vacuum hoses (Chapters 2 and 4).
5. Burned valves or incorrect valve timing (Chapter 2).

16 Pinging or knocking engine sounds during acceleration or uphill

1. Incorrect grade of fuel.
2. Problem with the engine control system (Chapter 6).
3. Fuel injection system faulty (Chapter 4).
4. Improper or damaged spark plugs or wires (Chapter 1).
5. EGR valve not functioning (Chapter 6).
6. Vacuum leak (Chapters 2 and 4).

17 Engine runs with oil pressure light on

1. Low oil level (Chapter 1).
2. Idle rpm below specification (Chapter 1).
3. Short in wiring circuit (Chapter 12).
4. Faulty oil pressure sender (Chapter 2).
5. Worn engine bearings and/or oil pump (Chapter 2).

18 Engine diesels (continues to run) after switching off

1. Idle speed too high (Chapter 1).
2. Excessive engine operating temperature (Chapter 3).
3. Excessive carbon deposits on valves and pistons (see Chapter 2).

Engine electrical system

19 Battery will not hold a charge

1. Alternator drivebelt defective or not adjusted properly (Chapter 1).
2. Battery electrolyte level low (Chapter 1).
3. Battery terminals loose or corroded (Chapter 1).
4. Alternator not charging properly (Chapter 5).
5. Loose, broken or faulty wiring in the charging circuit (Chapter 5).
6. Short in vehicle wiring (Chapter 12).
7. Internally defective battery (Chapters 1 and 5).

20 Alternator light fails to go out

1. Faulty alternator or charging circuit (Chapter 5).
2. Alternator drivebelt defective or out of adjustment (Chapter 1).
3. Alternator voltage regulator inoperative (Chapter 5).

21 Alternator light fails to come on when key is turned on

1. **Warning** light bulb defective (Chapter 12).
2. Fault in the printed circuit, dash wiring or bulb holder (Chapter 12).

Fuel system

22 Excessive fuel consumption

1. Dirty or clogged air filter element (Chapter 1).
2. Emissions system not functioning properly (Chapter 6).
3. Fuel injection system not functioning properly (Chapter 4).
4. Low tire pressure or incorrect tire size (Chapter 1).

23 Fuel leakage and/or fuel odor

1. Leaking fuel feed or return line (Chapters 1 and 4).
2. Tank overfilled.
3. Evaporative canister filter clogged (Chapters 1 and 6).
4. Problem with fuel injection system (Chapter 4).

Cooling system

24 Overheating

1. Insufficient coolant in system (Chapter 1).
2. Water pump defective (Chapter 3).
3. Radiator core blocked or grille restricted (Chapter 3).
4. Thermostat faulty (Chapter 3).
5. Electric coolant fan inoperative or blades broken (Chapter 3).
6. Radiator cap not maintaining proper pressure (Chapter 3).
7. Condenser fan inoperative (Chapter 3).

25 Overcooling

1. Faulty thermostat (Chapter 3).
2. Inaccurate temperature gauge sending unit (Chapter 3).

Troubleshooting

26 External coolant leakage

1. Deteriorated/damaged hoses; loose clamps (Chapters 1 and 3).
2. Water pump defective (Chapter 3).
3. Leak in coolant bypass hose or water pump outlet (Chapter 3).
4. Leakage from radiator core or coolant reservoir bottle (Chapter 3).
5. Engine drain or water jacket core plugs leaking (Chapter 2).

27 Internal coolant leakage

1. Leaking cylinder head gasket (Chapter 2).
2. Cracked cylinder bore or cylinder head (Chapter 2).

28 Coolant loss

1. Too much coolant in system (Chapter 1).
2. Coolant boiling away because of overheating (Chapter 3).
3. Internal or external leakage (Chapter 3).
4. Faulty pressure cap (Chapter 3).

29 Poor coolant circulation

1. Inoperative water pump (Chapter 3).
2. Restriction in cooling system (Chapters 1 and 3).
3. Thermostat sticking (Chapter 3).

Clutch

30 Pedal travels to floor - no pressure or very little resistance

1. Broken clutch cable (Chapter 8).
2. Broken release lever shaft, release fork or release bearing (Chapter 8).

31 Unable to select gears

1. Faulty transaxle (Chapter 7).
2. Faulty clutch disc or pressure plate (Chapter 8).
3. Faulty release lever shaft, release fork or release bearing (Chapter 8).
4. Faulty shift lever assembly or shift rod (Chapter 8).

32 Clutch slips (engine speed increases with no increase in vehicle speed)

1. Clutch plate worn (Chapter 8).
2. Clutch plate is oil soaked by leaking rear main seal (Chapter 8).
3. Clutch plate not seated (Chapter 8).
4. Warped pressure plate or flywheel (Chapter 8).
5. Weak diaphragm springs (Chapter 8).
6. Clutch plate overheated. Allow to cool.

33 Grabbing (chattering) as clutch is engaged

1. Oil on clutch plate lining, burned or glazed facings (Chapter 8).
2. Worn or loose engine or transaxle mounts (Chapters 2 and 7).
3. Worn splines on clutch plate hub (Chapter 8).
4. Warped pressure plate or flywheel (Chapter 8).
5. Burned or smeared resin on flywheel or pressure plate (Chapter 8).

34 Transaxle rattling (clicking)

1. Release fork loose (Chapter 8).
2. Low engine idle speed (Chapter 1).

35 Noise in clutch area

Faulty bearing (Chapter 8).

36 Clutch pedal stays on floor

1. Broken clutch cable (Chapter 8).
2. Broken release lever shaft, release fork or release bearing (Chapter 8).

37 High pedal effort

1. Clutch cable kinked or binding (see Chapter 8).
2. Pressure plate faulty (Chapter 8).

Manual transaxle

38 Knocking noise at low speeds

1. Worn driveaxle constant velocity (CV) joints (Chapter 8).
2. Worn side gear shaft counterbore in differential case (Chapter 7A).*

39 Noise most pronounced when turning

Differential gear noise (Chapter 7A).*

40 Clunk on acceleration or deceleration

1. Loose engine or transaxle mounts (Chapters 2 and 7A).
2. Worn differential pinion shaft in case.*
3. Worn side gear shaft counterbore in differential case (Chapter 7A).*
4. Worn or damaged driveaxle inboard CV joints (Chapter 8).

41 Clicking noise in turns

Worn or damaged outboard CV joint (Chapter 8).

42 Vibration

1. Rough wheel bearing (Chapters 1 and 10).
2. Damaged driveaxle (Chapter 8).
3. Out of round tires (Chapter 1).
4. Tire out of balance (Chapters 1 and 10).
5. Worn CV joint (Chapter 8).

Troubleshooting 0-25

43 Noisy in neutral with engine running

1 Damaged input gear bearing (Chapter 7A).*
2 Damaged clutch release bearing (Chapter 8).

44 Noisy in one particular gear

1 Damaged or worn constant mesh gears (Chapter 7A).*
2 Damaged or worn synchronizers (Chapter 7A).*
3 Bent reverse fork (Chapter 7A).*
4 Damaged fourth speed gear or output gear (Chapter 7A).*
5 Worn or damaged reverse idler gear or idler bushing (Chapter 7A).*

45 Noisy in all gears

1 Insufficient lubricant (Chapter 7A).
2 Damaged or worn bearings (Chapter 7A).*
3 Worn or damaged input gear shaft and/or output gear shaft (Chapter 7A).*

46 Slips out of gear

1 Worn or improperly adjusted linkage (Chapter 7A).
2 Transaxle loose on engine (Chapter 7A).
3 Shift linkage does not work freely, binds (Chapter 7A).
4 Input gear bearing retainer broken or loose (Chapter 7A).*
5 Foreign material between clutch cover and engine housing (Chapter 7A).
6 Worn shift fork (Chapter 7A).*

47 Leaks lubricant

1 Driveshaft seals worn (Chapter 7A).
2 Excessive amount of lubricant in transaxle (Chapters 1 and 7A).
3 Loose or broken input gear shaft bearing retainer (Chapter 7A).*
4 Input gear bearing retainer O-ring and/or lip seal damaged (Chapter 7A).*
5 Vehicle speed sensor O-ring leaking (Chapter 7A).

48 Hard to shift

Shift linkage loose or worn (Chapter 7A).
* Although the corrective action necessary to remedy the symptoms described is beyond the scope of this manual, the above information should be helpful in isolating the cause of the condition so that the owner can communicate clearly with a professional mechanic.

Automatic transaxle

Note: *Due to the complexity of the automatic transaxle, it is difficult for the home mechanic to properly diagnose and service this component. For problems other than the following, the vehicle should be taken to a dealer or transaxle shop.*

49 Fluid leakage

1 Automatic transaxle fluid is a deep red color. Fluid leaks should not be confused with engine oil, which can easily be blown onto the transaxle by air flow.
2 To pinpoint a leak, first remove all built-up dirt and grime from the transaxle housing with degreasing agents and/or steam cleaning. Then drive the vehicle at low speeds so air flow will not blow the leak far from its source. Raise the vehicle and determine where the leak is coming from. Common areas of leakage are:
   a) Pan (Chapters 1 and 7)
   b) Dipstick tube (Chapters 1 and 7)
   c) Transaxle oil lines (Chapter 7)
   d) Speedometer driven gear assembly or speed sensor (Chapter 7)
   e) Driveaxle oil seals (Chapter 7).

50 Transaxle fluid brown or has a burned smell

Transaxle fluid overheated (Chapter 1).

51 General shift mechanism problems

1 Chapter 7, Part B, deals with checking and adjusting the shift linkage on automatic transaxles. Common problems which may be attributed to poorly adjusted linkage are:
   a) Engine starting in gears other than Park or Neutral.
   b) Indicator on shifter pointing to a gear other than the one actually being used.
   c) Vehicle moves when in Park.
2 Refer to Chapter 7B for the shift linkage adjustment procedure.

52 Transaxle will not downshift with accelerator pedal pressed to the floor

The transaxle is electronically controlled. This type of problem - which is caused by a malfunction in the control unit, a sensor or solenoid, or the circuit itself - is beyond the scope of this book. Take the vehicle to a dealer service department or a competent automatic transmission shop.

53 Engine will start in gears other than Park or Neutral

Neutral start switch out of adjustment or malfunctioning (Chapter 7B).

54 Transaxle slips, shifts roughly, is noisy or has no drive in forward or reverse gears

There are many probable causes for the above problems, but the home mechanic should be concerned with only one possibility - fluid level. Before taking the vehicle to a repair shop, check the level and condition of the fluid as described in Chapter 1. Correct the fluid level as necessary or change the fluid and filter if needed. If the problem persists, have a professional diagnose the cause.

Driveaxles

55 Clicking noise in turns

Worn or damaged outboard CV joint (Chapter 8).

56 Shudder or vibration during acceleration

1. Excessive toe-in (Chapter 10).
2. Incorrect spring heights (Chapter 10).
3. Worn or damaged inboard or outboard CV joints (Chapter 8).
4. Sticking inboard CV joint assembly (Chapter 8).

57 Vibration at highway speeds

1. Out of balance front wheels and/or tires (Chapters 1 and 10).
2. Out of round front tires (Chapters 1 and 10).
3. Worn CV joint(s) (Chapter 8).

Brakes

Note: *Before assuming that a brake problem exists, make sure that:*
  a) *The tires are in good condition and properly inflated (Chapter 1).*
  b) *The front end alignment is correct (Chapter 10).*
  c) *The vehicle is not loaded with weight in an unequal manner.*

58 Vehicle pulls to one side during braking

1. Incorrect tire pressures (Chapter 1).
2. Front end out of alignment (have the front end aligned).
3. Front, or rear, tire sizes not matched to one another.
4. Restricted brake lines or hoses (Chapter 9).
5. Malfunctioning drum brake or caliper assembly (Chapter 9).
6. Loose suspension parts (Chapter 10).
7. Loose calipers (Chapter 9).
8. Excessive wear of brake shoe or pad material or disc/drum on one side.

59 Noise (high-pitched squeal when the brakes are applied)

Front and/or rear disc brake pads worn out. The noise comes from the wear sensor rubbing against the disc (does not apply to all vehicles). Replace pads with new ones immediately (Chapter 9).

60 Brake roughness or chatter (pedal pulsates)

1. Excessive lateral runout (Chapter 9).
2. Uneven pad wear (Chapter 9).
3. Defective disc (Chapter 9).

61 Excessive brake pedal effort required to stop vehicle

1. Malfunctioning power brake booster (Chapter 9).
2. Partial system failure (Chapter 9).
3. Excessively worn pads or shoes (Chapter 9).
4. Piston in caliper or wheel cylinder stuck or sluggish (Chapter 9).
5. Brake pads or shoes contaminated with oil or grease (Chapter 9).
6. Brake disc grooved and/or glazed (Chapter 1).
7. New pads or shoes installed and not yet seated. It will take a while for the new material to seat against the disc or drum.

62 Excessive brake pedal travel

1. Partial brake system failure (Chapter 9).

2. Insufficient fluid in master cylinder (Chapters 1 and 9).
3. Air trapped in system (Chapters 1 and 9).

63 Dragging brakes

1. Incorrect adjustment of brake light switch (Chapter 9).
2. Master cylinder pistons not returning correctly (Chapter 9).
3. Restricted brakes lines or hoses (Chapters 1 and 9).
4. Incorrect parking brake adjustment (Chapter 9).

64 Grabbing or uneven braking action

1. Malfunction of proportioning valve (Chapter 9).
2. Malfunction of power brake booster unit (Chapter 9).
3. Binding brake pedal mechanism (Chapter 9).

65 Brake pedal feels spongy when depressed

1. Air in hydraulic lines (Chapter 9).
2. Master cylinder mounting bolts loose (Chapter 9).
3. Master cylinder defective (Chapter 9).

66 Brake pedal travels to the floor with little resistance

1. Little or no fluid in the master cylinder reservoir caused by leaking caliper piston(s) (Chapter 9).
2. Loose, damaged or disconnected brake lines (Chapter 9).

67 Parking brake does not hold

Parking brake linkage improperly adjusted (Chapters 1 and 9).

Suspension and steering systems

Note: *Before attempting to diagnose the suspension and steering systems, perform the following preliminary checks:*
  a) *Tires for wrong pressure and uneven wear.*
  b) *Steering universal joints from the column to the rack and pinion for loose connectors or wear.*
  c) *Front and rear suspension and the rack and pinion assembly for loose or damaged parts.*
  d) *Out-of-round or out-of-balance tires, bent rims and loose and/or rough wheel bearings.*

68 Vehicle pulls to one side

1. Mismatched or uneven tires (Chapter 10).
2. Broken or sagging springs (Chapter 10).
3. Wheel alignment out-of-specifications (Chapter 10).
4. Front brake dragging (Chapter 9).

69 Abnormal or excessive tire wear

1. Wheel alignment out-of-specifications (Chapter 10).
2. Sagging or broken springs (Chapter 10).
3. Tire out-of-balance (Chapter 10).
4. Worn strut damper (Chapter 10).
5. Overloaded vehicle.
6. Tires not rotated regularly.

Troubleshooting 0-27

70 Wheel makes a thumping noise

1. Blister or bump on tire (Chapter 10).
2. Improper strut damper action (Chapter 10).

71 Shimmy, shake or vibration

1. Tire or wheel out-of-balance or out-of-round (Chapter 10).
2. Loose or worn wheel bearings (Chapters 1, 8 and 10).
3. Worn tie-rod ends (Chapter 10).
4. Worn lower balljoints (Chapters 1 and 10).
5. Excessive wheel runout (Chapter 10).
6. Blister or bump on tire (Chapter 10).

72 Hard steering

1. Lack of lubrication at balljoints, tie-rod ends and rack and pinion assembly (Chapter 10).
2. Front wheel alignment out-of-specifications (Chapter 10).
3. Low tire pressure(s) (Chapters 1 and 10).

73 Poor returnability of steering to center

1. Lack of lubrication at balljoints and tie-rod ends (Chapter 10).
2. Binding in balljoints (Chapter 10).
3. Binding in steering column (Chapter 10).
4. Lack of lubricant in steering gear assembly (Chapter 10).
5. Front wheel alignment out-of-specifications (Chapter 10).

74 Abnormal noise at the front end

1. Lack of lubrication at balljoints and tie-rod ends (Chapters 1 and 10).
2. Damaged strut mounting (Chapter 10).
3. Worn control arm bushings or tie-rod ends (Chapter 10).
4. Loose stabilizer bar (Chapter 10).
5. Loose wheel nuts (Chapters 1 and 10).
6. Loose suspension bolts (Chapter 10)

75 Wander or poor steering stability

1. Mismatched or uneven tires (Chapter 10).
2. Lack of lubrication at balljoints and tie-rod ends (Chapters 1 and 10).
3. Worn strut assemblies (Chapter 10).
4. Loose stabilizer bar (Chapter 10).
5. Broken or sagging springs (Chapter 10).
6. Wheels out of alignment (Chapter 10).

76 Erratic steering when braking

1. Wheel bearings worn (Chapter 10).
2. Broken or sagging springs (Chapter 10).
3. Leaking wheel cylinder or caliper (Chapter 10).
4. Warped rotors or drums (Chapter 10).

77 Excessive pitching and/or rolling around corners or during braking

1. Loose stabilizer bar (Chapter 10).
2. Worn strut dampers or mountings (Chapter 10).
3. Broken or sagging springs (Chapter 10).
4. Overloaded vehicle.

78 Suspension bottoms

1. Overloaded vehicle.
2. Worn strut dampers (Chapter 10).
3. Incorrect, broken or sagging springs (Chapter 10).

79 Cupped tires

1. Front wheel or rear wheel alignment out-of-specifications (Chapter 10).
2. Worn strut dampers (Chapter 10).
3. Wheel bearings worn (Chapter 10).
4. Excessive tire or wheel runout (Chapter 10).
5. Worn balljoints (Chapter 10).

80 Excessive tire wear on outside edge

1. Inflation pressures incorrect (Chapter 1).
2. Excessive speed in turns.
3. Front end alignment incorrect (excessive toe-in). Have professionally aligned.
4. Suspension arm bent or twisted (Chapter 10).

81 Excessive tire wear on inside edge

1. Inflation pressures incorrect (Chapter 1).
2. Front end alignment incorrect (toe-out). Have professionally aligned.
3. Loose or damaged steering components (Chapter 10).

82 Tire tread worn in one place

1. Tires out-of-balance.
2. Damaged or buckled wheel. Inspect and replace if necessary.
3. Defective tire (Chapter 1).

83 Excessive play or looseness in steering system

1. Wheel bearing(s) worn (Chapter 10).
2. Tie-rod end loose (Chapter 10).
3. Steering gear loose (Chapter 10).
4. Worn or loose steering intermediate shaft (Chapter 10).

84 Rattling or clicking noise in steering gear

1. Steering gear loose (Chapter 10).
2. Steering gear defective.

Notes

Chapter 1
Tune-up and routine maintenance

Contents

	Section		Section
Air filter replacement	22	Maintenance schedule	1
Automatic transaxle fluid and filter change	30	Manual transaxle lubricant change	33
Automatic transaxle fluid level check	8	Manual transaxle lubricant level check	18
Battery check, maintenance and charging	10	Positive Crankcase Ventilation (PCV) valve and	
Brake check	17	hose check and replacement	23
Clutch pedal height and freeplay - check and adjustment	12	Power steering fluid level check	7
Cooling system check	14	Rear wheel bearing check, repack and adjustment	31
Cooling system servicing (draining, flushing and refilling)	25	Spark plug check and replacement	20
Driveaxle boot check	27	Spark plug wire, distributor cap and rotor	
Drivebelt check, adjustment and replacement	19	check and replacement	21
Engine oil and oil filter change	6	Steering and suspension check	26
Evaporative emissions control system check	24	Tire and tire pressure checks	5
Exhaust system check	29	Tire rotation	11
Fluid level checks	4	Tune-up - general information	3
Fuel filter replacement	32	Underhood hose check and replacement	13
Fuel system check	28	Valve clearance check and adjustment (carbureted engines)	16
Idle speed check and adjustment	15	Windshield wiper blade inspection and replacement	9
Introduction	2		

Specifications

Recommended lubricants and fluids

Note: *Listed here are manufacturer recommendations at the time this manual was written. Manufacturers occasionally upgrade their fluid and lubricant specifications, so check with your local auto parts store for current recommendations.*

Engine oil
  Type.. API "certified for gasoline engines"
  Viscosity... See accompanying chart
  Fuel... Unleaded gasoline, 87 octane or higher

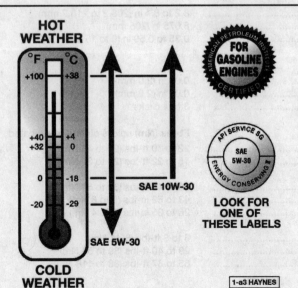

Engine oil viscosity chart - for best fuel economy and cold starting, select the lowest SAE viscosity grade for the expected temperature range

Chapter 1 Tune-up and routine maintenance

Recommended lubricants and fluids (continued)
Engine coolant	50/50 mixture of ethylene glycol based antifreeze and water
Automatic transaxle fluid	MERCON automatic transmission fluid
Manual transaxle lubricant	MERCON automatic transmission fluid
Brake fluid	DOT 3 brake fluid
Power steering fluid	Type F automatic transmission fluid
Chassis grease	SAE NLGI no. 2 chassis grease

Capacities*
	US qts (liters)
Engine oil (including filter)	3.6 qts (3.4 liters)
Engine coolant	
1993 and earlier	5.3 qts (5.0 liters)
1994 and later with A/C	6.3 qts (6.0 liters)
1994 and later without A/C	5.8 qts (5.5 liters)
Manual transaxle	2.6 qts (2.5 liters)
Automatic transaxle	6.0 qts (5.7 liters)

All capacities approximate. Add as necessary to bring up to appropriate level.

Ignition system
Spark plug type	Motorcraft AGS32C or equivalent
Spark plug gap	0.039 to 0.043 in (1.0 to 1.1 mm)
Engine firing order	1-3-4-2
Distributor rotation	Counterclockwise

Valve clearance (engine hot)
Intake valve	0.012 in (0.30 mm)
Exhaust valve	0.012 in (0.30 mm)

Cylinder location and distributor rotation
The blackened terminal shown on the distributor cap indicates the Number One spark plug wire position

Idle speed
1993 and earlier	
Manual transaxle	680 to 720 rpm*
Automatic transaxle	830 to 870 rpm*
1994 and later	
Manual transaxle	650 to 750 rpm**
Automatic transaxle	700 to 800 rpm**

Note: *Use the information printed on the Vehicle Emissions Control Information label, if different than the Specifications listed here.*
* With the set timing connector grounded.
** With a jumper wire connected between terminals STI and GND of the engine compartment service connector.

Accessory drivebelt deflection
Used drive belts	0.35 to 0.39 in (9 to 10 mm)
New drive belts	0.31 to 0.35 in (8 to 9 mm)

Clutch pedal
Height (including carpet)	
1993 and earlier	8.2 to 8.4 in (208.2 to 213.2 mm)
1994 and later	8.075 in (205 mm)
Freeplay	0.35 to 0.59 in (9 to 15 mm)

Brakes
Disc brake pad lining thickness (minimum)	0.08 in (2.0 mm)
Drum brake shoe lining thickness (minimum)	0.08 in (2.0 mm)
Parking brake adjustment	3 to 4 clicks

Torque specifications
	Ft-lbs (Nm) unless otherwise indicated
Engine oil drain plug	22 to 30 ft-lbs (29 to 41 Nm)
Spark plugs	15 to 22 ft-lbs (20 to 30 Nm)
Automatic transaxle	
Drain plug	29 to 40 ft-lbs (39 to 54 Nm)
Oil pan bolts	43 to 69 in-lbs (5 to 8 Nm)
Oil strainer/filter bolts	26 to 35 in-lbs (3 to 4 Nm)
Manual transaxle	
Speedometer driven gear retaining bolt	6 to 8 ft-lbs (8-11 Nm)
Drain plug	29 to 40 ft-lbs (39 to 54 Nm)
Wheel lug nuts	65 to 87 ft-lbs (88 to 118 Nm)

Chapter 1 Tune-up and routine maintenance

Typical engine compartment components (1993 and earlier)

1 Air cleaner assembly	5 Fusible links	9 Radiator cap	13 Engine oil dipstick
2 Fuel filter	6 Coolant reservoir	10 Distributor	14 Windshield washer fluid
3 Brake fluid reservoir	7 Battery	11 Oil filler cap	reservoir
4 Set timing connector	8 Upper radiator hose	12 Spark plug	15 PCV valve

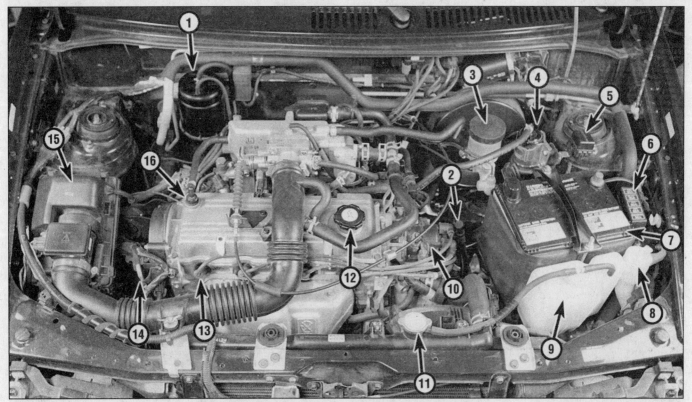

Typical engine compartment components (1994 and later)

1 EVAP canister	4 Fuel filter	8 Windshield washer fluid	12 Oil filler cap
2 Automatic transaxle	5 Data link connector	reservoir	13 Spark plug
dipstick	6 Engine compartment fuse	9 Coolant reservoir	14 Engine oil dipstick
3 Brake fluid reservoir	box	10 Distributor	15 Air cleaner assembly
	7 Battery	11 Radiator cap	16 PCV valve

Chapter 1 Tune-up and routine maintenance

Typical engine compartment underside components

1. Sway bar bushing
2. Lower radiator hose
3. Radiator drain fitting
4. Sway bar/lower control arm bushing
5. Driveaxle boot
6. Automatic transaxle drain plug
7. Exhaust system
8. Oil filter
9. Engine oil drain plug
10. Lower control arm bushing
11. Front disc brake caliper
12. Drivebelt

Typical rear underside components

1. Muffler
2. Suspension strut
3. Exhaust pipe
4. Rear brake hose
5. Gas tank
6. Rear brake assembly

1 Maintenance schedule

The maintenance intervals in this manual are provided with the assumption that you, not the dealer, will be doing the work. These are the minimum maintenance intervals recommended by the factory for vehicles that are driven daily. If you wish to keep your vehicle in peak condition at all times, you may wish to perform some of these procedures even more often. Because frequent maintenance enhances the efficiency, performance and resale value of your car, we encourage you to do so. If you drive in dusty areas, tow a trailer, idle or drive at low speeds for extended periods or drive for short distances (less than four miles) in below freezing temperatures, shorter intervals are also recommended.

When your vehicle is new, it should be serviced by a factory authorized dealer service department to protect the factory warranty. In many cases, the initial maintenance check is done at no cost to the owner.

Every 250 miles or weekly, whichever comes first

Check the engine oil level (Section 4)
Check the engine coolant level (Section 4)
Check the windshield washer fluid level (Section 4)
Check the brake fluid level (Section 4)
Check the tires and tire pressures (Section 5)

Every 3000 miles or 3 months, whichever comes first

All items listed above plus:
Change the engine oil and oil filter (Section 6)
Check the power steering fluid level (Section 7)
Check the automatic transaxle fluid level (Section 8)

Every 6000 miles or 6 months, whichever comes first

All items listed above plus:
Inspect and replace if necessary the windshield wiper blades (Section 9)
Check and service the battery (Section 10)
Rotate the tires (Section 11)

Every 15,000 miles or 12 months, whichever comes first

All items listed above plus:
Check the clutch pedal for proper height and freeplay (Section 12)
Inspect and replace if necessary all underhood hoses (Section 13)
Check the cooling system (Section 14)
Check and adjust if necessary the engine idle speed (Section 15)
Check and adjust valve clearance (carbureted engines) (Section 16)
Inspect the brake system (Section 17)*
Check the manual transaxle lubricant level (Section 18)
Check the seat belt operation (Chapter 11)

Every 30,000 miles or 24 months, whichever comes first

All items listed above plus:
Check and adjust if necessary the engine drivebelts (Section 19)*
Replace the spark plugs (Section 20)*
Inspect and replace if necessary the spark plug wires, distributor cap and rotor (Section 21)*
Replace the air filter (Section 22)*
Check and replace if necessary the PCV valve (Section 23)*
Inspect the fuel evaporative emissions control system (Section 24)*
Service the cooling system (drain, flush and refill) (Section 25)*
Inspect the steering and suspension components (Section 26)*
Check the driveaxle boots (Section 27)*
Inspect the fuel system (Section 28)*
Inspect the exhaust system (Section 29)
Change the automatic transaxle fluid (Section 30)**
Check and adjust if necessary the ignition timing (Chapter 5)*

Every 60,000 miles or 48 months, whichever comes first

Check and repack the front wheel bearings (Chapter 10)
Check and repack the rear wheel bearings (Section 31)
Replace the fuel filter (Section 32)*
Replace the timing belt (Chapter 2A)
Change the manual transaxle lubricant (Section 33)**

*This item is affected by "severe" operating conditions as described below. If your vehicle is operated under "severe" conditions, perform all maintenance indicated with an asterisk (*) at 15,000 mile/12 month intervals. Severe conditions are indicated if you mainly operate your vehicle under one or more of the following conditions:

Operating in dusty areas
Towing a trailer
Idling for extended periods and/or low speed operation
Operating when outside temperatures remain below freezing and when most trips are less than four miles

**If operated under one or more of the following conditions, change the manual or automatic transaxle fluid and differential lubricant every 15,000 miles:

In heavy city traffic where the outside temperature regularly reaches 90-degrees F (32-degrees C) or higher
In hilly or mountainous terrain
Frequent trailer pulling

2 Introduction

This Chapter is designed to help the home mechanic maintain the Festiva or Aspire for peak performance, economy, safety and long life.

Included is a master maintenance schedule, followed by procedures dealing specifically with each item on the schedule. Visual checks, adjustments, component replacement and other helpful items are included. Refer to the **accompanying illustrations** of the engine compartment and the underside of the vehicle for the locations of various components.

Servicing the vehicle, in accordance with the mileage/time maintenance schedule and the step-by-step procedures will result in a planned maintenance program that should produce a long and reliable service life. Keep in mind that it is a comprehensive plan, so maintaining some items but not others at the specified intervals will not produce the same results.

As you service the vehicle, you will discover that many of the procedures can - and should - be grouped together because of the nature of the particular procedure you're performing or because of the close proximity of two otherwise unrelated components to one another.

For example, if the vehicle is raised for chassis lubrication, you should inspect the exhaust, suspension, steering and fuel systems while you're under the vehicle. When you're rotating the tires, it makes good sense to check the brakes since the wheels are already removed. Finally, let's suppose you have to borrow or rent a torque wrench. Even if you only need it to tighten the spark plugs, you might as well check the torque of as many critical fasteners as time allows.

The first step in this maintenance program is to prepare yourself before the actual work begins. Read through all the procedures you're planning to do, then gather up all the parts and tools needed. If it looks like you might run into problems during a particular job, seek advice from a mechanic or an experienced do-it-yourselfer.

3 Tune-up - general information

The term tune-up is used in this manual to represent a combination of individual operations rather than one specific procedure.

If, from the time the vehicle is new, the routine maintenance schedule is followed closely and frequent checks are made of fluid levels and high wear items, as suggested throughout this manual, the engine will be kept in relatively good running condition and the need for additional work will be minimized.

More likely than not, however, there will be times when the engine is running poorly due to lack of regular maintenance. This is even more likely if a used vehicle, which has not received regular and frequent maintenance checks, is purchased. In such cases, an engine tune-up will be needed outside of the regular routine maintenance intervals.

The first step in any tune-up or engine diagnosis to help correct a poor running engine would be a cylinder compression check. A check of the engine compression (Chapter 2, Part B) will give valuable information regarding the overall performance of many internal components and should be used as a basis for tune-up and repair procedures. If, for instance, a compression check indicates serious internal engine wear, a conventional tune-up will not help the running condition of the engine and would be a waste of time and money.

The following series of operations are those most often needed to bring a generally poor running engine back into a proper state of tune.

Minor tune-up

Check all engine related fluids (Section 4)
Clean, inspect and test the battery (Section 10)
Check all underhood hoses (Section 13)
Check the cooling system (Section 14)
Check and adjust the idle speed (Section 15)
Check and adjust the valve clearances if necessary (Section 16)
Replace the spark plugs (Section 20)
Inspect the spark plug wires (Section 21)
Check the air filter (Section 22)
Check the PCV valve (Section 23)

Major tune-up

All items listed under Minor tune-up plus . . .
Check and adjust the drivebelts (Section 19)
Replace the spark plug wires, distributor cap and rotor (Section 21)
Replace the air filter (Section 22)
Replace the PCV valve (Section 23)
Check the fuel system (Section 28)
Replace the fuel filter (Section 32)
Check and adjust the ignition timing (Chapter 5)
Check the charging system (Chapter 5)

4 Fluid level checks (every 250 miles or weekly)

Note: *The following are fluid level checks to be done on a 250 mile or weekly basis. Additional fluid level checks can be found in specific maintenance procedures which follow. Regardless of intervals, be alert to fluid leaks under the vehicle which would indicate a problem to be corrected immediately.*

1 Fluids are an essential part of the lubrication, cooling, brake and windshield washer systems. Because the fluids gradually become depleted and/or contaminated during normal operation of the vehicle, they must be periodically replenished. See Recommended lubricants and fluids at the beginning of this Chapter before adding fluid to any of the following components. **Note:** *The vehicle must be on level ground when fluid levels are checked.*

Engine oil

Refer to illustrations 4.2, 4.4 and 4.6

2 The engine oil level is checked with a dipstick located on the right side of the engine compartment at the front of the engine **(see illustration)**. The dipstick extends through a metal tube from which it protrudes down into the engine oil pan.

3 The oil level should be checked before the vehicle has been driven, or about 15 minutes after the engine has been shut off. If the oil is checked immediately after driving the vehicle, some of the oil will remain in the upper engine components, producing an inaccurate reading on the dipstick.

4 Pull the dipstick from the tube and wipe all the oil from the end with a clean rag or paper towel. Insert the clean dipstick all the way back into its metal tube and pull it out again. Observe the oil at the end of the dipstick. At its highest point, the level should be between the L (Low) and F (Full) marks **(see illustration)**.

4.2 The oil dipstick (arrow) is located at the right front corner of the engine

Chapter 1 Tune-up and routine maintenance

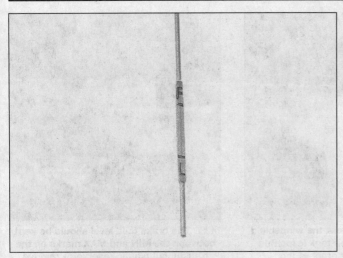

4.4 The oil level should be at or near the F mark - if it isn't, add enough oil to bring the level near to the F mark (it takes approximately 1 quart to raise the level from the L to the F mark)

4.6 The threaded oil filler cap (arrow) is located on the valve cover - always make sure the area around the opening is clean before unscrewing the cap to prevent dirt from contaminating the engine

5   It takes approximately 1.06 qts (1.0 liter) of oil to raise the level from the L mark to the F mark on the dipstick. Do not allow the level to drop below the L mark or oil starvation may cause engine damage. Conversely, overfilling the engine (adding oil above the F mark) may cause oil fouled spark plugs, oil leaks or oil seal failures.

6   Remove the threaded cap from the valve cover to add oil (see illustration). Use a funnel to prevent spills. After adding the oil, install the filler cap hand tight. Start the engine and look carefully for any small leaks around the oil filter or drain plug. Stop the engine and check the oil level again after it has had sufficient time to drain from the upper block and cylinder head galleys.

7   Checking the oil level is an important preventive maintenance step. A continually dropping oil level indicates oil leakage through damaged seals, from loose connections, or past worn rings or valve guides. If the oil looks milky in color or has water droplets in it, a cylinder head gasket may be blown. The engine should be checked immediately. The condition of the oil should also be checked. Each time you check the oil level, slide your thumb and index finger up the dipstick before wiping off the oil. If you see small dirt or metal particles clinging to the dipstick, the oil should be changed (Section 6).

Engine coolant

Refer to illustration 4.8

Warning: *Do not allow antifreeze to come in contact with your skin or painted surfaces of the vehicle. Flush contaminated areas immediately with plenty of water. Don't store new coolant or leave old coolant lying around where it's accessible to children or pets - they're attracted by its sweet smell and may drink it. Ingestion of even a small amount of coolant can be fatal! Wipe up garage floor and drip pan spills immediately. Keep antifreeze containers covered and repair cooling system leaks as soon as they're noticed.*

8   All vehicles covered by this manual are equipped with a pressurized coolant recovery system. A white plastic coolant reservoir located in the engine compartment is connected by a hose to the radiator filler neck (see illustration). If the engine overheats, coolant escapes through a valve in the radiator cap and travels through the hose into the reservoir. As the engine cools, the coolant is automatically drawn back into the cooling system to maintain the correct level.

9   The coolant level in the reservoir should be checked regularly. **Warning:** *Do not remove the radiator cap to check the coolant level when the engine is warm.* The level in the reservoir varies with the temperature of the engine. When the engine is cold, the coolant level should be at or slightly above the lower mark on the reservoir. Once the engine has warmed up, the level should be at or near the upper mark. If it isn't, allow the engine to cool, then remove the cap from the reservoir and add a 50/50 mixture of ethylene glycol-based antifreeze and water.

10 If the coolant level drops within a short time after replenishment, there may be a leak in the system. Inspect the radiator, hoses, engine coolant filler cap, drain plugs, air bleeder plugs and water pump. If no leak is evident, have the radiator cap pressure tested by your dealer. **Warning:** *Never remove the radiator cap or the coolant recovery reservoir cap when the engine is running or has just been shut down, because the cooling system is hot. Escaping steam and scalding liquid could cause serious injury.*

11 If it is necessary to open the radiator cap, wait until the system has cooled completely, then wrap a thick cloth around the cap and turn it to the first stop. If any steam escapes, wait until the system has cooled further, then remove the cap.

12 When checking the coolant level, always note its condition. It should be relatively clear. If it is brown or rust colored, the system should be drained, flushed and refilled. Even if the coolant appears to be normal, the corrosion inhibitors wear out with use, so it must be replaced at the specified intervals.

13 Do not allow antifreeze to come in contact with your skin or painted surfaces of the vehicle. Flush contacted areas immediately with plenty of water.

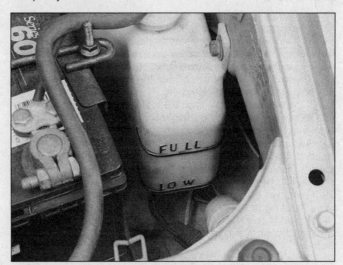

4.8 The coolant reservoir is located in the left front corner of the engine compartment - make sure the level is between the F and L marks on the side of the reservoir

Chapter 1 Tune-up and routine maintenance

4.14a On Festiva models, the windshield washer fluid reservoir tank is located in the right front corner of the engine compartment

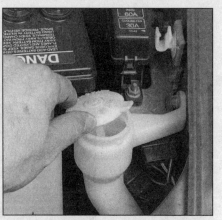

4.14b On Aspire models, the windshield washer fluid reservoir tank is located in the left front corner of the engine compartment

4.17 The brake fluid level should be kept between the MIN and MAX marks on the translucent plastic reservoir - remove the cap to add fluid

Windshield/window washer fluid

Refer to illustrations 4.14a and 14.4b

14 Fluid for the windshield washer system is stored in a plastic reservoir which is located at the front of the engine compartment **(see illustrations)**. Fluid for the rear hatchback window washer system is stored in a plastic reservoir located in the right rear corner of the cargo space. In milder climates, plain water can be used to top up the reservoir, but the reservoir should be kept no more than two-thirds full to allow for expansion should the water freeze. In colder climates, the use of a specially designed windshield washer fluid, available at your dealer and any auto parts store, will help lower the freezing point of the fluid. Mix the solution with water in accordance with the manufacturer's directions on the container. Do not use regular antifreeze. It will damage the vehicle's paint.

Battery electrolyte

15 On models not equipped with a sealed battery, unscrew the filler/vent cap and check the electrolyte level. It must be between the upper and lower levels. If the level is low, add distilled water. Install and securely retighten the cap. **Caution:** *Overfilling the cells may cause electrolyte to spill over during periods of heavy charging, causing corrosion or damage.*

Brake fluid

Refer to illustration 4.17

16 The brake fluid level is checked by looking through the plastic reservoir mounted on the master cylinder. The master cylinder is mounted on the front of the power booster unit in the (driver's side) rear corner of the engine compartment.
17 The fluid level should be between the MAX and MIN lines on the side of the reservoir **(see illustration)**.
18 If the fluid level is low, wipe the top of the reservoir and the cap with a clean rag to prevent contamination of the system as the cap is unscrewed.
19 Add only the specified brake fluid to the reservoir (refer to *Recommended lubricants and fluids* at the front of this Chapter or your owner's manual). Mixing different types of brake fluid can damage the system. Fill the reservoir to the MAX line. **Warning:** *Brake fluid can harm your eyes and damage painted surfaces, so use extreme caution when handling or pouring it. Do not use brake fluid that has been standing open or is more than one year old. Brake fluid absorbs moisture from the air, which can cause a dangerous loss of braking effectiveness.*
20 While the reservoir cap is off, check the master cylinder reservoir for contamination. If rust deposits, dirt particles or water droplets are present, the system should be drained and refilled by a dealer service department or repair shop.

21 After filling the reservoir to the proper level, make sure the cap is seated to prevent fluid leakage and/or contamination.
22 The fluid level in the master cylinder will drop slightly as the brake shoes or pads at each wheel wear down during normal operation. If the brake fluid level drops consistently, check the entire system for leaks immediately. Examine all brake lines, hoses and connections, along with the calipers, wheel cylinders and master cylinder (see Section 17).
23 When checking the fluid level, if you discover one or both reservoirs empty or nearly empty, the brake system should be bled (see Chapter 9).

5 Tire and tire pressure checks (every 250 miles or weekly)

Refer to illustrations 5.2, 5.3, 5.4a, 5.4b and 5.8

1 Periodic inspection of the tires may spare you from the inconvenience of being stranded with a flat tire. It can also provide you with vital information regarding possible problems in the steering and suspension systems before major damage occurs.
2 Normal tread wear can be monitored with a simple, inexpensive device known as a tread depth indicator **(see illustration)**. When the tread depth reaches the specified minimum, replace the tire(s).
3 Note any abnormal tread wear **(see illustration)**. Tread pattern irregularities such as cupping, flat spots and more wear on one side

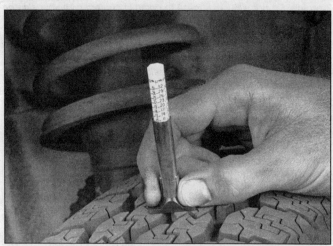

5.2 A tire tread depth indicator should be used to monitor tire wear - they are available at auto parts stores and service stations and cost very little

Chapter 1 Tune-up and routine maintenance

UNDERINFLATION

CUPPING

Cupping may be caused by:
- Underinflation and/or mechanical irregularities such as out-of-balance condition of wheel and/or tire, and bent or damaged wheel.
- Loose or worn steering tie-rod or steering idler arm.
- Loose, damaged or worn front suspension parts.

OVERINFLATION

INCORRECT TOE-IN OR EXTREME CAMBER

FEATHERING DUE TO MISALIGNMENT

5.3 This chart will help you determine the condition of your tires, the probable cause(s) of abnormal wear and the corrective action necessary

than the other are indications of front end alignment and/or balance problems. If any of these conditions are noted, take the vehicle to a tire shop or service station to correct the problem.

4 Look closely for cuts, punctures and embedded nails or tacks. Sometimes a tire will hold its air pressure for a short time or leak down very slowly even after a nail has embedded itself into the tread. If a slow leak persists, check the valve stem core to make sure it is tight **(see illustration)**. Examine the tread for an object that may have embedded itself into the tire or for a "plug" that may have begun to leak (radial tire punctures are repaired with a plug that is installed in a puncture). If a puncture is suspected, it can be easily verified by spraying a solution of soapy water onto the puncture area **(see illustration)**. The soapy solution will bubble if there is a leak. Unless the puncture is inordinately large, a tire shop or gas station can usually repair the punctured tire.

5 Carefully inspect the inner sidewall of each tire for evidence of brake fluid leakage. If you see any, inspect the brakes immediately.

5.4a If a tire loses air on a steady basis, check the valve core first to make sure it's snug (special inexpensive wrenches are commonly available at auto parts stores)

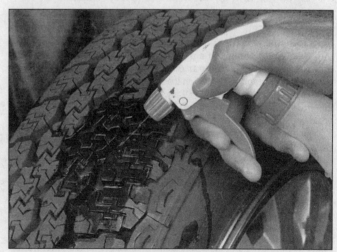
5.4b If the valve core is tight, raise the corner of the vehicle with the low tire and spray a soapy water solution onto the tread as the tire is turned slowly - slow leaks will cause small bubbles to appear

5.8 To extend the life of your tires, check the air pressure at least once a week with an accurate gauge (don't forget the spare!)

6   Correct tire air pressure adds miles to the lifespan of the tires, improves mileage and enhances overall ride quality. Tire pressure cannot be accurately estimated by looking at a tire, particularly if it is a radial. A tire pressure gauge is therefore essential. Keep an accurate gauge in the glovebox. The pressure gauges fitted to the nozzles of air hoses at gas stations are often inaccurate.

7   Always check tire pressure when the tires are cold. "Cold," in this case, means the vehicle has not been driven over a mile in the three hours preceding a tire pressure check. A pressure rise of four to eight pounds is not uncommon once the tires are warm.

8   Unscrew the valve cap protruding from the wheel or hubcap and push the gauge firmly onto the valve **(see illustration)**. Note the reading on the gauge and compare this figure to the recommended tire pressure shown on the tire placard on the driver's door frame. Be sure to reinstall the valve cap to keep dirt and moisture out of the valve stem mechanism. Check all four tires and, if necessary, add enough air to bring them up to the recommended pressure levels.

9   Don't forget to keep the spare tire inflated to the specified pressure (consult your owner's manual). Note that the air pressure specified for the compact spare is significantly higher than the pressure of the regular tires.

6   Engine oil and oil filter change (every 3000 miles or 3 months)

Refer to illustrations 6.2, 6.7, 6.12 and 6.14

1   Frequent oil changes are the best preventive maintenance the home mechanic can give the engine, because aging oil becomes diluted and contaminated, which leads to premature engine wear.

2   Make sure that you have all the necessary tools before you begin this procedure **(see illustration)**. You should also have plenty of rags or newspapers handy for mopping up any spills.

3   Access to the underside of the vehicle is greatly improved if the vehicle can be lifted on a hoist, driven onto ramps or supported by jackstands. **Warning:** *Do not work under a vehicle which is supported only by a bumper, hydraulic or scissors-type jack.*

4   If this is your first oil change, get under the vehicle and familiarize yourself with the location of the oil drain plug. The engine and exhaust components will be warm during the actual work, so try to anticipate any potential problems before the engine and accessories are hot.

5   Park the vehicle on a level spot. Start the engine and allow it to reach its normal operating temperature (the needle on the temperature gauge should be at least above the bottom mark). Warm oil and sludge will flow out more easily. Turn off the engine when it's warmed up. Remove the filler cap in the valve cover.

6   Raise the vehicle and support it on jackstands. **Warning:** *To avoid*

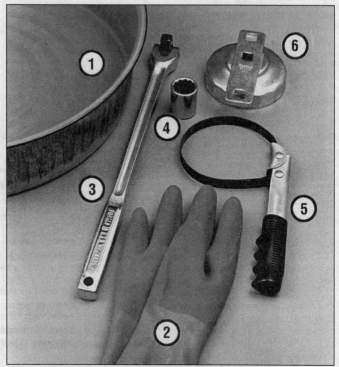

6.2 These tools are required when changing the engine oil and filter

1   **Drain pan -** *It should be fairly shallow in depth, but wide in order to prevent spills*
2   **Rubber gloves -** *When removing the drain plug and filter, it is inevitable that you will get oil on your hands (the gloves will prevent burns)*
3   **Breaker bar -** *Sometimes the oil drain plug is pretty tight and a long breaker bar is needed to loosen it*
4   **Socket –** *To be used with the breaker bar or a ratchet (must be the correct size to fit the drain plug)*
5   **Filter wrench -** *This is a metal band-type wrench, which requires clearance around the filter to be effective*
6   **Filter wrench -** *This type fits on the bottom of the filter and can be turned with a ratchet or beaker bar (different size wrenches are available for different types of filters)*

personal injury, never get beneath the vehicle when it is supported by only by a jack. The jack provided with your vehicle is designed solely for raising the vehicle to remove and replace the wheels. Always use jackstands to support the vehicle when it becomes necessary to place your body underneath the vehicle.

7   Being careful not to touch the hot exhaust components, place the drain pan under the drain plug in the bottom of the pan and remove the plug **(see illustration)**. You may want to wear gloves while unscrewing the plug the final few turns if the engine is really hot.

8   Allow the old oil to drain into the pan. It may be necessary to move the pan farther under the engine as the oil flow slows to a trickle. Inspect the old oil for the presence of metal shavings and chips.

9   After all the oil has drained, wipe off the drain plug with a clean rag. Even minute metal particles clinging to the plug would immediately contaminate the new oil.

10   Clean the area around the drain plug opening, reinstall the plug and tighten it securely, but do not strip the threads.

11   Move the drain pan into position under the oil filter.

12   Loosen the oil filter **(see illustration)** by turning it counterclockwise with the filter wrench. Any standard filter wrench should work. Once the filter is loose, use your hands to unscrew it from the block. Just as the filter is detached from the block, immediately tilt the open end up to prevent the oil inside the filter from spilling out. **Warning:** *The engine exhaust components may still be hot, so be careful.*

Chapter 1 Tune-up and routine maintenance

6.7 Use a proper size box-end wrench or socket to remove the oil drain plug and avoid rounding it off

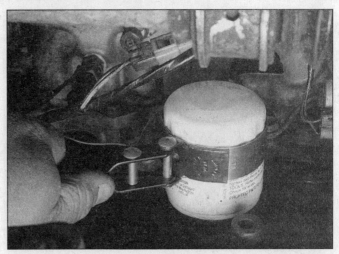

6.12 The oil filter is usually on very tight, so you'll need a special wrench for removal - DO NOT use the wrench to tighten the new filter

13   With a clean rag, wipe off the mounting surface on the block. If a residue of old oil is allowed to remain, it will smoke when the block is heated up. It will also prevent the new filter from seating properly. Also make sure that the none of the old gasket remains stuck to the mounting surface. It can be removed with a scraper if necessary.

14   Compare the old filter with the new one to make sure they are the same type. Smear some engine oil on the rubber gasket of the new filter and screw it into place **(see illustration)**. Because overtightening the filter will damage the gasket, do not use a filter wrench to tighten the filter. Tighten it by hand until the gasket contacts the seating surface. Then seat the filter by giving it an additional 3/4-turn.

15   Remove all tools, rags, etc. from under the vehicle, being careful not to spill the oil in the drain pan, then lower the vehicle.

16   Add new oil to the engine through the oil filler cap in the valve cover. Use a spout or funnel to prevent oil from spilling onto the top of the engine. Pour three quarts of fresh oil into the engine. Wait a few minutes to allow the oil to drain into the pan, then check the level on the oil dipstick (see Section 4 if necessary). If the oil level is at or near the F mark, install the filler cap hand tight, start the engine and allow the new oil to circulate.

17   Allow the engine to run for about a minute. While the engine is running, look under the vehicle and check for leaks at the oil pan drain plug and around the oil filter. If either is leaking, stop the engine and tighten the plug or filter slightly.

18   Wait a few minutes to allow the oil to trickle down into the pan, then recheck the level on the dipstick and, if necessary, add enough oil to bring the level to the F mark.

19   During the first few trips after an oil change, make it a point to check frequently for leaks and proper oil level.

20   The old oil drained from the engine cannot be reused in its present state and should be discarded. Check with your local refuse disposal company, disposal facility or environmental agency to see if they will accept the oil for recycling. Don't pour used oil into drains or onto the ground. After the oil has cooled, it can be drained into a suitable container (capped plastic jugs, topped bottles, milk cartons, etc.) for transport to one of these disposal sites.

7 Power steering fluid level check (every 3000 miles or 3 months)

Refer to illustrations 7.2, 7.4 and 7.5

1   Unlike manual steering, the power steering system relies on fluid which may, over a period of time, require replenishing.

2   The fluid reservoir for the power steering pump is located in the right rear corner of the engine compartment **(see illustration)**.

3   Run the engine until it reaches its normal operating temperature. Then make sure the front wheels are pointed straight ahead and shut off the engine.

6.14 Lubricate the oil filter gasket with clean engine oil before installing the filter on the engine

7.2 The power steering fluid reservoir is located in the right rear corner of the engine compartment

Chapter 1 Tune-up and routine maintenance

7.4 On earlier models, the power steering fluid level is checked by removing the cap and observing the level on the dipstick

7.5 On later models, the power steering reservoir is translucent - the fluid level can be checked by looking at the F and L lines on the side of the reservoir

4   On earlier models, use a clean rag to wipe off the reservoir cap and the area around the cap. This will help prevent any foreign matter from entering the reservoir during the check. Twist off the cap and wipe off the fluid with a clean rag, reinsert the dipstick, then withdraw it and read the fluid level **(see illustration)**.
5   On later models, simply look at the F (Full) and L (Low) lines on the reservoir **(see illustration)**. The fluid level should be between the F and L lines.
6   If additional fluid is required, pour the specified type directly into the reservoir, using a funnel to prevent spills. Fill the reservoir to the F line on later models or to the H mark of the dipstick on earlier models.
7   If the reservoir requires frequent fluid additions, all power steering hoses, hose connections, the power steering pump and the rack and pinion assembly should be carefully checked for leaks.

8   Automatic transaxle fluid level check (every 3000 miles or 3 months)

Refer to illustrations 8.4a and 8.4b

1   The level of the automatic transaxle fluid should be carefully maintained. Low fluid level can lead to slipping or loss of drive, while overfilling can cause foaming, loss of fluid and transaxle damage.
2   The transaxle fluid level should only be checked when the transaxle is hot (at its normal operating temperature). If the vehicle has just been driven over 10 miles (15 miles in a frigid climate), and the fluid temperature is about 150-degrees F (65-degrees C), the transaxle is hot. **Caution:** *If the vehicle has just been driven for a long time at high speed or in city traffic in hot weather, or if it has been pulling a trailer, an accurate fluid level reading cannot be obtained. Allow the fluid to cool down for about 30 minutes.*
3   If the vehicle has not just been driven, park the vehicle on level ground, set the parking brake and start the engine. While the engine is idling, depress the brake pedal and move the selector lever through all the gear ranges, beginning and ending in Park.
4   With the engine still idling, remove the dipstick from its tube **(see illustration)**. Check the level of the fluid on the dipstick **(see illustration)** and note its condition.
5   Wipe the fluid from the dipstick with a clean rag and reinsert it back into the filler tube until the cap seats.
6   Pull the dipstick out again and note the fluid level. If the level is below the L (low) line, add the specified automatic transmission fluid through the dipstick tube with a funnel.
7   Add just enough of the recommended fluid to fill the transaxle to the proper level. It takes about one pint to raise the level from the low line to the full line on the dipstick when the fluid is hot, so add the fluid a little at a time and keep checking the level until it is correct.
8   The condition of the fluid should also be checked along with the level. If the fluid at the end of the dipstick is black or a dark reddish brown color, or if it emits a burned smell, the fluid should be changed (see Section 30). If you are in doubt about the condition of the fluid, purchase some new fluid and compare the two for color and smell.

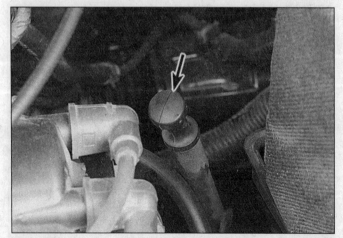

8.4a The automatic transaxle dipstick (arrow) is located on the left side of the engine compartment

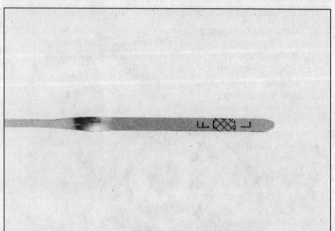

8.4b Check the fluid with the transaxle at normal operating temperature - the level should be kept in the cross-hatched area (don't add fluid if the level is anywhere in the cross-hatched area)

Chapter 1 Tune-up and routine maintenance

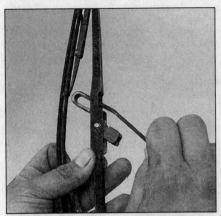

9.5 Push on the release lever and slide the wiper assembly down out of the hook in the end of the wiper arm

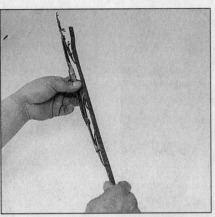

9.6 After detaching the end of the element, slide it out of the end of the frame

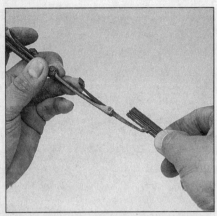

9.7 Insert the end of the element with the protrusions in first

9 Windshield wiper blade inspection and replacement (every 6000 miles or 6 months)

Refer to illustrations 9.5, 9.6 and 9.7

1 The windshield wiper and blade assembly should be inspected periodically for damage, loose components and cracked or worn blade elements.
2 Road film can build up on the wiper blades and affect their efficiency, so they should be washed regularly with a mild detergent solution.
3 The action of the wiping mechanism can loosen bolts, nuts and fasteners, so they should be checked and tightened, as necessary, at the same time the wiper blades are checked.
4 If the wiper blade elements are cracked, worn or warped, or no longer clean adequately, they should be replaced with new ones.
5 Remove the wiper blade assembly from the arm by pushing on the release lever, then sliding the assembly down and out of the hook in the end of the arm **(see illustration)**.
6 Detach the blade insert element and pull it out of the right end of the wiper frame **(see illustration)**.
7 Insert the new element end with the small protrusions into the right side of the wiper frame **(see illustration)**. Slide the element fully into place, then seat the protrusions in the end of the frames to secure it.

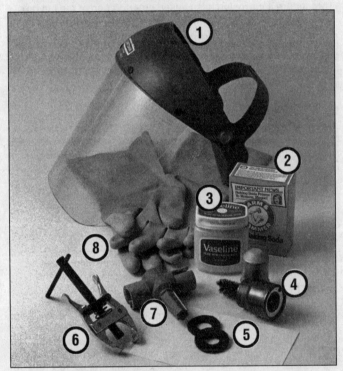

10.1 Tools and materials required for battery maintenance

1 **Face shield/safety goggles** - When removing corrosion with a brush, the acidic particles can easily fly up into your eyes
2 **Baking soda** - A solution of baking soda and water can be used to neutralize corrosion
3 **Petroleum jelly** - A layer of this on the battery posts will help prevent corrosion
4 **Battery post/cable cleaner** - This wire brush cleaning tool will remove all traces of corrosion from the battery posts and cable clamps
5 **Treated felt washers** - Placing one of these on each post, directly under the cable clamps, will help prevent corrosion
6 **Puller** - Sometimes the cable clamps are very difficult to pull off the posts, even after the nut/bolt has been completely loosened. This tool pulls the clamp straight up and off the post without damage
7 **Battery post/cable cleaner** - Here is another cleaning tool which is a slightly different version of number 4 above, but it does the same thing
8 **Rubber gloves** - Another safety item to consider when servicing the battery; remember that's acid inside the battery

10 Battery check, maintenance and charging (every 6000 miles or 6 months)

Refer to illustrations 10.1, 10.6a, 10.6b, 10.7a, 10.7b and 10.8
Warning: *Certain precautions must be followed when checking and servicing the battery. Hydrogen gas, which is highly flammable, is always present in the battery cells, so keep lighted tobacco and all other open flames and sparks away from the battery. The electrolyte inside the battery is actually dilute sulfuric acid, which will cause injury if splashed on your skin or in your eyes. It will also ruin clothes and painted surfaces. When removing the battery cables, always detach the negative cable first and hook it up last!*

1 A routine preventive maintenance program for the battery in your vehicle is the only way to ensure quick and reliable starts. But before performing any battery maintenance, make sure that you have the proper equipment necessary to work safely around the battery **(see illustration)**.
2 There are also several precautions that should be taken whenever battery maintenance is performed. Before servicing the battery, always turn the engine and all accessories off and disconnect the cable from the negative terminal of the battery.
3 The battery produces hydrogen gas, which is both flammable and explosive. Never create a spark, smoke or light a match around the battery. Always charge the battery in a ventilated area.

10.6a Battery terminal corrosion usually appears as light, fluffy powder

10.6b Removing a cable from the battery post with a wrench - sometimes a pair of special battery pliers are required for this procedure if corrosion has caused deterioration of the nut hex (always remove the ground (-) cable first and hook it up last!)

4   Electrolyte contains poisonous and corrosive sulfuric acid. Do not allow it to get in your eyes, on your skin on your clothes. Never ingest it. Wear protective safety glasses when working near the battery. Keep children away from the battery.

5   Note the external condition of the battery. If the positive terminal and cable clamp on your vehicle's battery is equipped with a rubber protector, make sure that it's not torn or damaged. It should completely cover the terminal. Look for any corroded or loose connections, cracks in the case or cover or loose hold-down clamps. Also check the entire length of each cable for cracks and frayed conductors.

6   If corrosion, which looks like white, fluffy deposits (see illustration) is evident, particularly around the terminals, the battery should be removed for cleaning. Loosen the cable clamp bolts with a wrench, being careful to remove the ground cable first, and slide them off the terminals (see illustration). Then disconnect the hold-down clamp bolt and nut, remove the clamp and lift the battery from the engine compartment.

7   Clean the cable clamps thoroughly with a battery brush or a terminal cleaner and a solution of warm water and baking soda (see illustration). Wash the terminals and the top of the battery case with the same solution but make sure that the solution doesn't get into the battery. When cleaning the cables, terminals and battery top, wear safety goggles and rubber gloves to prevent any solution from coming in contact with your eyes or hands. Wear old clothes too - even diluted, sulfuric acid splashed onto clothes will burn holes in them. If the terminals have been extensively corroded, clean them up with a terminal cleaner (see illustration). Thoroughly wash all cleaned areas with plain water.

8   Make sure that the battery tray is in good condition and the hold-down nut and bolt are tight (see illustration). If the battery is removed from the tray, make sure no parts remain in the bottom of the tray

10.7a When cleaning the cable clamps, all corrosion must be removed (the inside of the clamp is tapered to match the taper on the post, so don't remove too much material)

when the battery is reinstalled. When reinstalling the hold-down clamp bolt or nut, do not overtighten it.

9   Information on removing and installing the battery can be found in Chapter 5. Information on jump starting can be found at the front of this manual. For more detailed battery checking procedures, refer to the *Haynes Automotive Electrical Manual*.

10.7b Regardless of the type of tool used to clean the battery posts, a clean, shiny surface should be the result

10.8 Make sure the battery clamp retaining nuts (arrows) are tight

Chapter 1 Tune-up and routine maintenance

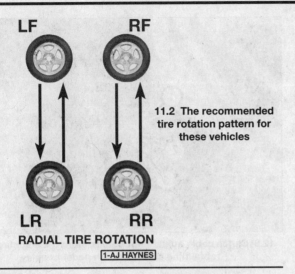

11.2 The recommended tire rotation pattern for these vehicles

hardest on the battery and can damage batteries not in good condition. This type of charging should only be used in emergency situations.

14 The average time necessary to charge a battery should be listed in the instructions that come with the charger. As a general rule, a trickle charger will charge a battery in 12 to 16 hours.

11 Tire rotation (every 6000 miles or 6 months)

Refer to illustration 11.2

1 The tires should be rotated at the specified intervals and whenever uneven wear is noticed. Since the vehicle will be raised and the tires removed anyway, check the brakes (see Section 17) at this time.
2 Radial tires must be rotated in a specific pattern **(see illustration)**.
3 Refer to the information in *Jacking and towing* at the front of this manual for the proper procedures to follow when raising the vehicle and changing a tire. If the brakes are to be checked, do not apply the parking brake as stated. Make sure the tires are blocked to prevent the vehicle from rolling.
4 Preferably, the entire vehicle should be raised at the same time. This can be done on a hoist or by jacking up each corner and then lowering the vehicle onto jackstands placed under the frame rails. Always use four jackstands and make sure the vehicle is firmly supported.
5 After rotation, check and adjust the tire pressures as necessary and be sure to check the lug nut tightness.
6 For further information on the wheels and tires, refer to Chapter 10.

Cleaning

10 Corrosion on the hold-down components, battery case and surrounding areas can be removed with a solution of water and baking soda. Thoroughly rinse all cleaned areas with plain water.
11 Any metal parts of the vehicle damaged by corrosion should be covered with a zinc-based primer, then painted.

Charging

Warning: *When batteries are being charged, hydrogen gas, which is very explosive and flammable, is produced. Do not smoke or allow open flames near a charging or a recently charged battery. Wear eye protection when near the battery during charging. Also, make sure the charger is unplugged before connecting or disconnecting the battery from the charger.*

12 Slow-rate charging is the best way to restore a battery that's discharged to the point where it will not start the engine. It's also a good way to maintain the battery charge in a vehicle that's only driven a few miles between starts. Maintaining the battery charge is particularly important in the winter when the battery must work harder to start the engine and electrical accessories that drain the battery are in greater use.
13 It's best to use a one or two-amp battery charger (sometimes called a "trickle" charger). They are the safest and put the least strain on the battery. They are also the least expensive. For a faster charge, you can use a higher amperage charger, but don't use one rated more than 1/10th the amp/hour rating of the battery. Rapid boost charges that claim to restore the power of the battery in one to two hours are

12 Clutch pedal height and freeplay - check and adjustment (every 15,000 miles or 12 months)

Refer to illustrations 12.1, 12.3, 12.6 and 12.8

1 To check the clutch pedal height, measure the distance from the center of the clutch pedal surface to the carpet or pad on the firewall **(see illustration)**. The height should be within the limits listed in this Chapter's Specifications. If it isn't, it must be adjusted.
2 To adjust the clutch pedal height, disconnect the clutch switch electrical connector.
3 Loosen the switch locknut **(see illustration)**.
4 Turn the clutch switch in or out until the pedal height is correct.
5 Tighten the locknut and recheck the pedal height to verify it is correct. **Note:** *Whenever the pedal height is adjusted it will most likely be necessary to adjust the freeplay because increasing or decreasing pedal height will cause a similar change in pedal freeplay.*

12.1 Pedal height is the distance between the pedal pad and the floor

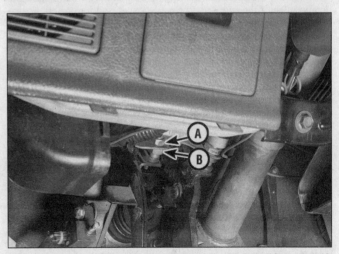

12.3 Loosen the locknut (A), then rotate the clutch pedal position switch (B) to adjust the pedal height

12.6 Pedal freeplay is the distance from the natural resting point of the pedal to the point at which resistance is felt

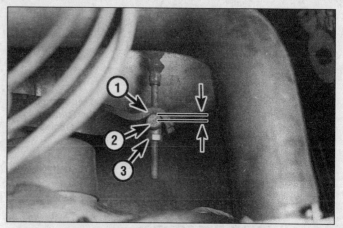

12.8 Clutch cable adjustment details - turn the adjusting nut to obtain the correct clutch pedal freeplay

1 Clutch release lever
2 Clutch cable pivot
3 Clutch cable adjusting nut

6 Check the clutch pedal freeplay by lightly pushing the clutch pedal down and measuring the distance that it moves freely before the clutch resistance is felt **(see illustration)**. The freeplay should be within the limits listed in this Chapter's Specifications. If it isn't, it must be adjusted.
7 Adjustment of the clutch pedal freeplay is obtained by loosening or tightening the clutch cable adjusting nut.
8 Working in the engine compartment, pull the clutch release lever back and measure the clearance between the release lever and the cable pivot **(see illustration)**. A clearance of 0.060 to 0.100 in (1.5 to 2.5 mm) between the release lever and the cable pivot should give the correct pedal freeplay.
9 Recheck the pedal freeplay to verify it is correct. If the pedal freeplay is still incorrect refer to Chapter 8 and inspect the clutch release components for damage.

13 Underhood hose check and replacement (every 15,000 miles or 12 months)

Caution: *Replacement of air conditioning hoses must be left to a dealer service department or air conditioning shop that has the equipment to depressurize the system safely. Never remove air conditioning components or hoses until the system has been depressurized.*

General

1 High temperatures in the engine compartment can cause the deterioration of the rubber and plastic hoses used for engine, accessory and emission systems operation. Periodic inspection should be made for cracks, loose clamps, material hardening and leaks.
2 Information specific to the cooling system hoses can be found in Section 14.
3 Some, but not all, hoses are secured to the fittings with clamps. Where clamps are used, check to be sure they haven't lost their tension, allowing the hose to leak. If clamps aren't used, make sure the hose has not expanded and/or hardened where it slips over the fitting, allowing it to leak.

Vacuum hoses

4 It's quite common for vacuum hoses, especially those in the emissions system, to be color coded or identified by colored stripes molded into them. Various systems require hoses with different wall thickness, collapse resistance and temperature resistance. When replacing hoses, be sure the new ones are made of the same material.
5 Often the only effective way to check a hose is to remove it completely from the vehicle. If more than one hose is removed, be sure to label the hoses and fittings to ensure correct installation.
6 When checking vacuum hoses, be sure to include any plastic T-fittings in the check. Inspect the fittings for cracks and the hose where it fits over the fitting for distortion, which could cause leakage.
7 A small piece of vacuum hose (1/4-inch inside diameter) can be used as a stethoscope to detect vacuum leaks. Hold one end of the hose to your ear and probe around vacuum hoses and fittings, listening for the "hissing" sound characteristic of a vacuum leak. **Warning:** *When probing with the vacuum hose stethoscope, be very careful not to come into contact with moving engine components such as the drivebelts, cooling fan, etc.*

Fuel hose

Warning: *There are certain precautions which must be taken when inspecting or servicing fuel system components. Work in a well ventilated area and do not allow open flames (cigarettes, appliance pilot lights, etc.) or bare light bulbs near the work area. Mop up any spills immediately and do not store fuel soaked rags where they could ignite.*
8 Check all rubber fuel lines for deterioration and chafing. Check especially for cracks in areas where the hose bends and just before fittings, such as where a hose attaches to the fuel filter.
9 High quality fuel line, specifically designed for fuel injection systems, must be used for fuel line replacement. **Warning:** *Never use anything other than the proper fuel line for fuel line replacement.*
10 Spring-type clamps are commonly used on fuel lines. These clamps often lose their tension over a period of time, and can be "sprung" during removal. Replace all spring-type clamps with screw clamps whenever a hose is replaced.

Metal lines

11 Sections of metal line are often used for fuel line between the fuel pump and fuel injection unit. Check carefully to be sure the line has not been bent or crimped and that cracks have not started in the line.
12 If a section of metal fuel line must be replaced, only seamless steel tubing should be used, since copper and aluminum tubing don't have the strength necessary to withstand normal engine vibration.
13 Check the metal brake lines where they enter the master cylinder and brake proportioning unit (if used) for cracks in the lines or loose fittings. Any sign of brake fluid leakage calls for an immediate thorough inspection of the brake system.

14 Cooling system check (every 15,000 miles or 12 months)

Refer to illustration 14.4
1 Many major engine failures can be attributed to a faulty cooling system. If the vehicle is equipped with an automatic transaxle, the

Chapter 1 Tune-up and routine maintenance 1-17

Check for a chafed area that could fail prematurely.

Check for a soft area indicating the hose has deteriorated inside.

Overtightening the clamp on a hardened hose will damage the hose and cause a leak.

Check each hose for swelling and oil-soaked ends. Cracks and breaks can be located by squeezing the hose.

14.4 Hoses, like drivebelts, have a habit of failing at the worst possible time - to prevent the inconvenience of a blown radiator or heater hose, inspect them carefully as shown here

3 Remove the radiator cap by turning it to the left until it reaches a stop. If you hear a hissing sound (indicating there is still pressure in the system), wait until it stops. Now press down on the cap with the palm of your hand and continue turning to the left until the cap can be removed. Thoroughly clean the cap, inside and out, with clean water. Also clean the filler neck on the radiator. All traces of corrosion should be removed. The coolant inside the radiator should be relatively transparent. If it's rust colored, the system should be drained and refilled (see Section 25). If the coolant level isn't up to the top, add additional antifreeze/coolant mixture (see Section 4).

4 Carefully check the large upper and lower radiator hoses along with the smaller diameter heater hoses which run from the engine to the firewall. Inspect each hose along its entire length, replacing any hose which is cracked, swollen or shows signs of deterioration. Cracks may become more apparent if the hose is squeezed (see illustration). Regardless of condition, it's a good idea to replace hoses with new ones every two years.

5 Make sure that all hose connections are tight. A leak in the cooling system will usually show up as white or rust colored deposits on the areas adjoining the leak. If wire-type clamps are used at the ends of the hoses, it may be a good idea to replace them with more secure screw-type clamps.

6 Use compressed air or a soft brush to remove bugs, leaves, etc. from the front of the radiator or air conditioning condenser. Be careful not to damage the delicate cooling fins or cut yourself on them.

7 Every other inspection, or at the first indication of cooling syst problems, have the cap and system pressure tested. If you don't ha a pressure tester, most gas stations and repair shops will do this fo minimal charge.

15 Idle speed check and adjustment (every 15,000 miles or 12 months)

Refer to illustrations 15.5a, 15.5b, 15.6a and 15.6b

Note: *Check and adjust the ignition timing as described in Chapter before attempting to the adjust idle speed.*

1 Engine idle speed is the speed at which the engine operates whe no throttle pedal pressure is applied. The idle speed is critical to th performance of the engine itself, as well as many engine sub-systems.

2 A hand-held tachometer must be used when adjusting idle speed to get an accurate reading. The exact hook-up for these meters varies with the manufacturer, so follow the particular directions included.

3 Set the parking brake and block the wheels. Be sure the transmission is in Neutral (manual transaxle) or Park (automatic transaxle).

4 Turn off the headlights and any other accessories during this procedure.

5 Attach a jumper wire from the set timing connector to ground (see illustrations).

cooling system also cools the transaxle fluid and thus plays an important role in prolonging transaxle life.

2 The cooling system should be checked with the engine cold. Do this before the vehicle is driven for the day or after the engine has been shut off for at least three hours.

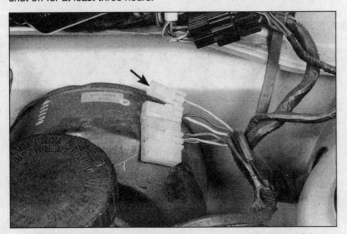
15.5a On 1993 and earlier models, connect a jumper wire between the test connector terminal (arrow) and body ground before checking and adjusting the idle speed

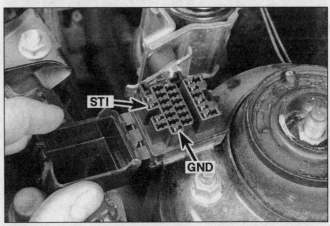
15.5b On 1994 and later models, connect a jumper wire between terminals STI and GND of the test connector before checking and adjusting the idle speed

Chapter 1 Tune-up and routine maintenance

15.6a On carbureted models, the idle speed screw (arrow) is located at the rear of the carburetor

15.6b On fuel injected models, the idle speed screw (arrow) is located on top of the throttle body

6 The idle speed is set by turning an adjustment screw which is located next to the throttle lever **(see illustrations)**. This screw changes the amount the throttle plate is held open by the throttle linkage.

7 For all applications, the engine must be completely warmed-up to operating temperature, which will automatically render the choke and fast idle systems inoperative.

8 Once you have located the idle screw, experiment with different length screwdrivers until the adjustment can be easily made without coming into contact with hot or moving engine components.

9 If the air cleaner is removed, the vacuum hose to the snorkel should be plugged.

10 Check the engine idle speed with the tachometer and compare it to the specifications. If the idle speed is incorrect, turn the idle speed adjusting screw until the idle speed is correct.

11 Most models have a tune-up decal or Vehicle Emission Control Information (VECI) label located in the engine compartment with the specifications for setting idle speed. If no VECI label is found refer to the Specifications Section at the beginning of this Chapter.

16 Valve clearance check and adjustment (carbureted engines) (every 15,000 miles or 12 months)

Refer to illustrations 16.3, 16.4 and 16.6

Note: *The manufacturer recommends adjusting the valve clearance at the specified interval only if the valve train is making excessive noise.*

1 Make sure the engine is warmed up to operating temperature before beginning this procedure.

2 Refer to Chapter 2A and remove the valve cover, then use the rotor and timing cover marks, to position the number one piston at TDC on the compression stroke (see Chapter 2A, Section 3). **Note:** *Verify TDC for the number one piston by observing that both valves are closed and the rocker arms are loose.*

3 With the crankshaft at number one TDC, measure the clearance of the indicated valves **(see illustration)**. Insert a feeler gauge of the specified thickness (see this Chapter's Specifications) between the valve stem tip and the rocker arm. The feeler gauge should slip between the valve stem tip and rocker arm with a slight amount of drag.

4 If the clearance is incorrect (too loose or too tight), loosen the locknut and turn the adjusting screw slowly until you can feel a slight drag on the feeler gauge as you withdraw it from between the valve stem tip and the rocker arm **(see illustration)**.

5 Once the clearance is adjusted, hold the adjusting screw with a screwdriver (to keep it from turning) and tighten the locknut to the torque listed in this Chapter's Specifications. Recheck the clearance to make sure it hasn't changed after tightening the locknut.

6 Make a mark on the crankshaft pulley and an adjacent mark on the front cover. Rotate the crankshaft one complete revolution (360-degrees) and realign the marks. Check the clearance of the

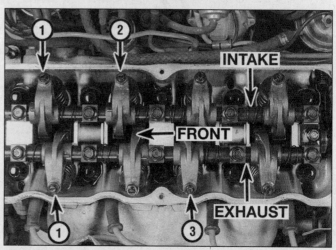

16.3 When the no. 1 piston is at TDC on the compression stroke, the valve clearance for the no. 1 and no. 3 cylinder exhaust valves and the no. 1 and no. 2 cylinder intake valves can be measured

remaining valves **(see illustration)**.

7 If necessary, repeat the adjustment procedure described in Steps 3, 4 and 5 until all the valves are adjusted to specifications.

8 Install the valve cover and the air cleaner assembly. Check for oil leaks and proper engine operation.

16.4 Measure the clearance for each valve with a feeler gauge of the specified thickness - if the clearance is correct, you should feel a slight drag on the gauge as you pull it out

Chapter 1 Tune-up and routine maintenance

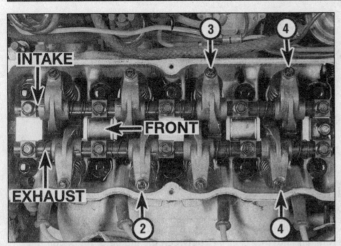

16.6 Rotate the crankshaft 360-degrees from no. 1 TDC, then check and adjust the valve clearances for the no. 2 and no. 4 cylinder exhaust valves and the no. 3 and no. 4 cylinder intake valves

17 Brake check (every 15,000 miles or 12 months)

Warning: *The dust created by the brake system may contain asbestos, which is harmful to your health. Never blow it out with compressed air and don't inhale any of it. An approved filtering mask should be worn when working on the brakes. Do not, under any circumstances, use petroleum-based solvents to clean brake parts. Use brake system cleaner only!* **Note:** *For detailed photographs of the brake system, refer to Chapter 9.*

1 In addition to the specified intervals, the brakes should be inspected every time the wheels are removed or whenever a defect is suspected. Any of the following symptoms could indicate a potential brake system defect: The vehicle pulls to one side when the brake pedal is depressed; the brakes make squealing or dragging noises when applied; brake pedal travel is excessive; the pedal pulsates; brake fluid leaks, usually onto the inside of the tire or wheel.
2 Loosen the wheel lug nuts.
3 Raise the vehicle and place it securely on jackstands.
4 Remove the wheels (see *Jacking and towing* at the front of this book, or your owner's manual, if necessary).

Disc brakes

Refer to illustrations 17.5 and 17.10

5 There are two pads (an outer and an inner) in each caliper. The pads are visible with the wheels removed **(see illustration)**.
6 If the lining material is less than the thickness listed in this Chapter's Specifications, replace the pads. **Note:** *Keep in mind that the lining material is riveted or bonded to a metal backing plate and the metal portion is not included in this measurement.*
7 If it is difficult to determine the exact thickness of the remaining pad material by the above method, or if you are at all concerned about the condition of the pads, remove the pads from the calipers for further inspection (refer to Chapter 9).
8 Once the pads are removed from the calipers, clean them with brake cleaner and re-measure them with a ruler or a vernier caliper.
9 Measure the disc thickness with a micrometer to make sure that it still has service life remaining. If any disc is thinner than the specified minimum thickness, replace it (refer to Chapter 9). Even if the disc has service life remaining, check its condition. Look for scoring, gouging and burned spots. If these conditions exist, remove the disc and have it resurfaced (see Chapter 9).
10 Before installing the wheels, check all brake lines and hoses for damage, wear, deformation, cracks, corrosion, leakage, bends and twists, particularly in the vicinity of the rubber hoses at the calipers **(see illustration)**. Check the clamps for tightness and the connections for leakage. Make sure that all hoses and lines are clear of sharp edges, moving parts and the exhaust system. If any of the above conditions are noted, repair, reroute or replace the lines and/or fittings as necessary (see Chapter 9).

Drum brakes

Refer to illustrations 17.12 and 17.14

11 Refer to Chapter 9 and remove the rear brake drums.
12 Note the thickness of the lining material on the rear brake shoes **(see illustration)** and look for signs of contamination by brake fluid and grease. If the lining material is within 1/16-inch of the recessed rivets or metal shoes, replace the brake shoes with new ones. The shoes should also be replaced if they are cracked, glazed (shiny lining surfaces) or contaminated with brake fluid or grease. See Chapter 9 for the replacement procedure.
13 Check the shoe return and hold-down springs and the adjusting mechanism to make sure they're installed correctly and in good condition. Deteriorated or distorted springs, if not replaced, could allow the linings to drag and wear prematurely.

17.5 With the wheels removed, the brake pad lining (arrow) can be inspected

17.10 Check for any sign of brake fluid leakage at the line fittings and the brake hoses (arrow)

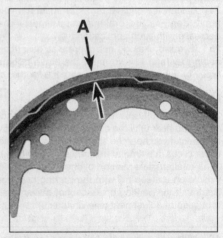

17.12 If the lining is bonded to the brake shoe, measure the lining thickness from the outer surface to the metal shoe, as shown here; if the lining is riveted to the shoe, measure from the lining outer surface to the rivet head

17.14 Carefully peel back the wheel cylinder boot and check for leaking fluid indicating that the cylinder must be replaced or rebuilt

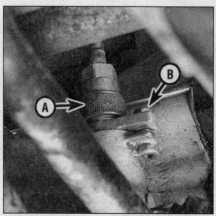

18.3 Detach the speedometer cable by unscrewing the cable retaining nut (A), then remove the retaining bolt (B) securing the drive gear assembly to the transaxle - remove the drive gear assembly from the transaxle to check the lubricant level

18.6 The lubricant level should be between the indicator marks on the side of the drive gear assembly

14 Check the wheel cylinders for leakage by carefully peeling back the rubber boots **(see illustration)**. If brake fluid is noted behind the boots, the wheel cylinders must be replaced (see Chapter 9).
15 Check the drums for cracks, score marks, deep scratches and hard spots, which will appear as small discolored areas. If imperfections cannot be removed with emery cloth, the drums must be resurfaced by an automotive machine shop (see Chapter 9 for more detailed information).
16 Refer to Chapter 9 and install the brake drums.
17 Install the wheels and snug the wheel lug nuts finger tight.
18 Remove the jackstands and lower the vehicle.
19 Tighten the wheel lug nuts to the torque listed in this Chapter's Specifications.

Brake booster check

20 Sit in the driver's seat and perform the following sequence of tests.
21 With the engine stopped, depress the brake pedal several times - the travel distance should not change.
22 With the brake fully depressed, start the engine - the pedal should move down a little when the engine starts.
23 Depress the brake, stop the engine and hold the pedal in for about 30 seconds - the pedal should neither sink nor rise.
24 Restart the engine, run it for about a minute and turn it off. Then firmly depress the brake several times - the pedal travel should decrease with each application.
25 If your brakes do not operate as described above when the preceding tests are performed, the brake booster is either in need of repair or has failed. Refer to Chapter 9 for the removal procedure.

Parking brake

26 Slowly pull up on the parking brake and count the number of clicks you hear until the handle is up as far as it will go. The adjustment is correct if you hear the specified number of clicks. If you hear more or fewer clicks, it's time to adjust the parking brake (refer to Chapter 9).
27 An alternative method of checking the parking brake is to park the vehicle on a steep hill with the parking brake set and the transaxle in Neutral. If the parking brake cannot prevent the vehicle from rolling, it is in need of adjustment (see Chapter 9).

18 Manual transaxle lubricant level check (every 15,000 miles or 12 months)

Refer to illustrations 18.3 and 18.6

1 Park the vehicle on level ground and set the parking brake firmly. Turn the engine off.
2 The oil level is checked by removing the speedometer cable and driven gear from the transaxle housing.
3 Disconnect the speedometer cable by turning the knurled nut securing it to the driven gear assembly **(see illustration)**.
4 Remove the bolt securing the driven gear assembly to the transaxle and slowly pull the driven gear assembly from the transaxle **(see illustration 18.3)**.
5 Wipe the driven gear clean and reinsert the assembly in the transaxle.
6 Pull it out again. The oil level should be between L (Low) and F (Full) **(see illustration)**.
7 If the oil level is low, add oil through the speedometer gear hole until it is at the proper level. **Warning:** *Do not overfill.*
8 Inspect the O-ring seal on the driven gear. Replace it if it appears damaged, flattened or age hardened. Re-install the driven gear in the transaxle and tighten the retaining bolt to the torque listed in this Chapter's Specifications. Then re-install the speedometer cable.
9 Drive the vehicle a short distance, then check carefully for leaks.

19 Drivebelt check, adjustment and replacement (every 30,000 miles or 24 months)

Refer to illustrations 19.3a, 19.3b and 19.4

Check

1 The alternator/water pump and power steering pump/air conditioning compressor drivebelts are located at the front of the engine. Because of their composition and the high stresses to which they are subjected, drivebelts stretch and deteriorate as they get older. They must therefore be periodically inspected. The good condition and proper adjustment of the alternator/water pump drivebelt is especially critical because it effects the operation of the engine.
2 Vehicles equipped with power steering and air conditioning have a second drivebelt dedicated to these accessories. This accessory drivebelt is mounted outboard of the alternator/water pump drivebelt on the crankshaft pulley.
3 With the engine off, open the hood and locate the drivebelts. With a flashlight visually check the belts. Look for cracking, fraying, separation, tears and glazing, which gives the belt a shiny appearance **(see illustrations)**. Both sides of the belt should be inspected, which means you will have to twist the belt to check the underside. Use your fingers to feel the belt where you can't see it. If any of the above conditions are evident, replace the belt (go to Step 8).
4 To check the tension of each belt in accordance with factory

Chapter 1 Tune-up and routine maintenance

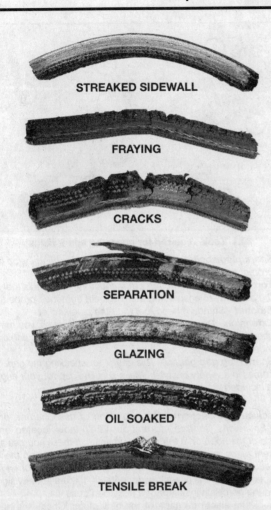

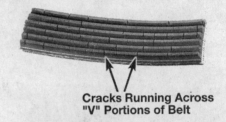

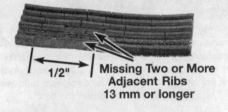

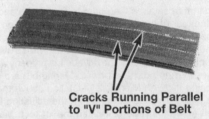

19.3a Here are some of the more common problems associated with drivebelts (check belts very carefully to prevent an untimely breakdown)

19.3b Small cracks in the underside of a V-ribbed belt are acceptable - lengthwise cracks, or missing pieces that cause the belt to make noise, are cause for replacement

specifications, apply moderate pressure (22 pounds) midway between the specified pulleys. Measure the deflection **(see illustration)** and compare your measurement to the specified drivebelt deflection for either a used or new belt. **Note:** *A "used" belt is defined as any belt which has been operated more than five minutes on the engine; a "new" belt is one that has been used for less than five minutes.*

Adjustment

Refer to illustrations 19.5 and 19.7

5   If the alternator/water pump belt must be adjusted, loosen the alternator mounting bolt located under the alternator. Loosen the adjusting bolt on the top of the alternator and lever the alternator away from the engine to tension the belt **(see illustration)**. Tighten the mounting and adjusting bolt. Measure the belt deflection in accordance with the above method. Repeat this step until the drivebelt is properly adjusted.

6   Adjust the power steering pump belt by loosening the upper

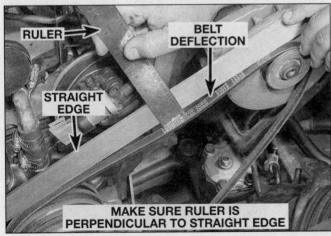

19.4 Measuring drivebelt deflection with a straightedge and ruler

19.5 After loosening the lower pivot bolt and the adjusting bolt (arrow), lever the alternator away from the engine to tension the drivebelt

Chapter 1 Tune-up and routine maintenance

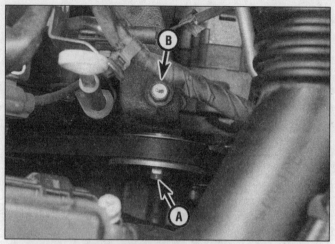

19.7 On vehicles equipped with air conditioning that do not have power steering, loosen the idler pulley locknut (A) then turn the adjusting bolt (B) counterclockwise to loosen or clockwise to tighten the belt

mounting bolt that secures the pump to the engine and the adjusting locknut at the rear of the pump. Adjust the belt tension by turning the adjusting bolt. Tighten the adjusting locknut and the pump mounting bolt. Measure the belt deflection in accordance with the above method. Repeat this step until the drivebelt is properly adjusted.

7 Vehicles that do not have power steering but are equipped with air conditioning have an idler pulley installed above the compressor. Loosen the idler pulley locknut and turn the adjusting bolt to tension the drivebelt **(see illustration)**. Tighten the locknut. Measure the belt deflection in accordance with the above method. Repeat this step until the drivebelt is properly adjusted.

Replacement

8 To replace a belt, follow the above procedures to loosen the drivebelt enough to slip the belt off the crankshaft pulley and remove it. If you are replacing the alternator/water pump belt, you will have to remove the power steering and/or air conditioning belt first because of the way they are arranged on the crankshaft pulley. Because of this and because belts tend to wear out more or less together, it is a good idea to replace both belts at the same time. Mark each belt and its appropriate pulley groove so the replacement belts can be installed in their proper positions.
9 Take the old belts to the parts store in order to make a direct comparison for length, width and design.
10 After replacing the drivebelt, make sure that it fits properly. When installing a multi-ribbed belt, make sure that it is centered - it must not overlap either edge of the pulley.
11 Adjust the drivebelt(s) in accordance with the procedure outlined above.

20 Spark plug check and replacement (every 30,000 miles or 24 months)

Refer to illustrations 20.1, 20.4a and 20.4b
1 Spark plug replacement requires a spark plug socket, a ratchet and, in most cases, an extension. This socket is lined with a rubber grommet to protect the porcelain insulator of the spark plug and to hold the plug while you insert it into the spark plug hole. You will also need a wire-type feeler gauge to check and adjust the spark plug gap and a torque wrench to tighten the new plugs to the specified torque **(see illustration)**.
2 If you are replacing the plugs, purchase the new plugs, adjust them to the proper gap and then replace each plug one at a time. **Note:** *When buying new spark plugs, it's essential that you obtain the correct plugs for your specific vehicle. This information can be found in*

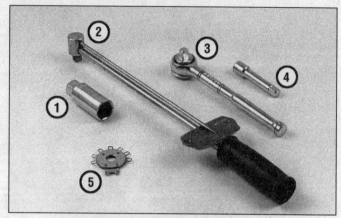

20.1 Tools required for changing spark plugs

1 **Spark plug socket** - This will have special padding inside to protect the spark plug porcelain insulator
2 **Torque wrench** - Although not mandatory, use of this tool is the best way to ensure that the plugs are tightened properly
3 **Ratchet** - Standard hand tool to fit the plug socket
4 **Extension** - Depending on model and accessories, you may need special extensions and universal joints to reach one or more of the plugs
5 **Spark plug gap gauge** - This gauge for checking the gap comes in a variety of styles. Make sure the gap for your engine is included

the Specifications Section at the beginning of this Chapter, on the Vehicle Emissions Control Information (VECI) label located on the underside of the hood or in the owner's manual. If these sources specify different plugs, purchase the spark plug type specified on the VECI label because that information is provided specifically for your engine.
3 Inspect each of the new plugs for defects. If there are any signs of cracks in the porcelain insulator of a plug, don't use it.
4 Check the electrode gaps of the new plugs. Check the gap by inserting the wire gauge of the proper thickness between the electrodes at the tip of the plug **(see illustration)**. The gap between the electrodes should be identical to that listed in this Chapter's Specifications or on the VECI label. If the gap is incorrect, use the notched adjuster on the feeler gauge body to bend the curved side electrode slightly **(see illustration)**.
5 If the side electrode is not exactly over the center electrode, use the notched adjuster to align them. **Caution:** *If the gap of a new plug must be adjusted, bend only the base of the ground electrode – do not touch the tip.*

20.4a Spark plug manufacturers recommend using a wire-type gauge when checking the gap - if the wire does not slide between the electrodes with a slight drag, adjustment is required

Chapter 1 Tune-up and routine maintenance   1-23

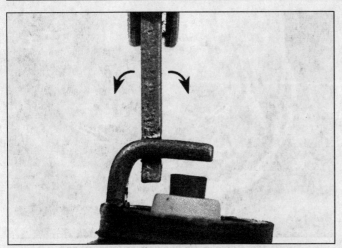

20.4b To change the gap, bend the side electrode only, as indicated by the arrows, and be very careful not to crack or chip the porcelain insulator surrounding the center electrode

20.6 When removing the spark plug wires, pull only on the boot and use a twisting/pulling motion

Removal

Refer to illustrations 20.6 and 20.8

6   To prevent the possibility of mixing up spark plug wires, work on one spark plug at a time. Remove the wire and boot from one spark plug. Grasp the boot - not the cable - as shown, give it a half twisting motion and pull straight up **(see illustration)**.
7   If compressed air is available, blow any dirt or foreign material away from the spark plug area before proceeding (a common bicycle pump will also work).
8   Remove the spark plug **(see illustration)**.
9   Whether you are replacing the plugs at this time or intend to reuse the old plugs, compare each old spark plug with the chart on the inside back cover of this manual to determine the overall running condition of the engine.

Installation

Refer to illustrations 20.10a and 20.10b

10 Prior to installation, it's a good idea to coat the spark plug threads with anti-seize compound **(see illustration)**. Also, it's often difficult to insert spark plugs into their holes without cross-threading them. To avoid this possibility, fit a short piece of 3/8-inch ID rubber hose over the end of the spark plug **(see illustration)**. The flexible hose acts as a universal joint to help align the plug with the plug hole. Should the plug begin to cross-thread, the hose will slip on the spark plug, preventing thread damage. Tighten the plug to the torque listed in this Chapter's Specifications.

11 Attach the plug wire to the new spark plug, again using a twisting motion on the boot until it is firmly seated on the end of the spark plug.
12 Follow the above procedure for the remaining spark plugs, replacing them one at a time to prevent mixing up the spark plug wires.

21 Spark plug wire, distributor cap and rotor check and replacement (every 30,000 miles or 24 months)

Refer to illustrations 21.8, 21.11a, 21.11b, 21.11c and 21.12

1   The spark plug wires should be checked whenever new spark plugs are installed.
2   Begin this procedure by making a visual check of the spark plug wires while the engine is running. In a darkened garage (make sure there is ventilation) start the engine and observe each plug wire. Be careful not to come into contact with any moving engine parts. If there is a break in the wire, you will see arcing or a small spark at the damaged area. If arcing is noticed, make a note to obtain new wires, then allow the engine to cool and check the distributor cap and rotor.
3   The spark plug wires should be inspected one at a time to prevent mixing up the order, which is essential for proper engine operation. Each original plug wire should be numbered to help identify its location. If the number is illegible, a piece of tape can be marked with the correct number and wrapped around the plug wire.
4   Disconnect the plug wire from the spark plug. A removal tool can be used for this purpose or you can grasp the rubber boot, twist the boot half a turn and pull the boot free. Do not pull on the wire itself.

20.8 Use a spark plug socket with a long extension to unscrew the spark plug

20.10a Apply a thin coat of anti-seize compound to the spark plug threads

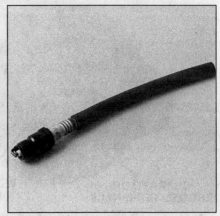

20.10b A length of 3/8-inch ID rubber hose will save time and prevent damaged threads when installing the spark plugs

21.8 Pull only on the boot when removing ignition wires from the distributor

21.11a Remove the two screws (arrows) and detach the distributor cap

5 Check inside the boot for corrosion, which will look like a white crusty powder.

6 Push the wire and boot back onto the end of the spark plug. It should fit tightly onto the end of the plug. If it doesn't, remove the wire and use pliers to carefully crimp the metal connector inside the wire boot until the fit is snug.

7 Using a clean rag, wipe the entire length of the wire to remove built-up dirt and grease. Once the wire is clean, check for burns, cracks and other damage. Do not bend the wire sharply, because the conductor might break.

8 Disconnect the wire from the distributor cap. Again, pull only on the boot **(see illustration)**. Check for corrosion and a tight fit. Replace the wire in the distributor cap.

9 Inspect the remaining spark plug wires, making sure that each one is securely fastened at the distributor and spark plug when the check is complete.

10 If new spark plug wires are required, purchase a set for your specific engine model. Pre-cut wire sets with the boots already installed are available. Remove and replace the wires one at a time to avoid mix-ups in the firing order.

11 Detach the distributor cap by removing the two retaining screws **(see illustration)**. Then inspect it for cracks, carbon tracks and worn, burned or loose contacts **(see illustrations)**.

12 Pull the rotor off the distributor shaft and examine it for cracks and carbon tracks **(see illustration)**. Replace the cap and rotor if any damage or defects are noted.

13 It is common practice to install a new cap and rotor whenever new spark plug wires are installed, but if you wish to continue using the old cap, clean the terminals first.

14 When installing a new cap, remove the wires from the old cap one at a time and attach them to the new cap in the exact same location – do not simultaneously remove all the wires from the old cap or firing order mix-ups may occur. **Note:** *If an accidental mix-up occurs, refer to the firing order diagrams at the beginning of this Chapter.*

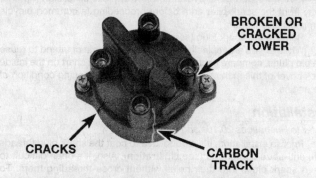

21.11b Inspect the outside of the distributor cap for cracks, carbon tracking and broken or cracked terminals (if in doubt about its condition, install a new one)

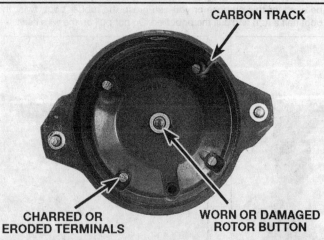

21.11c Inspect the inside of the distributor cap for charred or eroded terminals and other damage (if in doubt about its condition, install a new one)

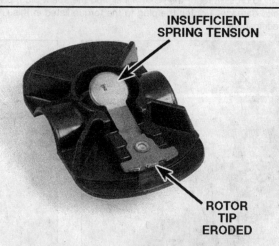

21.12 Check the rotor for wear and corrosion as indicated here (if in doubt about its condition, buy a new one)

Chapter 1 Tune-up and routine maintenance   1-25

22.3 On carbureted models, unscrew the wingnut from the center of the air cleaner cover, then detach the clips (arrows)

22.4 Remove the cover to access the air filter element (A) and the PCV filter (B) - if either filter is dirty it should be replaced

22 Air filter replacement (every 30,000 miles or 24 months)

Caution: *Never drive the vehicle with the air cleaner removed. Excessive engine wear could result and backfiring could even cause a fire under the hood.*

1   At the specified intervals, the air filter should be replaced with a new one. A thorough program of preventive maintenance would also call for the filter to be inspected periodically between changes, especially if the vehicle is often driven in dusty conditions.

2   The air filter is located inside the air cleaner housing, which is mounted on top of the carburetor or at the right front corner of the engine compartment on fuel injected models.

Carbureted models

Refer to illustrations 22.3 and 22.4

3   Remove the wing nut that holds the top plate to the air cleaner body, release the clips and lift the top plate off the air cleaner housing **(see illustration)**.

4   Lift the air filter out of the housing. If it's covered with dirt, it should be replaced. The PCV filter should also be replaced at this time **(see illustration)**.

5   Wipe the inside of the air cleaner housing with a rag.

6   Place the old filter (if in good condition) or the new filter (if replacement is necessary) into the air cleaner housing.

7   Reinstall the top plate on the air cleaner and tighten the wing nut, then snap the clips into place.

Fuel-injected models

Refer to illustrations 22.8a and 22.8b

8   To remove the air filter, loosen the intake air duct band and remove the duct/hose. then detach the electrical connector from the mass airflow sensor. Remove the four bolts or clips attaching the air cleaner cover to the air cleaner housing, then lift the cover up and remove the air filter element **(see illustrations)**.

9   Inspect the outer surface of the filter element. If it is dirty, replace it. If it is only moderately dusty, it can be reused by blowing it clean from the back to the front surface with compressed air. Because it is a pleated paper type filter, it cannot be washed or oiled. If it cannot be cleaned satisfactorily with compressed air, discard and replace it. While the cover is off, be careful not to drop anything down into the housing.

10   Wipe out the inside of the air cleaner housing with a damp cloth.

11   Place the new filter into the air cleaner housing, making sure it seats properly.

12   Installation of the cover is the reverse of removal.

22.8a On fuel-injected models, detach the Intake air duct (A), the mass air flow sensor electrical connector (B) and the cover retaining clips or bolts (C)

22.8b Remove the cover and lift the element out

23.2 Grasp the hose securely and pull the PCV valve out of the cover

24.2a On 1993 and earlier models, the EVAP canister is mounted next to the battery

24.2b On 1994 and later models, the EVAP canister is mounted on the firewall

23 Positive Crankcase Ventilation (PCV) valve and hose check and replacement (every 30,000 miles or 24 months)

Refer to illustration 23.2

1 The Positive Crankcase Ventilation (PCV) system directs blowby gases from the crankcase through the PCV valve and hose back into the intake manifold so they can be burned in the engine. The system consists of a hose leading from the rocker arm cover to the intake manifold and a fresh air hose between the air cleaner assembly and the rocker arm cover.
2 The PCV valve and hose is located in the valve cover **(see illustration)**.
3 With the engine idling at normal operating temperature, pull the valve (with hose attached) from the valve cover.
4 Place your finger over the valve opening or hose. If there's no vacuum, check for a plugged hose, manifold port, or the valve itself. Replace any plugged or deteriorated hoses.
5 Turn off the engine and shake the PCV valve, listening for a rattle. If the valve doesn't rattle, replace it with a new one.
6 To replace the valve, pull it from the end of the hose, noting its installed position.
7 When purchasing a replacement PCV valve, make sure it's for your particular vehicle and engine size. Compare the old valve with the new one to make sure they're the same.
8 Push the valve into the end of the hose until it's seated.
9 Inspect all rubber hoses and grommets for damage and hardening. Replace them, if necessary.
10 Press the PCV valve and hose securely into position. For further information on the PCV system refer to Chapter 6.

24 Evaporative emissions control system check (every 30,000 miles or 24 months)

Refer to illustrations 24.2a and 24.2b

1 The function of the evaporative emissions control system is to draw fuel vapors from the gas tank and fuel system, store them in a charcoal canister and then burn them during normal engine operation.
2 The most common symptom of a fault in the evaporative emissions system is a strong fuel odor in the engine compartment. If a fuel odor is detected, inspect the charcoal canister, located at the rear of the engine compartment. Check the canister and all hoses for damage and deterioration **(see illustrations)**.
3 The evaporative emissions control system is explained in more detail in Chapter 6.

25 Cooling system servicing (draining, flushing and refilling) (every 30,000 miles or 24 months)

Warning: *Do not allow engine coolant (antifreeze) to come in contact with your skin or painted surfaces of the vehicle. Rinse off spills immediately with plenty of water. Antifreeze is highly toxic if ingested. Never leave antifreeze laying around in an open container or in puddles on the floor; children and pets are attracted by it's sweet smell and may drink it. Check with local authorities about disposing of used antifreeze. Many communities have collection centers which will see that antifreeze is disposed of safely.*

1 Periodically, the cooling system should be drained, flushed and refilled to replenish the antifreeze mixture and prevent formation of rust and corrosion, which can impair the performance of the cooling system and cause engine damage. When the cooling system is serviced, all hoses and the radiator cap should be checked and replaced if necessary.

Draining

Refer to illustrations 25.4a and 25.4b

2 Apply the parking brake and block the wheels. If the vehicle has just been driven, wait several hours to allow the engine to cool down before beginning this procedure.
3 Once the engine is completely cool, remove the radiator cap.
4 Move a large container under the radiator drain to catch the coolant. Attach a 3/8-inch inner diameter hose to the drain fitting to direct the coolant into the container (some models are already equipped with a hose), then open the drain fitting (a pair of pliers may be required to turn it) **(see illustrations)**.
5 While the coolant is draining, check the condition of the radiator hoses, heater hoses and clamps (refer to Section 14 if necessary).
6 Replace any damaged clamps or hoses (see Chapter 3).

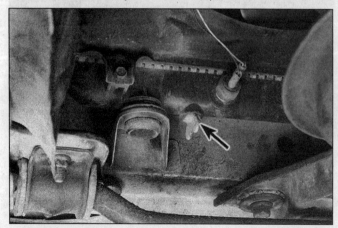

25.4a Radiator drain location - 1993 and earlier vehicles

Chapter 1 Tune-up and routine maintenance   1-27

25.4b Radiator drain location - 1994 and later vehicles

Flushing

7   Once the system is completely drained, flush the radiator with fresh water from a garden hose until water runs clear at the drain. The flushing action of the water will remove sediments from the radiator but will not remove rust and scale from the engine and cooling tube surfaces.

8   These deposits can be removed by the chemical action of a cleaner. Follow the procedure outlined in the manufacturer's instructions. If the radiator is severely corroded, damaged or leaking, it should be removed (see Chapter 3) and taken to a radiator repair shop.

9   Remove the overflow hose from the coolant recovery reservoir. Drain the reservoir and flush it with clean water, then reconnect the hose.

Refilling

10 Close and tighten the radiator drain.
11 Place the heater temperature control in the maximum heat position.
12 Slowly add new coolant (a 50/50 mixture of water and antifreeze) to the radiator until it's full. Add coolant to the reservoir up to the lower mark.
13 Leave the radiator cap off and run the engine in a well-ventilated area until the thermostat opens (coolant will begin flowing through the radiator and the upper radiator hose will become hot).
14 Turn the engine off and let it cool. Add more coolant mixture to bring the level back up to the lip on the radiator filler neck.
15 Squeeze the upper radiator hose to expel air, then add more coolant mixture if necessary. Replace the radiator cap.
16 Start the engine, allow it to reach normal operating temperature and check for leaks.

26 Steering and suspension check (every 30,000 miles or 24 months)

Note: *The steering linkage and suspension components should be checked periodically. Worn or damaged suspension and steering linkage components can result in excessive and abnormal tire wear, poor ride quality and vehicle handling and reduced fuel economy. For detailed illustrations of the steering and suspension components, refer to Chapter 10.*

Shock absorber check

1   Park the vehicle on level ground, turn the engine off and set the parking brake. Check the tire pressures.
2   Push down at one corner of the vehicle, then release it while noting the movement of the body. It should stop moving and come to rest in a level position within one or two bounces.
3   If the vehicle continues to move up-and-down or if it fails to return to its original position, a worn or weak shock absorber is probably the reason.
4   Repeat the above check at each of the three remaining corners of the vehicle.
5   Raise the vehicle and support it securely on jackstands.
6   Check the shock absorbers and front struts for evidence of fluid leakage. A light film of fluid is no cause for concern. Make sure that any fluid noted is from the shocks and not from some other source. If leakage is noted, replace the shocks or struts as a set.
7   Check the shocks and struts to be sure that they are securely mounted and undamaged. Check the upper mounts for damage and wear. If damage or wear is noted, replace the shocks or struts as a set (front or rear).
8   If the shocks or struts must be replaced, refer to Chapter 10 for the procedure.

Steering and suspension check

Refer to illustrations 26.9a, 26.9b and 26.9c

9   Visually inspect the steering and suspension system components for damage and distortion. Look for damaged seals, boots and bushings and leaks of any kind **(see illustrations)**.
10 Grasp each front tire at the front and rear edges, push in at the front, pull out at the rear and feel for play in the steering system components. If any freeplay is noted, check the steering gear mounts and the tie-rod ends for looseness.
11 Clean the lower end of the steering knuckle. Have an assistant grasp the lower edge of the tire and move the wheel in-and-out while you look for movement at the steering knuckle-to-control arm balljoint. If there is any movement the suspension balljoint(s) must be replaced.
12 Additional steering and suspension system information and illustrations can be found in Chapter 10.

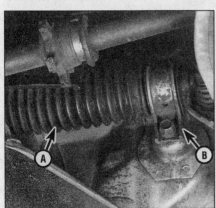

26.9a Check the steering gear boots (A) for cracks and leaking steering fluid, then check the steering gear mounts (B) for looseness

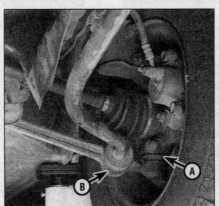

26.9b Check the lower balljoints (A) for torn grease seals and the lower stabilizer bar bushings (B) for deterioration

26.9c Also check the upper stabilizer bar bushings (arrow) for deterioration

27 Driveaxle boot check (every 30,000 miles or 24 months)

Refer to illustration 27.2

1 The driveaxle boots are very important because they prevent dirt, water and foreign material from entering and damaging the constant velocity (CV) joints. Oil and grease can cause the boot material to deteriorate prematurely, so it's a good idea to wash the boots with soap and water. Because it constantly pivots back and forth following the steering action of the front hub, the outer CV boot wears out sooner and should be inspected regularly.
2 Inspect the boots for tears and cracks as well as loose clamps **(see illustration)**. If there is any evidence of cracks or leaking lubricant, they must be replaced as described in Chapter 8.

28 Fuel system check (every 30,000 miles or 24 months)

Refer to illustration and 28.6

Warning: *Certain precautions should be observed when inspecting or servicing the fuel system components. Work in a well ventilated area and do not allow open flames (cigarettes, appliance pilot lights, etc.) near the work area. Mop up spills immediately and do not store fuel soaked rags where they could ignite. It is a good idea to keep a dry chemical (Class B) fire extinguisher near the work area any time the fuel system is being serviced.*

1 If you smell gasoline while driving or after the vehicle has been sitting in the sun, inspect the fuel system immediately.
2 Remove the gas filler cap and inspect if for damage and corrosion. The gasket should have an unbroken sealing imprint. If the gasket is damaged or corroded, remove it and install a new one.
3 Inspect the fuel feed and return lines for cracks. Make sure that the threaded flare-nut type connectors which secure the metal fuel lines to the fuel injection system are tight.
4 Since some components of the fuel system - the fuel tank and part of the fuel feed and return lines, for example - are underneath the vehicle, they can be inspected more easily with the vehicle raised on a hoist. If that's not possible, raise the vehicle and support it securely on jackstands.
5 With the vehicle raised and safely supported, inspect the gas tank and filler neck for punctures, cracks and other damage. The hose connecting the filler neck to the tank is particularly critical. Sometimes this hose will leak because of loose clamps or deteriorated rubber. These are problems a home mechanic can usually rectify. **Warning:** *Do not, under any circumstances, try to repair a fuel tank (except rubber components). A welding torch or any open flame can easily cause fuel vapors inside the tank to explode.*
6 Carefully check all rubber hoses and metal lines leading away from the fuel tank. Check for loose connections, deteriorated hoses, crimped lines and other damage **(see illustration)**. Carefully inspect the lines from the tank to the fuel injection system. Repair or replace damaged sections as necessary (see Chapter 4).

27.2 Flex the driveaxle boots by hand to check for cracks and/or leaking grease

29 Exhaust system check (every 30,000 miles or 24 months)

Refer to illustrations 29.2a and 29.2b

1 With the engine cold (at least three hours after the vehicle has been driven), check the complete exhaust system from the engine to the end of the tailpipe. Ideally, the inspection should be done with the vehicle on a hoist to permit unrestricted access. If a hoist isn't available, raise the vehicle and support it securely on jackstands.
2 Check the exhaust pipes and connections for evidence of leaks, severe corrosion and damage. Make sure that all brackets and hangers are in good condition and tight **(see illustrations)**.
3 At the same time, inspect the underside of the body for holes, corrosion, open seams, etc. which may allow exhaust gases to enter the passenger compartment. Seal all body openings with silicone or body putty.
4 Rattles and other noises can often be traced to the exhaust system, especially the mounts and hangers. Try to move the pipes, muffler and catalytic converter. If the components can come in contact with the body or suspension parts, secure the exhaust system with new mounts.
5 Check the running condition of the engine by inspecting inside the end of the tailpipe. The exhaust deposits here are an indication of engine state-of-tune. If the pipe is black and sooty or coated with white deposits, the engine may need a tune-up, including a thorough fuel system inspection and adjustment.

28.6 Carefully inspect all fuel lines (arrow) leading from the gas tank to the engine - check the hoses for cracks and make sure the clamps are tight

29.2a Check the flange connections (arrow) for exhaust leaks - also check that the retaining nuts are securely tightened

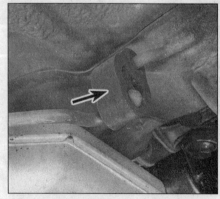

29.2b Check the exhaust system hangers (arrow) for damage and cracks

Chapter 1 Tune-up and routine maintenance

30.6 Remove the transaxle drain plug

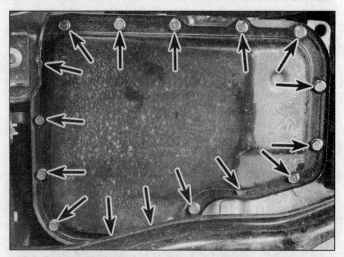

30.7 Remove the bolts (arrows) at the rear of the pan first, then loosen the bolts at the front of the pan a few turns - pry the pan free and let it hang down so the remaining fluid can drain

30 Automatic transaxle fluid and filter change (every 30,000 miles or 24 months)

Refer to illustrations 30.6, 30.7, 30.11 and 30.13

1   At the specified time intervals, the automatic transaxle fluid should be drained and replaced.
2   Before beginning work, purchase the specified transmission fluid (see *Recommended fluids and lubricants* at the front of this Chapter).
3   Other tools necessary for this job include jackstands to support the vehicle in a raised position, wrenches, drain pan capable of holding at least eight quarts, newspapers and clean rags.
4   The fluid should be drained immediately after the vehicle has been driven. Hot fluid is more effective than cold fluid at removing built up sediment. **Warning:** *Fluid temperature can exceed 350-degrees F in a hot transaxle. Wear protective gloves.*
5   After the vehicle has been driven to warm up the fluid, raise it and support it securely on jackstands.
6   Position a drain pan under the transaxle drain plug and remove the plug **(see illustration)**. Allow the oil to completely drain, then re-install the plug and tighten it to the torque listed in this Chapter's Specifications.
7   Move the drain pan under the transaxle pan and remove the rear and side pan mounting bolts **(see illustration)**.
8   Loosen the front pan bolts approximately four turns.
9   Carefully pry the transmission pan loose with a screwdriver, allowing the fluid to drain. Once the fluid has drained, remove the remaining bolts, pan and gasket.
10 Carefully clean the gasket surface of the transmission to remove all traces of the old gasket and sealant.
11 Remove the oil strainer/filter retaining bolts and lower the filter from the transaxle **(see illustration)**. Be careful when lowering the filter as it contains residual fluid.
12 Place the new filter in position, and install the bolts. Tighten the bolts to the torque listed in this Chapter's Specifications.
13 Carefully clean the gasket surfaces of the fluid pan, removing all traces of old gasket material. Wash the pan in clean solvent and dry it with compressed air. Be sure to clean the metal filings from the magnet **(see illustration)**.
14 Install a new gasket, place the fluid pan in position and install the bolts in their original positions. Tighten the bolts to the torque listed in this Chapter's Specifications.
15 Lower the vehicle.
16 With the engine off, add new fluid (approximately 3 qts) to the transaxle through the dipstick tube (see *Recommended fluids and lubricants* for the recommended fluid type and capacity). Use a funnel to prevent spills. It is best to add a little fluid at a time, continually checking the level with the dipstick (see Section 8). Allow the fluid time to drain into the pan.
17 Start the engine and shift the selector into all positions from P through L, then shift into P and apply the parking brake.
18 With the engine at operating temperature and idling, check the fluid level. Add fluid until the level is between the F and L marks on the dipstick. DO NOT OVERFILL.

30.11 Remove the filter bolts (arrows) and lower the filter (be careful, there will be some residual fluid)

30.13 Wash the pan in clean solvent and remove metal filings from the magnet (arrow)

1-30   Chapter 1   Tune-up and routine maintenance

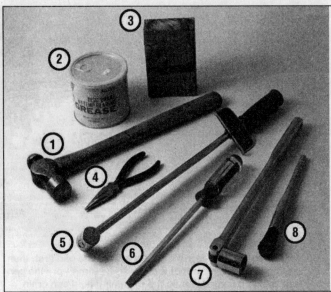

31.1 Tools and materials needed for front wheel bearing maintenance

1. **Hammer** - A common hammer will do just fine
2. **Grease** - High-temperature grease that is formulated specially for front wheel bearings should be used
3. **Wood block** - If you have a scrap piece of 2x4, it can be used to drive the new seal into the hub
4. **Needle-nose pliers** - Used to straighten and remove the cotter pin in the spindle
5. **Torque wrench** - This is very important in this procedure; if the bearing is too tight, the wheel won't turn freely - if it's too loose, the wheel will "wobble" on the spindle. Either way, it could mean extensive damage
6. **Screwdriver** - Used to remove the seal from the hub (a long screwdriver is preferred)
7. **Socket/breaker bar** - Needed to loosen the nut on the spindle if it's extremely tight
8. **Brush** - Together with some clean solvent, this will be used to remove old grease from the hub and spindle

31   Rear wheel bearing check, repack and adjustment (every 30,000 miles or 24 months)

Check and repack

Refer to illustrations 31.1, 31.6, 31.7, 31.8a, 31.8b, 31.8c, 31.11 and 31.15

1   In most cases the rear wheel bearings will not need servicing until the brake shoes or pads are changed. However, the bearings should be checked whenever the rear of the vehicle is raised for any reason. Several items, including a torque wrench and special grease, are required for this procedure **(see illustration)**.

2   With the vehicle securely supported on jackstands, spin each wheel and check for noise, rolling resistance and freeplay.

3   Grasp the top of each tire with one hand and the bottom with the other. Move the wheel in and out on the spindle. If there's any noticeable movement, the bearings should be checked and then repacked with grease, or replaced if necessary.

4   Engage the parking brake and remove the wheel.

5   Disengage the parking brake and remove the brake drum (see Chapter 9).

6   Dislodge the dust cap from the hub/drum assembly using a screwdriver or hammer and chisel **(see illustration)**.

7   If equipped with a cotter pin, straighten the bent ends of the cotter pin, then pull the cotter pin out of the nut lock **(see illustration)**. Discard the cotter pin and use a new one during reassembly. If the nut is staked in place, use a dull chisel to unstake the nut.

8   Remove the nut lock, nut and washer from the end of the spindle **(see illustrations)**. **Caution:** *On some models, the right-side wheel bearing retaining nut may be left-hand thread - turn clockwise to loosen.*

9   Pull the hub/drum assembly out slightly, then push it back into its original position. This should force the outer bearing off the spindle enough so it can be removed

10   Pull the drum assembly off the spindle.

11   Use a screwdriver or a seal puller tool to pry the seal out of the rear of the rotor **(see illustration)**. Note how the seal is installed.

12   Remove the inner wheel bearing from the hub/drum assembly.

13   Use solvent to remove all traces of the old grease from the bearings, hub and spindle. A small brush may prove helpful; however make sure no bristles from the brush embed themselves inside the bearing rollers. Allow the parts to air dry.

14   Carefully inspect the bearings for cracks, heat discoloration, worn rollers, etc. Check the bearing races inside the hub for wear and damage. If the bearing races are defective, the hubs should be taken to a machine shop with the facilities to remove the old races and press new ones in. Note that the bearings and races come as matched sets and old bearings should never be installed on new races.

15   Use high-temperature wheel bearing grease to pack the bearings.

31.6 Dislodge the dust cap by working around the outer circumference with a hammer and chisel

31.7 Remove the cotter pin and discard it - use a new one when the hub/drum assembly is reinstalled

31.8a Remove the nut lock (if equipped)

Chapter 1 Tune-up and routine maintenance

1-31

31.8b Remove the wheel bearing retaining nut. Caution: *On some models, the right-side wheel bearing retaining nut may be left-hand thread - turn clockwise to loosen*

31.8c Remove the retaining washer

Work the grease completely into the bearings, forcing it between the rollers, cone and cage from the back side **(see illustration)**.

16 Apply a thin coat of grease to the spindle at the outer bearing seat, inner bearing seat, shoulder and seal seat.

17 Put a small quantity of grease inboard of each bearing race inside the hub. Using your finger, form a dam at these points to provide extra grease availability and to keep thinned grease from flowing out of the bearing.

18 Place the grease-packed inner bearing into the rear of the hub and put a little more grease outboard of the bearing.

19 Place a new seal over the inner bearing and tap the seal evenly into place until it's flush with the hub/drum assembly.

20 Carefully place the hub/drum assembly onto the spindle and push the grease-packed outer bearing into position.

Adjustment

21 Install the washer and spindle nut. Tighten the nut only slightly (no more than 12 ft-lbs of torque).

22 Spin the hub/drum assembly in a forward direction while tightening the spindle nut to approximately 18-22 ft-lbs to seat the bearings and remove any grease or burrs which could cause excessive bearing play later.

23 Loosen the spindle nut 1/4-turn, then using your hand (not a wrench of any kind), tighten the nut until a slight drag is felt (approximately 3-7 in lbs). Install a new cotter pin (if equipped) through the hole in the spindle and the slots in the locknut. If the slots in the locknut don't line up with the hole in the spindle, turn the nut slightly until they do.

24 Check that the hub/disc assembly spins freely with no noticeable freeplay. If freeplay exists repeat Steps 22 and 23 until proper adjustment is obtained.

25 If equipped with a cotter pin, bend the ends of the cotter pin until they're flat against the nut. Cut off any extra length which could interfere with the dust cap. If the nut was staked in place, use a dull chisel to stake the nut.

26 Install the dust cap, lightly tapping it into place with a hammer.

27 Install the wheel on the hub/disc assembly and tighten the lug nuts.

28 Lower the vehicle.

32 Fuel filter replacement (every 60,000 miles or 48 months)

Warning: *Gasoline is extremely flammable, so take extra precautions when you work on any part of the fuel system. Don't smoke or allow open flames or bare light bulbs near the work area, and don't work in a garage where a natural gas-type appliance (such as a water heater or clothes dryer) with a pilot light is present. Since gasoline is carcinogenic, wear latex gloves when there's a possibility of being exposed to fuel, and, if you spill any fuel on your skin, rinse it off immediately with soap and water. Mop up any spills immediately and do not store fuel-soaked rags where they could ignite. When you perform any kind of work on the fuel system, wear safety glasses and have a Class B Type fire extinguisher on hand.*

1 Fuel filters often become clogged due to excessive dirt or foreign material in the fuel system and must be replaced according to the

31.11 Use a seal puller or large screwdriver to remove the inner grease seal - note the seal installed position

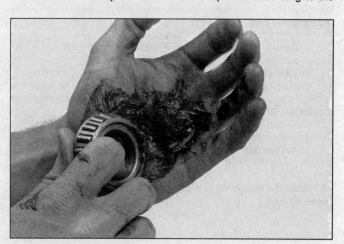

31.15 Work the grease completely into the bearing rollers

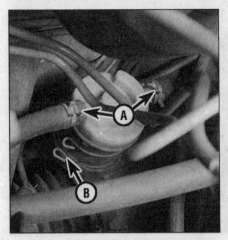

32.3 On carbureted models, detach the hose clamps (A) and the hoses, then release the retaining clips (B) to remove the fuel filter

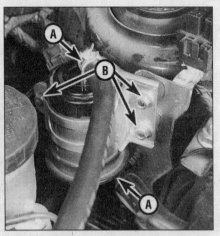

32.9 On fuel injected models, relieve the fuel system pressure as described in Chapter 4 - detach the hose clamps (A) and the hoses, then remove the retaining nuts (B) from the filter support bracket to remove the fuel filter

33.8 The manual transaxle drain plug (arrow) can accessed through the opening in the engine crossmember

maintenance interval suggestions.
2   Considering the volatile nature of gasoline, precautions should be exercised when a fuel filter is replaced. The job should be done with the engine cold or after sitting at least three hours. Be prepared to immediately clean up any spilled gasoline. In addition, you need to obtain a replacement filter for your specific vehicle.

Carbureted models

Refer to illustration 32.3
3   The canister filter is mounted in a bracket on the firewall on the left side of the engine compartment **(see illustration)**.
4   To replace the filter, release the hose clamps and slide them down the hoses, past the fittings on the filter.
5   Carefully twist and pull on the hoses to separate them from the filter. If the hoses are in bad shape, now would be a good time to replace them with new ones. Slide off the old clamps and install new ones.
6   Pull the filter out of the clip and install the new one, then hook up the hoses and reposition the clamps. Note that the arrow on the filter must point in the direction of fuel flow (toward the carburetor). Start the engine and check carefully for leaks at the filter hose connections.

Fuel-injected models

Refer to illustration 32.9
Warning: *Refer to Chapter 4 and depressurize the fuel system before removing the filter!*
7   Disconnect the negative battery cable.
8   The canister filter is mounted in a bracket on the left side of the engine compartment.
9   Disconnect the fuel hoses from the fuel filter **(see illustration)**.
10 Detach the retaining nuts from the filter support bracket and remove the old filter and the filter retaining bracket assembly **(see illustration 32.9)**.
11 Note that the inlet and outlet lines are clearly labeled and that the flanged end of the filter faces down.
12 Install the new filter and bracket assembly and tighten the retaining nuts securely. Make sure that the new filter is installed flanged end down.
13 Start the engine and check carefully for leaks at the filter hose connections.

33 Manual transaxle lubricant change (every 60,000 miles or 48 months)

Refer to illustration 33.8
1   At the specified time intervals, the manual transaxle lubricant should be drained and replaced.
2   Before beginning work, purchase the specified lubricant (see *Recommended fluids and lubricants* at the front of this Chapter) and a new drain plug washer/seal.
3   Other tools necessary for this job include jackstands to support the vehicle in a raised position, wrenches, drain pan capable of holding at least four quarts, newspapers and clean rags.
4   The oil should be drained immediately after the vehicle has been driven. Hot oil is more effective than cold oil at removing built up sediment. **Warning:** *Oil temperature can exceed 350-degrees F in a hot transaxle. Wear protective gloves.*
5   After the vehicle has been driven to warm up the oil, disconnect the speedometer cable and remove the speedometer driven gear assembly as described in Section 18.
6   Raise the vehicle and support it securely on jackstands. Make sure it is safely supported and as level as possible.
7   Move the necessary equipment under the vehicle, being careful not to touch any of the hot exhaust components.
8   Place the drain pan under the transaxle drain plug and loosen the drain plug **(see illustration)**.
9   Carefully unscrew the drain plug and washer with your fingers. Be careful not to burn yourself on the oil.
10 Allow the oil to drain completely. Clean the drain plug then reinstall it with a new washer. Tighten the drain plug to the torque listed in this Chapter's Specifications.
11 Lower the vehicle.
12 With the engine off, add new oil to the transaxle through the speedometer driven gear case hole (see *Recommended fluids and lubricants* at the front of this Chapter for oil type and transaxle capacity). Use a funnel to prevent spills. It is best to add a little oil at a time, continually checking the level (see Section 18).
13 Install the speedometer driven gear assembly and reconnect the speedometer cable.

Chapter 2 Part A Engine

Contents

	Section
Camshaft oil seal - replacement	8
Camshaft - removal, inspection and installation	10
CHECK ENGINE light	See Chapter 6
Crankshaft front oil seal - replacement	15
Cylinder compression check	See Chapter 2B
Cylinder head - removal and installation	12
Drivebelt - check, adjustment and replacement	See Chapter 1
Engine - removal and installation	See Chapter 2B
Engine mounts - check and replacement	18
Engine oil and filter change	See Chapter 1
Engine overhaul - general information	See Chapter 2B
Exhaust manifold - removal and installation	6
Flywheel/driveplate - removal and installation	16
Oil pressure check	See Chapter 2B
Oil pump - removal, inspection and installation	14

	Section
General information	1
Intake manifold - removal and installation	5
Oil pan - removal and installation	13
Rear main oil seal - replacement	17
Repair operations possible with the engine in the vehicle	2
Rocker arm assembly and lash adjusters - removal, inspection and installation	9
Spark plug replacement	See Chapter 1
Timing belt and sprockets - removal, inspection and installation	7
Top Dead Center (TDC) - locating	3
Valve cover - removal and installation	4
Valve springs, retainers and seals - replacement	11
Valves - servicing	See Chapter 2B
Water pump - removal and installation	See Chapter 3

Specifications

General
Firing order	1-3-4-2
Displacement	79.3 cubic inches (1.3 liters)

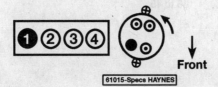

Cylinder location and distributor rotation

The blackened terminal shown on the distributor cap indicates the Number One spark plug wire position

Camshaft

Lobe height - intake and exhaust	
1988 and 1989	1.4185 to 1.4224 inches (36.029 to 36.129 mm)
Wear limit	1.4126 inches (35.879 mm) minimum
1990 to 1993	1.4331 to 1.4371 inches (36.403 to 36.503 mm)
Wear limit	1.4272 inches (36.253 mm) minimum
1994 to 1997	
Intake	1.4222 inches (36.124 mm)
Wear limit	1.4025 inches (35.624 mm) minimum
Exhaust	1.4332 inches (36.404 mm)
Wear limit	1.4135 inches (35.904 mm) minimum
Journal diameter	
Front and rear journals	1.7103 to 1.7112 inches (43.440 to 43.465 mm)
Center journal	1.7091 to 1.7100 inches (43.410 to 43.435 mm)
Out-of-round limit	0.002 inch (0.05 mm) maximum
Journal oil clearance	0.006 inch (0.15 mm) maximum
Endplay	0.008 inch (0.20 mm) maximum

Oil pump
Rotor tip clearance	0.0078 inch (0.198 mm) maximum
Outer rotor-to-housing clearance	0.0087 inch (0.22 mm) maximum
Endplay	0.005 inch (0.14 mm) maximum

Rocker arm and shaft
Oil clearance	0.004 inch (0.10 mm) maximum

Valve clearance
1988 and 1989 (intake and exhaust, warm)	0.012 inch (0.3 mm)

Torque specifications
	Ft-lbs (unless otherwise indicated)
Camshaft sprocket bolt	36 to 45
Camshaft thrust plate bolt	69 to 95 in-lbs
Crankshaft pulley bolts	109 to 152 in-lbs
Crankshaft sprocket bolt	80 to 85
Cylinder head bolts	
Step 1	35 to 40
Step 2	56 to 60
Engine mount bolts/nuts	
Transaxle crossmember-to-chassis bolts	47 to 66
Transaxle mount through-bolts	39 to 47
Front transaxle mount to crossmember nuts	27 to 38
Rear transaxle mount to crossmember nut	21 to 34
Right engine mount throughbolt	48 to 69
Right engine mount to engine bracket nuts	39 to 47
Exhaust manifold bolts/nuts	144 to 204 in-lbs
Exhaust manifold heat shield bolts/nuts	144 to 204 in-lbs
Exhaust pipe to manifold nuts	23 to 34
Flywheel cover bolts	61 to 87 in-lbs
Flywheel/driveplate bolts	71 to 76
Intake manifold bolts	14 to 20
Intake manifold brace bolts	22 to 34
Oil pan bolts	69 to 78 in-lbs
Oil pump bolts	14 to 19
Oil pump strainer	71 to 97 in-lbs
Rocker arm shaft bolts	16 to 21
Tensioner pulley bolt	14 to 19
Timing belt cover bolts	69 to 95 in-lbs
Valve cover bolts	44 to 80 in-lbs
Water pump pulley	36 to 45

1 General information

This Part of Chapter 2 is devoted to in-vehicle repair procedures for all engines. All information concerning engine removal and installation and engine block and cylinder head overhaul can be found in Part B of this Chapter.

The following repair procedures are based on the assumption that the engine is installed in the vehicle. If the engine has been removed from the vehicle and mounted on a stand, many of the Steps outlined in this Part of Chapter 2 will not apply.

The Specifications included in this Part of Chapter 2 apply only to the procedures contained in this Part. Part B of Chapter 2 contains the Specifications necessary for cylinder head and engine block rebuilding.

2 Repair operations possible with the engine in the vehicle

Many major repair operations can be accomplished without removing the engine from the vehicle.

Clean the engine compartment and the exterior of the engine with some type of degreaser before any work is done. It will make the job easier and help keep dirt out of the internal areas of the engine.

Depending on the components involved, it may be helpful to remove the hood to improve access to the engine as repairs are performed (refer to Chapter 11 if necessary). Cover the fenders to prevent damage to the paint. Special pads are available, but a substitute such as a thick bedspread or blanket will also work.

If vacuum, exhaust, oil or coolant leaks develop, indicating a need

Chapter 2 Part A Engine

3.6 Mark the distributor where the rotor is pointing at the number one cylinder position

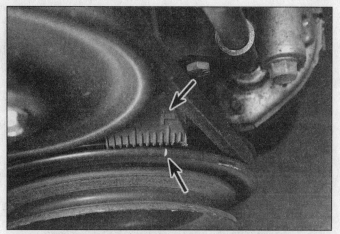

3.8 On 1988 and 1989 engines, align the crankshaft pulley notch with the "T" on the timing cover scale (arrows)

for gasket or seal replacement, the repairs can generally be made with the engine in the vehicle. The intake and exhaust manifold gaskets, oil pan gasket, crankshaft oil seals and cylinder head gasket are all accessible with the engine in place.

Exterior engine components, such as the intake and exhaust manifolds, the oil pan, the oil pump, the water pump, the starter motor, the alternator, the distributor and the fuel system components can be removed for repair with the engine in place.

Since the cylinder head can be removed without pulling the engine, camshaft and valve component servicing can also be accomplished with the engine in the vehicle. Replacement of the timing belt and sprockets is also possible with the engine in the vehicle.

In extreme cases caused by a lack of necessary equipment, repair or replacement of piston rings, pistons, connecting rods and rod bearings is possible with the engine in the vehicle. However, this practice is not recommended because of the engine cleaning and other preparation work, such as driveshaft, steering, stabilizer bar, engine mount members that may require partial disassembly or removal for access to the engine.

3 Top Dead Center (TDC) - locating

Refer to illustrations 3.6 and 3.8

Note: *The following procedure is based on the assumption that the distributor is correctly installed. If you are trying to locate TDC to install the distributor correctly, piston position must be determined by feeling for compression at the number one spark plug hole, then aligning the ignition timing marks as described in Step 8.*

1 Top Dead Center (TDC) is the highest point in the cylinder that each piston reaches traveling up-and-down as the crankshaft turns. Each piston reaches TDC on the compression stroke and again on the exhaust stroke, but TDC generally refers to piston position on the compression stroke.

2 Positioning the piston(s) at TDC is an essential part of many procedures such as camshaft and timing belt/sprocket removal and distributor removal.

3 Before beginning this procedure, be sure to place the transmission in Neutral and apply the parking brake or block the rear wheels. Also, disable the ignition system by detaching the primary (low voltage) wires from the coil (see Chapter 5). Remove the spark plugs (see Chapter 1).

4 In order to bring any piston to TDC, the crankshaft must be turned using one of the methods outlined below. When looking at the front of the engine, normal crankshaft rotation is clockwise.

  a) *The preferred method is to turn the crankshaft with a socket and ratchet attached to the bolt threaded into the front of the crankshaft.*

  b) *A remote starter switch, which may save some time, can also be used. Follow the instructions included with the switch. Once the piston is close to TDC, use a socket and ratchet as described in the previous paragraph.*

  c) *If an assistant is available to turn the ignition switch to the Start position in short bursts, you can get the piston close to TDC without a remote starter switch. Make sure your assistant is out of the vehicle, away from the ignition switch, then use a socket and ratchet as described in Paragraph a) to complete the procedure.*

5 Note the position of the terminal for the number one spark plug wire on the distributor cap. If the terminal is not marked, follow the plug wire from the number one cylinder spark plug to the cap.

6 Use a felt-tip pen or chalk to make a mark on the distributor body and the cap - directly at the terminal **(see illustration)**.

7 Detach the cap from the distributor and set it aside (see Chapter 1 if necessary).

8 Turn the crankshaft (see Step 3 above) until the notch in the crankshaft sprocket is aligned with the mark on the timing belt cover **(see illustration)**. **Note:** *On 1988 and 1989 engines, align the notch with the "T" mark on the timing cover. On later engines align the yellow notch on the pulley with the with the pointer (not the scale) on the timing cover. There is also a white timing mark (10 degrees) on the later-model crankshaft pulley, but this is for ignition timing only. Use the yellow mark for finding TDC.*

9 Look at the distributor rotor - it should be pointing directly at the mark you made on the distributor body.

10 If the rotor is 180-degrees off, the number one piston is at TDC on the exhaust stroke.

11 To get the piston to TDC on the compression stroke, turn the crankshaft one complete turn (360-degrees) clockwise. The rotor should now be pointing at the mark on the distributor. When the rotor is pointing at the number one spark plug wire terminal in the distributor cap and the ignition timing marks are aligned, the number one piston is at TDC on the compression stroke. **Note:** *If it is impossible to align the ignition timing marks when the rotor is pointing at the mark on the distributor body, the timing belt may have jumped the teeth on the sprockets or may have been installed incorrectly.*

12 After the number one piston has been positioned at TDC on the compression stroke, TDC for any of the remaining pistons can be located by turning the crankshaft and following the firing order. Mark the remaining spark plug wire terminal locations on the distributor body just like you did for the number one terminal, then number the marks to correspond with the cylinder numbers. As you turn the crankshaft, the rotor will also turn. When it's pointing directly at one of the marks on the distributor, the piston for that particular cylinder is at TDC on the compression stroke.

2A-4   Chapter 2 Part A   Engine

4.4 Remove the two upper timing belt cover bolts

4.6a Remove the valve cover bolts (arrows) (carbureted engine shown)

4   Valve cover - removal and installation

Removal

Refer to illustrations 4.4, 4.6a and 4.6b

1   Disconnect the negative cable from the battery.
2   Detach the Positive Crankcase Ventilation (PCV) valve and breather hoses from the valve cover.
3   Disconnect the spark plug wires from the clips.
4   Remove the two upper bolts from the upper timing belt cover **(see illustration)**. Loosen, but do not remove the lower timing belt cover bolts.
5   On fuel-injected engines, refer to Chapter 4 and remove the air cleaner tube from the throttle body and remove the throttle cable from its bracket on the valve cover.
6   Remove the valve cover bolts **(see illustrations)**.
7   Lift the valve cover from the cylinder head. If it sticks, gently tap it with a rubber mallet or a hammer and a block of wood. Don't pry between the sealing surfaces.
8   Visually check the valve cover gasket for damage and to ensure that has not hardened and is still flexible. Save it for reuse if satisfactory.

Installation

Refer to illustration 4.9

9   If a new valve cover gasket is being installed, clean the groove of the valve cover. Apply silicone sealant in the groove of the valve cover and press the new gasket into the groove **(see illustration)**.
10   If the valve cover gasket is being reused, remove the gasket, make sure the groove is clean and the gasket is clean. Apply silicone sealant in the groove of the valve cover, and reinstall the gasket.
11   Position the valve cover in place and insert the bolts by hand, starting the threads several turns before using a wrench.
12   Tighten the valve cover bolts/screws in several steps to the torque listed in this Chapter's Specifications.
13   Reinstallation of the remaining parts is the reverse of removal.
14   Run the engine and check for oil leaks.

5   Intake manifold - removal and installation

Removal

Refer to illustrations 5.6a, 5.6b, 5.9a and 5.9b

1   Disconnect the negative cable from the battery.
2   Drain the cooling system (see Chapter 1).
3   Refer to Chapter 4 and relieve the fuel system pressure (fuel-injected models only).
4   Disconnect the air intake hose from the throttle body, the electrical connector at the throttle position sensor, the electrical connectors at each fuel injector, and tag and disconnect all other hoses and wires connected to the fuel rail or intake manifold (see Chapter 4). **Note:** *On carbureted models, disconnect the fuel, electrical and vacuum connections at the carburetor (see Chapter 4).*

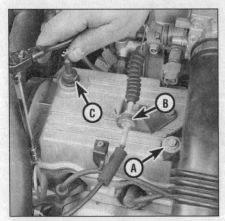

4.6b On fuel injected models, remove the air cleaner tube (A), the throttle linkage (B) and the PCV hose/valve (C), then remove the valve cover bolts

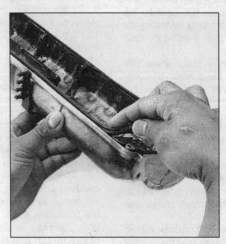

4.9 Press the valve cover gasket into the groove in the valve cover

5.6a Remove the bolts (arrows) holding the intake manifold brace (shown from below)

Chapter 2 Part A Engine

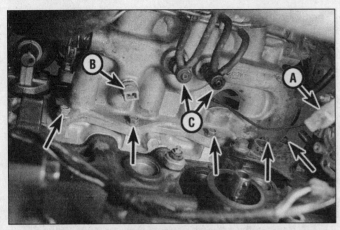

5.6b On carbureted models, disconnect the red wire connector (A), the temperature sensor (B), and the two vacuum connectors (C), then remove the lower bolts (arrows)

5.9a Upper intake manifold mounting bolts (arrows) - carbureted models

5 Disconnect the accelerator cable (see Chapter 4).
6 Raise the vehicle and support it securely on jackstands. Under the intake manifold, remove the intake manifold bracket, as applicable **(see illustration)**. On carbureted models, disconnect the vacuum hoses and coolant temperature sensor from below **(see illustration)**.
7 The intake manifold on carbureted models is one-piece. On fuel-injected models, it consists of an upper and lower section, with the upper section being the plenum. The intake manifold can be removed from the engine as a unit, complete with throttle body and injectors, or the plenum can be unbolted first and then the lower manifold removed from the cylinder head. **Note:** *On carbureted models, the carburetor can be left in place when unbolting the intake manifold.*
8 Remove the lower intake manifold mounting bolts, while supporting the intake manifold from above the engine
9 Lower the vehicle and remove the upper bolts and the intake manifold **(see illustrations)**.

Installation

10 Carefully use a gasket scraper to remove all traces of old gasket material and any sealant from the manifold and cylinder head, then clean the mating surfaces with gasket cleaner or solvent - be careful to not gouge the gasket surfaces when cleaning. If the gasket was leaking, have the manifold checked for warpage at an automotive machine shop and resurfaced if necessary.
11 Install a new gasket, then position the manifold on the head and install the nuts/bolts. **Note:** *If the plenum was separated from the intake manifold, clean the mating surfaces of the two sections, install a new gasket and bolt the plenum to the intake manifold and tighten to Specifications.*
12 Tighten the manifold-to-engine nuts/bolts in three or four equal steps to the torque listed in this Chapter's Specifications. Work from the center out towards the ends, while alternating upper to lower bolts/nuts to avoid warping the manifold.
13 Reinstall the remaining parts in the reverse order of removal.
14 Before starting the engine, check the throttle linkage for smooth operation.
15 Check coolant level. Run the engine and check for coolant and vacuum leaks.
16 Road test the vehicle and check for proper operation of all accessories, including the cruise control system (if equipped).

6 Exhaust manifold - removal and installation

Refer to illustrations 6.3, 6.6 and 6.7
Warning: *The engine must be completely cool before beginning this procedure.*

Removal

1 Disconnect the negative cable from the battery.
2 Unplug the oxygen sensor electrical connector from the exhaust manifold. If you are installing a new manifold, remove the sensor (see Chapter 6).
3 Remove the heat shield bolts and remove the heat shield from the manifold **(see illustration)**.

5.9b Upper intake manifold mounting bolts (arrows) - fuel-injected models

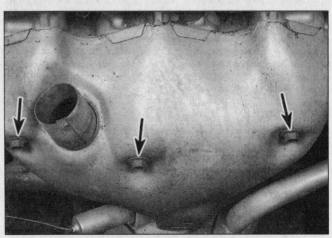

6.3 Remove the exhaust manifold heat shield bolts (arrows)

6.6 Remove the nuts and washers connecting the exhaust pipe to the manifold

6.7 Remove the exhaust manifold nuts and bolts (arrows)

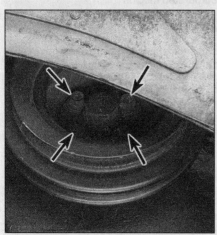

7.7 Remove the four bolts (arrows) and the crankshaft outer pulley retaining washer

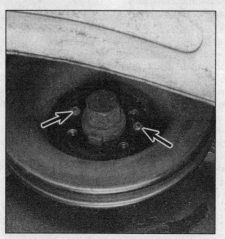

7.8 Remove these two screws to remove the inner pulley and its inner spacer

7.9 Timing belt cover bolt locations (arrows)

4   Apply penetrating oil to the exhaust manifold mounting bolts/nuts and to the exhaust pipe flange nuts.
5   Raise the vehicle and support it securely on jackstands (see Chapter 1).
6   Disconnect the exhaust pipe from the exhaust manifold **(see illustration)**. Lower the vehicle.
7   Remove the manifold bolts/nuts and detach the manifold from the cylinder head **(see illustration)**. **Note:** *If any bolts/nuts are difficult to remove, reapply penetrating oil to the bolts/nuts and let them soak for at least 15 minutes. If any bolts or studs break during removal, you may be able to use a vise-grip pliers after the manifold is removed to unscrew the broken bolt/stud. If unable to remove the broken bolt/stud, see your automotive parts store for stud removal tools. Replace any damaged parts with factory parts, or parts specifically designed for exhaust system application.*

Installation

8   Use a scraper to remove all traces of old gasket material and carbon deposits from the manifold and cylinder head mating surfaces. If the gasket was leaking, have the manifold checked for warpage at an automotive machine shop and resurfaced if necessary. **Caution:** *When scraping, be very careful not to gouge or scratch the delicate aluminum cylinder head manifold mounting surface.*
9   Position a new exhaust manifold gasket over the studs on the cylinder head.
10 Install the manifold and thread the mounting bolts/nuts into place.
11 Working from the center out, tighten the bolts/nuts to the torque listed in this Chapter's Specifications in several equal steps.

12 Reinstall the remaining parts in the reverse order of removal. If reinstalling the oxygen sensor, use a special anti-seize thread lubricant available at your automotive parts store.
13 Run the engine and check for exhaust leaks.

7   Timing belt and sprockets - removal, inspection and installation

Removal

Refer to illustrations 7.7, 7.8, 7.9, 7.10a, 7.10b, 7.10c and 7.12
1   Disconnect the negative cable from the battery
2   Block the rear wheels and set the parking brake.
3   Remove the splash shield from under the right-front fender (see Chapter 11).
4   Remove the power steering and air conditioner drivebelt, if applicable (see Chapter 1).
5   Remove the alternator drivebelt (see Chapter 1). Remove the spark plugs from all cylinders.
6   Refer to Chapter 3 and remove the water pump pulley bolts and remove the water pump pulley.
7   Hold the crankshaft pulley with a strap wrench to keep it from turning while removing the four retaining washer bolts **(see illustration)**. Also loosen the crankshaft sprocket bolt while holding the pulley. **Note:** *If necessary to prevent the crankshaft from turning, remove the starter motor (see Chapter 5) and carefully wedge a screwdriver between the ring gear teeth and the engine block. Do not damage the*

Chapter 2 Part A Engine

7.10a Install the crankshaft sprocket bolt and rotate the crankshaft with a socket and breaker bar

7.10b Timing belt camshaft sprocket alignment marks (arrows)

7.10c Timing belt crankshaft sprocket alignment marks (arrows)

7.12 To release the tension on the timing belt, loosen the timing belt tensioner pulley bolt, rotate the pulley away from the belt and retighten the bolt

7.15 Timing belt tensioner components

- A Tensioner
- B Spring
- C Spring cover

gear teeth when holding the flywheel/driveplate.

8 Remove the outer crankshaft pulley retainer and the outer pulley, then remove the two screws and the inner pulley and spacer **(see illustration)**.

9 Remove the timing belt upper cover (four bolts) and the (two bolts) lower cover **(see illustration)**. Remove the crankshaft sprocket bolt and the pulley hub, if equipped.

10 Temporarily install the crankshaft sprocket bolt (if removed) and using a socket and breaker bar, rotate the engine to align the crankshaft and camshaft sprocket timing marks **(see illustrations)**. The crankshaft sprocket pulley Woodruff key should be facing up (top).

11 If reusing the timing belt, paint match marks on the sprockets and belt and an arrow indicating direction of travel on the belt.

12 Loosen the timing belt tensioner **(see illustration)**, and temporarily tighten the tensioner with the spring fully extended. Remove the timing belt.

13 Remove the crankshaft sprocket bolt (if necessary).

14 If it is necessary to remove the camshaft sprocket for replacement, refer to Section 8 for the procedure. If it is necessary to remove the crankshaft sprocket, carefully pry it off with two prybars.

Inspection

Refer to illustrations 7.15 and 7.17

Caution: *Do not bend, twist or turn the timing belt inside out. Do not allow it to come in contact with oil, coolant or fuel. Do not use timing belt tension to keep the camshaft or crankshaft from turning when installing the sprocket bolt(s). Do not turn the crankshaft or camshaft more than a few degrees (necessary for tooth alignment) while the timing belt is removed.*

15 Remove the tensioner and check the bearing for smooth operation and excessive play. Inspect the spring for damage **(see illustration)**.

16 If the timing belt was broken during engine operation, the belt may have been fouled by debris or may have been damaged by a defective component in the area of the timing belt; check for belt material in the teeth of the sprockets. Any defective parts or debris in the sprockets must be cleaned out of all the sprockets before installing the new belt or the belt will not mesh properly when installed.

17 If the belt teeth are cracked or stripped off **(see illustration)**, the distributor, water pump, oil pump or camshaft(s) may have seized.

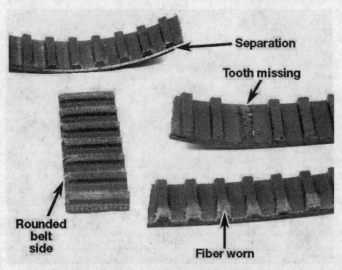

7.17 Check the timing belt for cracked and missing teeth

toward the oil pump body, and that the camshaft sprocket aligns with its dowel pin. Coat the threads of the crankshaft sprocket bolt with RTV sealant, and install the bolt. Tighten the crankshaft and camshaft sprocket bolts to the torque listed in this Chapter's Specifications.

24 Recheck the camshaft sprocket and crankshaft sprocket timing marks to be sure they are properly aligned **(see illustrations 7.10a and 7.10b)**. The crankshaft sprocket is aligned with the mark upwards. The camshaft sprocket has two marks, one at the 12 o'clock and one at the 3 o'clock position, when looking at the sprocket, which must be in alignment with the marks on the cylinder head. **Note:** *If necessary, rotate the camshaft sprocket slightly to achieve proper alignment.*

25 Slip the timing belt over the crankshaft sprocket and camshaft sprocket, and position the belt with no looseness on the side opposite the tensioner pulley. If the original belt is being reinstalled, align the marks made during removal with the marks on the sprockets, and be sure to install the timing belt so that it will rotate in the same direction as removed (the direction of rotation was marked during removal).

26 Install the timing belt guide (if equipped).

27 Loosen the tensioner pulley bolt to apply tension to the timing belt. **Note:** *The tensioner pulley spring applies the proper tension to the belt.*

28 Rotate the crankshaft two turns clockwise, align the crankshaft sprocket timing marks and verify the camshaft sprocket marks are aligned with the marks on the cylinder head **(see illustrations 7.10b and 7.10c)**. **Caution:** *If you feel resistance while rotating the engine by hand, do not continue. The valves may be contacting the pistons due to incorrect valve timing. Recheck the camshaft and crankshaft sprockets to be sure they are correctly aligned with their marks.*

29 Tighten the tensioner pulley bolt to Specifications.

30 Reinstall the remaining parts in the reverse order of removal **(see illustration)**. Run the engine and check for proper operation. **Caution:** *DO NOT start the engine until you are absolutely certain that the timing belt is installed correctly. Serious and costly engine damage could occur if the belt is improperly installed.*

18 If there is noticeable wear or cracks in the belt, check to see if there are nicks or burrs on the sprockets.

19 If there is wear or damage on only one side of the belt, check the belt guide and the alignment of all sprockets. Also check the oil seals at the front of the engine and replace them if they are leaking.

20 Replace the timing belt with a new one if obvious wear or damage is noted or if it is the least bit questionable. Correct any problems which contributed to belt failure prior to belt installation. **Note:** *We recommend replacing the belt whenever it is removed, since belt failure can lead to expensive engine damage.*

Installation

Refer to illustration 7.30

21 Remove all dirt and oil from the timing belt area at the front of the engine.

22 If it was removed, install the tensioner and spring. The tensioner should be pulled back against spring tension with the spring fully extended and the tensioner bolt temporarily tightened.

23 If they were removed, install the camshaft and crankshaft sprockets and the crankshaft pulley hub (if equipped). Make sure the crankshaft sprocket Woodruff key is installed with the tapered side

8 Camshaft oil seal - replacement

Refer to illustrations 8.4, 8.5, 8.7 and 8.8

1 Disconnect the cable from the negative battery terminal.

2 Block the rear wheels and set the parking brake. Refer to Section 7 and remove the timing belt.

3 Position cylinder number one to TDC on the compression stroke (see Section 3).

4 Remove the valve cover (see Section 4). Hold the camshaft's hex portion with a wrench while removing the camshaft sprocket bolt **(see illustration)**. Slip the camshaft sprocket from the camshaft.

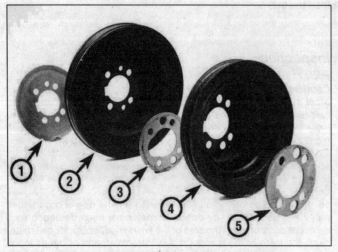

7.30 Crankshaft pulley assembly components

1 Timing belt guide
2 Inner crankshaft pulley
3 Inner pulley spacer
4 Outer pulley
5 Outer pulley retainer

8.4 Hold the camshaft from turning with a wrench (A) while removing the camshaft sprocket bolt (B)

Chapter 2 Part A Engine

8.5 Carefully pry the camshaft seal out of the bore - DO NOT nick or scratch the camshaft or seal bore

8.7 Drive the new seal in squarely with a large socket, to the depth of the original seal

8.8 Align the camshaft sprocket with the dowel (arrow) on the camshaft

5 Carefully pry the camshaft seal out of the bore using a thin screwdriver (see illustration). Another method is to use a deep socket the size of the seal and drive the seal into the cylinder head until it comes out on the other side. It will be loose over the camshaft. Cut one side of the seal with side-cutter pliers and remove the old seal. **Caution:** *DO NOT nick or scratch the camshaft or seal bore.*

6 Apply some clean engine oil to the lip of the new camshaft oil seal. Push the seal in slightly by hand.

7 Hold a short length of pipe or deep socket sized to fit the camshaft oil seal bore against the seal, and lightly tap the seal in, flush to the edge of the cylinder head or to the depth of the original seal **(see illustration)**. **Note:** *There is no shoulder or stop in the cylinder head for the seal to sit against. Do not drive the seal in beyond the point where it is flush with the front of the head and even all around.*

8 Reinstall the remaining parts in the reverse order of removal. **Note:** *When reinstalling the camshaft sprocket, make sure the hole in the sprocket lines up with the dowel pin on the camshaft* **(see illustration)**.

9 Run the engine and check for proper operation.

9 Rocker arm assembly and lash adjusters - removal, inspection and installation

Removal

Refer to illustrations 9.4, 9.5a, 9.5b and 9.6

1 Remove the valve cover (see Section 4).

2 Number or mark the components (rocker arms, spacers, shafts and rocker arm bolts seats) before removal to be sure that the parts will be reinstalled in the same location when reassembled. **Note:** *This engine uses rocker arms on two rocker shafts. On 1988 and 1989 models, the lash is mechanically adjusted, while later models use hydraulic lash adjusters. The hydraulic lash adjusters eliminate the need for valve lash adjusting screws or shims. After initial installation, no further valve adjustment is necessary.*

3 Push down on each rocker arm with your fingers. If the hydraulic lash adjusters allow movement, that lash adjuster should be replaced.

4 Loosen the rocker arm bolts in the sequence shown in two or three steps **(see illustration)**.

5 Remove the intake and exhaust rocker arms and rocker shaft assemblies. Mark or otherwise store the rocker arms and spacers (springs on 1988 and 1989 models) for later reinstallation in the same locations from which they were removed **(see illustrations)**. **Note:** *The*

9.4 Rocker arm/shaft bolt loosening/tightening sequence

9.5a Rocker arm shaft assembly - 1988 and 1989 models use springs between the rockers

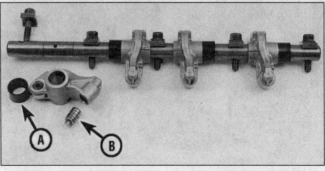

9.5b 1990 and later models use spacers (A) and hydraulic lash adjusters (B)

Chapter 2 Part A Engine

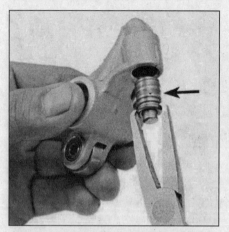

9.6 Gently pull the lash adjuster (arrow) from the rocker arm

9.8a Measure the rocker arm shaft outer diameter

9.8b Measure the rocker arm inside diameter - subtract the shaft diameter from the rocker inside diameter to determine the oil clearance

intake and exhaust assemblies are different. If the two shafts get mixed up, measure the distance between the last two (rear of the engine) oil holes in each shaft. The shaft with the closely-spaced holes is the exhaust shaft.

6 Pull the hydraulic lash adjusters from their rocker arms, keeping them identified as to which rocker arm they were originally installed (see illustration).

Inspection

Refer to illustrations 9.8a and 9.8b

7 Check each rocker arm for wear, cracks and other damage, especially where the camshaft contacts the rocker arm.

8 Clean and inspect the rocker arm shafts for signs of galling or excessive wear. Shiny spots where the rocker arms ride is normal. Slide your fingers along the oiled shaft to feel for significant wear. replace the shaft if there are noticeable ridges. Using a micrometer, measure the rocker shaft diameter, then the inside diameter of the rocker arms (see illustrations). The difference between the two is the shaft-to-rocker clearance. Compare to this Chapter's Specifications, and replace the shafts and rockers if the clearance is beyond Specifications.

9 Inspect the hydraulic lash adjusters for signs of abnormal wear or fractures. Make sure the O-ring is intact, with no cuts or burrs.

Installation

Refer to illustration 9.12

10 If the hydraulic lash adjusters were removed from the rocker arms, or are being replaced, fill the rocker arm cavity with clean engine oil, coat the lash adjusters with clean engine oil, and install them into the rocker arm, making certain not to damage or twist the O-rings.

11 Assemble and install the rocker arm assemblies as they were removed, with all components lubricated with clean engine oil. Make sure that the rocker shaft oil holes face downward. **Note:** *The intake and exhaust assemblies are different. If the two shafts get mixed up, measure the distance between the last two (rear of the engine) oil holes in each shaft. The shaft with the closely-spaced holes is the exhaust shaft.*

12 Tighten the rocker arm bolts to the torque listed in this Chapter's Specifications in several steps and in the recommended sequence (see illustration 9.4). On 1988 and 1989 models, use pliers to push the springs away from the rocker-shaft bolt saddles while tightening the rocker shaft mounting bolts (see illustration).

13 Install the camshaft sprocket (see Section 8) and tighten the bolt to the torque listed in this Chapter's Specifications.

14 Install the timing belt (see Section 7).

15 The remainder of installation is the reverse of the removal procedure. On 1988 and 1989 models, follow the procedure in Chapter 1 for adjusting the valve lash.

10 Camshaft - removal, inspection and installation

Removal

Refer to illustrations 10.5, 10.6 and 10.8

1 Remove the valve cover (see Section 4).

9.12 On 1988 and 1989 models, push back the springs for clearance when tightening the rocker-shaft bolts

10.5 Pry the camshaft back-and-forth to check the endplay (thrust clearance)

Chapter 2 Part A Engine

10.6 Remove the bolt and the camshaft thrust plate (arrow)

10.8 Pull the camshaft straight out of the cylinder head toward the left side of the vehicle

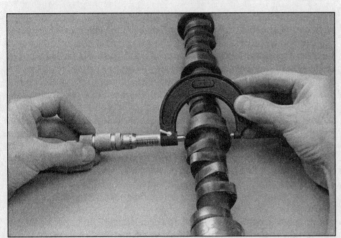

10.10a Measure each journal diameter with a micrometer (if any journal measures less than the specified limit, replace the camshaft)

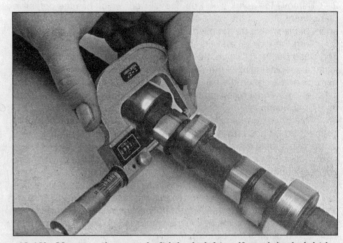

10.10b Measure the camshaft lobe heights - if any lobe height is less than the minimum listed in this Chapter's Specifications, replace the camshaft

2 Remove the distributor (see Chapter 5).
3 Remove the timing belt covers, timing belt, and camshaft sprocket (see Sections 7 and 8).
4 Refer to Section 9 and remove the rocker arms/shafts assembly.
5 Measure the thrust clearance (endplay) of the camshaft(s) with a dial indicator **(see illustration)**. If the clearance is greater than the value listed in this Chapter's Specifications, replace the camshaft thrust plate, camshaft, and/or the cylinder head (see Section 12).
6 Remove the bolt and the camshaft thrust plate **(see illustration)**.
7 Depending on the model, you may have to remove the air cleaner assembly (see Chapter 4) and/or the battery (see Chapter 5) for room to remove the camshaft from the back of the cylinder head (left side of vehicle).
8 Remove the camshaft oil seal (see Section 8), then pull the camshaft straight out of the cylinder head **(see illustration)**.

Inspection

Refer to illustrations 10.10a and 10.10b

9 Examine all parts, looking for signs of pitting, scoring or scuffing.
10 Measure the camshaft journals and camshaft lobes. Measure each camshaft bore inside diameter and subtract the camshaft journal outside diameter measurement to determine the camshaft journal oil clearance **(see illustrations)**. Compare your measurements to the values listed in this Chapter's Specifications. If you don't have access to the measuring tools required, they can be rented, or you can bring your components to an automotive machine shop for measurement.

Installation

11 Apply camshaft assembly lubricant to the camshaft lobes and bearing journals.
12 Install the camshaft in the cylinder head bolt the thrust plate in place.
13 Refer to Section 8 for installation of a new camshaft oil seal.
14 The remainder of installation is the reverse of the disassembly process. On 1988 and 1989 engines, refer to Chapter 1 for the procedure to adjust the valve clearances. On 1990 and later engines, no adjustment is necessary.
15 Start the engine and after it warms up, observe and listen for valvetrain noises, and oil or coolant leaks.

11 Valve springs, retainers and seals - replacement

Refer to illustrations 11.4, 11.8a, 11.8b, 11.9, 11.14 and 11.16
Note: *Broken valve springs and defective valve stem seals can be replaced without removing the cylinder head. Two special tools and a compressed air source are normally required to perform this operation, so read through this Section carefully and rent or buy the tools before beginning the job.*

1 Refer to Section 9 and remove the rocker arm assemblies. Keep all parts marked and in order so they may be reinstalled in their original location.

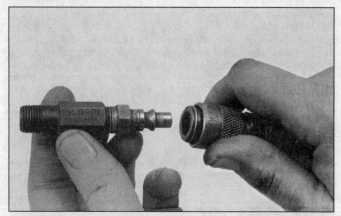

11.4 This is what the air hose adapter that threads into the spark plug hole looks like - they are commonly available from auto parts stores

11.8a Use a valve spring compressor to compress the springs - this is the lever type

2   Remove the spark plug from the cylinder which has the defective component. If all of the valve stem seals are being replaced, all of the spark plugs should be removed.

3   Turn the crankshaft until the piston in the affected cylinder is at Top Dead Center (TDC) on the compression stroke (refer to Section 3). If you are replacing all of the valve stem seals, begin with cylinder number one and work on the valves for one cylinder at a time. Move from cylinder-to-cylinder following the firing order sequence (see this Chapter's Specifications).

4   Thread an adapter into the spark plug hole **(see illustration)** and connect an air hose from a compressed air source. Most auto parts stores can supply the air hose adapter. **Note:** *Many cylinder compression gauges utilize a screw-in fitting that may work with your air hose quick-disconnect fitting.*

5   Apply compressed air to the cylinder. **Warning:** *The piston may be forced down by compressed air, causing the crankshaft to turn suddenly. If the wrench used when positioning the number one piston at TDC is still attached to the bolt in the crankshaft pulley end, damage or injury could occur if the crankshaft moves.*

6   The valves should now be held in place by the air pressure.

7   Use a large ratchet and socket to rotate the crankshaft in the normal direction of rotation (clockwise, when viewed from the timing belt end) until slight resistance is felt.

8   Stuff clean shop rags into any cylinder head holes above and below the valves to prevent parts and tools from falling into the engine, then use a valve-spring compressor to compress the spring. Remove the keepers with small needle-nose pliers or a magnet **(see illustrations)**. **Note:** *There are several types of valve spring compressors. The simplest type clamps to the spring and tightening a knob on top compresses the spring. A second type attaches with clips that go in opposite sides of the spring to hold it, while the spring is pushed down by a lever on top of the tool. Each will do the job, with the lever-type tool being faster but more expensive.*

9   Remove the spring retainer and valve spring (mark the top end of the valve spring for later reinstallation), then remove the spring seat (looks like a thinner version of the spring retainer, but fits between the bottom of the spring and the cylinder head) oil seal **(see illustration)**. **Note:** *If using air pressure to hold the valve(s) and this fails to hold the valve in the closed position during this operation, the valve face and/or seat is probably damaged. If so, the cylinder head will have to be removed (see Section 12).*

10 Wrap a rubber band or tape around the top of the valve stem so the valve won't fall into the combustion chamber, then release the air pressure.

11 Inspect the valve spring for cracks or damage and check that the free length is as listed in this Chapter's Specifications. Inspect the valve stem for damaged or rough face or an unevenly worn stem tip. Rotate the valve in its guide and check the end of the valve stem for eccentric movement, which would indicate that the valve is bent.

12 Move the valve up-and-down in the guide and make sure there is no binding. If the valve stem binds, either the valve is bent or the guide is damaged. Rock the valve stem side to side; clearance should not exceed the valve stem to valve guide clearance in this Chapter's Specifications. If the valve stem is defective, the cylinder head will have to be removed for repair of valves and/or valve guides.

11.8b With the spring compressed, remove the keepers from the valve stem with a magnet or small needle-nose pliers

11.9 Remove the valve guide oil seal with an oil seal removal tool or a pair of pliers

Chapter 2 Part A Engine

11.14 Gently tap the seal into place with a hammer and a deep socket

11.16 Apply a small dab of grease to each keeper as shown here before installation - it'll hold them in place on the valve stem as the spring is released

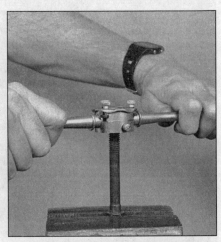

12.16 A die should be used to remove sealant and corrosion from the head bolt threads prior to installation

13 Reapply air pressure to the cylinder to retain the valve in the closed position, then remove the tape or rubber band from the valve stem.
14 Lubricate the valve stem with engine oil or engine assembly lubricant and install a new oil seal using an oil seal installer tool or deep socket **(see illustration)**, measuring the installed depth as necessary.
15 Install the spring seat and spring in position over the valve. Replace any springs not meeting the inspection above. Be sure the spring is installed as marked during removal (top of the spring is up) and also note that the end of the spring with a closer pitch is toward the cylinder head.
16 Install the valve spring retainer. Compress the valve spring and carefully position the keepers in the groove. Apply a small dab of grease to the inside of each keeper to hold it in place if necessary **(see illustration)**.
17 Remove the pressure from the spring tool and make sure the keepers are seated.
18 Disconnect the compressed air hose and remove the adapter from the spark plug hole.
19 Install the rest of the parts in the reverse order of the removal procedure.
20 Start and run the engine, then check for oil leaks and unusual sounds coming from the valve cover area.

12 Cylinder head - removal and installation

Caution: *The engine must be completely cool before beginning this procedure.*

Removal

1 Disconnect the negative cable from the battery and remove the spark plugs.
2 Drain the coolant from the engine block and radiator (see Chapter 1).
3 Drain the engine oil and remove the oil filter (see Chapter 1).
4 Remove the air intake tube assembly from the throttle body (see Chapter 4), then refer to Section 4 and remove the valve cover.
5 Refer to Section 5 and remove the intake manifold, and Section 6 to remove the exhaust manifold.
6 Remove all electrical connectors/wiring harness connections to the cylinder head. Mark the connectors for later reinstallation. Refer to Chapter 5 and remove the distributor.
7 Refer to Section 7 and remove the timing belt covers and the timing belt.
8 Refer to Chapter 3 and disconnect the upper radiator hose and the coolant bypass hose.

9 Check the cylinder head. Label and detach any remaining components that would interfere with cylinder head removal.
10 Using a breaker bar and the appropriate socket, loosen the cylinder head bolts in 1/4-turn increments, loosening in the reverse of the tightening sequence **(see illustration 12.22)** until they can be removed by hand.
11 Lift the cylinder head off the engine block. If it is stuck, very carefully pry up at the transaxle end, away from the head gasket surface.
12 Remove all external components from the head to allow for thorough cleaning and inspection. See Chapter 2, Part B, for cylinder head inspection and servicing procedures.

Installation

Refer to illustrations 12.16 and 12.22

13 The mating surfaces of the cylinder head and block must be perfectly clean when the head is installed.
14 Use a gasket scraper to remove all traces of carbon and old gasket material, then clean the mating surfaces with lacquer thinner or acetone. If any oil residue is on the mating surfaces when the head is installed, the gasket may not seal correctly and leaks could develop. When working on the block, stuff the cylinders with clean shop rags to prevent the entry of debris. Use a vacuum cleaner to remove material that falls into the cylinders. **Caution:** *Be careful not to gouge the soft aluminum of the cylinder head.*
15 Check the block and head mating surfaces for nicks, deep scratches and other damage. If damage is slight, it can be removed with a file; if it's excessive, machining may be the only alternative.
16 Use a thread die of the correct thread size to chase (clean up) the head bolt threads **(see illustration)**. **Note:** *Cleaning up the threads using a thread die should not cut any metal from the threads. If you observe any metal cuttings while chasing the threads, stop and replace the bolt with a new bolt from an automotive parts store or dealer. Make sure the replacement bolt is a OEM (Original Equipment Manufacturer) replacement cylinder head bolt specifically designed for this engine, and is the correct length and thread type. Discard the defective bolt.* Use a tap of the correct thread size to chase the threads in the head bolt holes, then clean the holes with compressed air - make sure that no residue such as dirt, corrosion, or sealant remains in the holes and the threads are not damaged as this will affect torque readings, which affects the quality of the head installation job. **Warning:** *Wear eye protection when using compressed air!*
17 Install the components that were removed from the head.
18 Position the new gasket over the dowel pins in the block. **Note:** *There is one corner of the head gasket which has a serrated or wavy edge. This end of the gasket goes toward the front (timing belt end) of the engine, with the serrated section to the left as viewed from the timing belt end. Aftermarket (non-factory) head gaskets may have another*

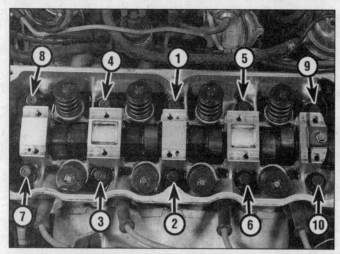

12.22 Cylinder head bolt TIGHTENING sequence

13.11 Areas to apply sealant include the oil-pump-to-block joints, and the rear main seal retainer plate-to-block joints

method of identifying the front or top of the gasket. Read the directions that come with the gasket.
19 Carefully set the head on the block without disturbing the gasket.
20 Before installing the head bolts, apply a small amount of clean engine oil to the threads.
21 Install the bolts and tighten them finger tight.
22 Tighten the bolts following the recommended sequence in several steps to the torque listed in this Chapter's Specifications **(see illustration)**.
23 The remaining installation steps are the reverse of removal.
24 Refill the cooling system, install a new oil filter and add oil to the engine (see Chapter 1).
25 Run the engine and check for leaks. Set the ignition timing (see Chapter 5) and road test the vehicle. On 1988 and 1989 models, readjust the valve clearances after the engine has been warmed up to normal operating temperatures.
26 Frequently recheck coolant level for the first few hundred miles to be sure that no leakage exists.

13 Oil pan - removal and installation

Removal

1 Disconnect the negative cable from the battery.
2 Set the parking brake and block the rear wheels. Raise the front of the vehicle and support it securely on jackstands.
3 Remove the under cover splash shield(s) from under the engine
4 Drain the engine oil and remove the oil filter (see Chapter 1).
5 Disconnect the exhaust pipe from the exhaust manifold and remove the pipe (see Chapter 4).
6 Remove the lower flywheel cover (see Chapter 8).
7 Remove the oil pan bolts and then remove the oil pan. If the oil pan is stuck, pry it loose carefully by inserting a screwdriver or putty knife at the corners of the oil pan. **Note:** *It may be necessary to rotate the crankshaft to reach the point of least interference between the pan and the crankshaft throws.* On 1988 through 1992 models, there are stiffener plates along the bottom of the oil pan flanges.

Installation

Refer to illustrations 13.11

8 Use a scraper to remove all traces of old gasket material and sealant from the block and oil pan. Clean the mating surfaces with gasket cleaner or equivalent solvent, available at automotive parts stores. **Caution:** *Be very careful not to scratch, bend, or otherwise damage the mating surfaces of the pan and block or oil leaks could develop.*
9 Make sure the threaded bolt holes in the block are clean. Visually check the condition of the oil strainer.
10 Check the oil pan flange for cracks or distortion, particularly at the bolting flange.
11 Apply silicone sealant to the indicated areas of the engine block **(see illustration)**.
12 The one-piece rubber gasket can be applied to the block, holding it in place long enough for it to stick to the sealant.
13 Carefully position the oil pan on the engine block and install the bolts/nuts. On models so equipped, install the stiffener plates before inserting any of the oil pan bolts. Working from the center out, tighten the pan bolts to the torque listed in this Chapter's Specifications in three or four steps.
14 The remainder of installation is the reverse of removal. Be sure to add oil and install a new oil filter.
15 After sufficient time for the silicone sealant to set up, run the engine and check for oil leaks.

14 Oil pump - removal, inspection and installation

Refer to illustrations 14.5, 14.6 and 14.7

Removal

1 Disconnect the negative battery cable. **Note:** *If the oil pump has high mileage, check the oil pressure performance (see Part B of this Chapter) and compare its pressure to this Chapter's Specifications.*
2 Remove the under cover splash shield.
3 Securely support the front of the vehicle on jackstands, and remove the right-front tire for access to the front cover. Refer to Chapter 11 for removal of the right fenderwell splash shield. Refer to Section 13 and remove the oil pan.
4 Refer to Section 7 and remove the timing belt and crankshaft sprocket.
5 Unbolt the oil pump pickup tube from the bottom of the pump **(see illustration)**.
6 Remove the oil pump bolts from the engine block **(see illustration)** and separate it from the engine block. You may have to pry carefully between the front main bearing cap and the pump housing with a screwdriver.
7 Place the oil pump on a workbench. Note how far the oil seal is seated in the bore. Using a seal removal tool or a screwdriver taped or wrapped with a rag to protect the pump bore, remove the oil seal from the housing **(see illustration)**. **Caution:** *Do not scratch the housing bore.* **Note:** *If you are replacing the front oil seal only and not removing, inspecting, repairing or replacing the oil pump, the front oil seal can be replaced without oil pump removal as described in Section 15.*

Chapter 2 Part A Engine

2A-15

14.5 Remove the oil pickup tube/screen assembly

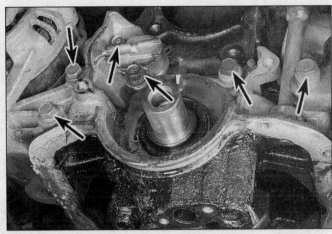

14.6 Remove the bolts (arrows) and separate the oil pump from the engine block

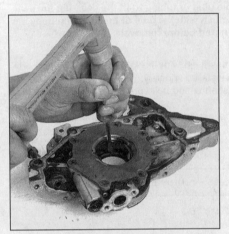

14.7 Remove the front oil seal using screwdriver or punch - wrap the tool tip with tape to protect the oil pump bore

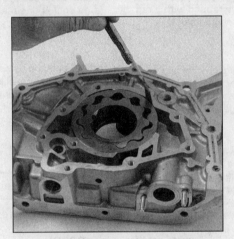

14.10a Measure the clearance between the oil pump driven rotor and the pump housing . . .

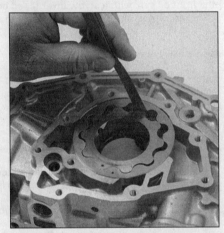

14.10b . . . then measure the clearance between the drive and driven rotor tips . . .

8   Remove the screws that hold the oil pump cover (slotted plate) to the front cover/oil pump housing. Inspect the oil pump cover for distortion or damage.

9   Remove the oil pressure relief valve. Note or mark the direction of the components as installed, then remove the oil pump inner and outer rotors from the housing.

Inspection

Refer to illustrations 14.10a, 14.10b and 14.10c

10   Reinstall the oil pump inner and outer rotors into the oil pump housing **(see illustrations)** and measure the clearance of:

  a) *The driven rotor-to-pump housing.*
  b) *The drive rotor tip-to-oil pump driven rotor tip.*
  c) *The rotor-to-oil pump housing endplay clearance.*

Compare your measurements to the clearance listed in this Chapter's Specifications.

11   Check the oil pressure relief valve for scoring, nicks or burns. Check the spring for damage. Replace the components if necessary.

12   Be sure the surfaces of the pump housing are clean and dry before reassembly.

13   Lightly coat the outer edge of a new oil seal with engine assembly lubricant or clean engine oil. Using a socket with an outside diameter slightly smaller than the outside diameter of the seal, carefully drive the new seal into place with a hammer. Make sure it's installed squarely and driven in to the same depth as the original. If a socket is not available, a short section of large diameter pipe will also work. Apply engine

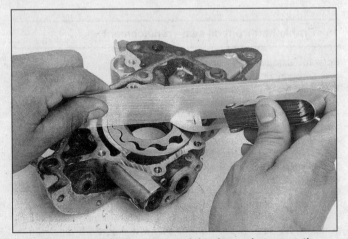

14.10c . . . and finally, use a straightedge and measure the endplay between the rotors and the pump housing

assembly lubricant to the seal lip surface that contacts the crankshaft.

14   Lubricate the oil pressure relief valve piston with clean engine oil and reinstall the valve components into the pump case.

15   Lubricate the rotor set with clean engine oil. Reinstall the rotors.

16   Install the cover, apply thread locking compound to the threads of the screws and tighten the screws securely. Install the spring retainer

2A-16   Chapter 2 Part A   Engine

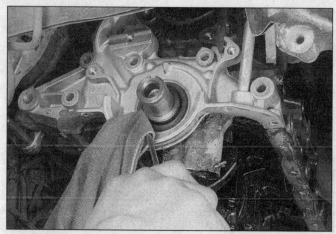

15.7 Before removal of the old oil seal, note how far the seal is seated in the bore and the direction the oil seal lip faces. Remove the front oil seal with a screwdriver taped or wrapped with a rag to protect the crankshaft surface and engine block

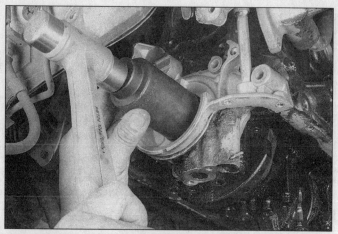

15.9 With the oil seal lips facing the correct direction as when removed, press the oil seal partially into place by hand. Using a large socket or suitable pipe or tubing to fit the seal, tap the seal into the bore until it is flush with the face of the oil pump body, as noted before removal

using a new cotter pin.
17   Inspect the screen at the end of the oil pick-up tube for any debris that might plug it. Either clean the tube and screen completely or replace it with a new one at this time. **Note:** *When replacing the oil pump pickup tube/screen, be sure to use a new gasket.*

Installation

18   Use a scraper to remove all traces of gasket and sealant from the pump and engine block, then clean the mating surfaces with solvent.
19   Install new gasket with a thin coat of silicone sealant on the front cover gasket surface. Reinstall the oil pump to the engine block. **Note:** *Be sure the sealant doesn't plug or cover any oil passages.*
20   Install the bolts, tightening them to the torque listed in this Chapter's Specifications.
21   The remainder of installation is the reverse of the removal procedure.
22   Add oil (see Chapter 1), start the engine and check for oil pressure and leaks.

15   Crankshaft front oil seal - replacement

Refer to illustrations 15.7 and 15.9

1   Disconnect the negative battery cable.
2   Remove the under cover splash shield.
3   Securely support the front of the vehicle on jackstands, and remove right-front tire for access to the front cover. Refer to Chapter 11 for removal of the inner splash shield from the fenderwell.
4   Remove the crankshaft pulley, timing belt covers, timing belt, and timing belt guide (see Section 7).
5   Remove the crankshaft sprocket bolt and slide the crankshaft sprocket off the crankshaft.
6   Cut the front oil seal lip with a razor knife.
7   Note how far the seal is seated in the bore and the direction the oil seal lip faces (the oil seal should be flush with the face of the oil pump body. Remove the front oil seal with a screwdriver taped or wrapped with a rag to protect the crankshaft surface and engine block **(see illustration).**
8   Clean the bore in the engine block and clean the crankshaft surface. Coat the outside of the new front oil seal with engine oil. Apply engine assembly lubricant or clean engine oil to the seal lip.
9   Press the oil seal in slightly by hand, with the oil seal lip facing the same direction as removed. Using a seal-driver or a socket with an outside diameter slightly smaller than the outside diameter of the front oil seal, carefully tap the new seal into place with a hammer **(see illus-**

tration) until the oil seal is flush with the face of the oil pump body. Make sure the oil seal is installed squarely.
10   Reinstall the crankshaft timing belt sprocket and timing belt (see Section 7).
11   The remainder of the installation is the reverse of the removal procedure.
12   Run the engine and check for oil leaks at the front oil seal.

16   Flywheel/driveplate - removal and installation

Refer to illustrations 16.1 and 16.3

Removal

1   Raise the vehicle and support it securely on jackstands, then refer to Chapter 7 and remove the transaxle. The engine should be supported by an overhead crane or engine support cradle that mounts to the body **(see illustration).**
2   If you're working on a model with a manual transaxle, remove the clutch cover and clutch disc (see Chapter 8). Now is a good time to check/replace the clutch components and pilot bearing.
3   Use a center-punch or paint to make alignment marks on the fly-

16.1 Use an engine support cradle (they can be rented) that mounts on the fenders and radiator support

16.3 Mark the flywheel/driveplate and the crankshaft with paint (arrows) so they can be reassembled in the same relative position

16.9 When installing an automatic-transaxle driveplate, be sure to install the backing ring (arrow) oriented the same way as when it was removed

17.4 The quick way to replace the rear main oil seal is to simply pry the old one out . . .

wheel/driveplate and crankshaft to ensure correct reinstallation and alignment **(see illustration)**.

4 Remove the bolts that secure the flywheel/driveplate to the crankshaft. If the crankshaft turns, wedge a screwdriver in the ring gear teeth to jam the flywheel.

5 Remove the flywheel/driveplate from the crankshaft. On the automatic transaxle models, also remove the driveplate backing plate and adapter, taking note of which sides of the driveplate the adapter plates are mounted for correct reinstallation later. Since the flywheel is fairly heavy, be sure to support it while removing the last bolt.

6 Clean the flywheel to remove grease and oil. Inspect the surface for cracks, rivet grooves, burned areas and score marks. Light scoring can be removed with emery cloth. Check for cracked and broken ring gear teeth. Lay the flywheel on a flat surface and use a straightedge to check for warpage. If necessary, take the flywheel to an automotive machine shop to have it resurfaced.

7 Clean and inspect the mating surfaces of the flywheel/driveplate and the crankshaft. If the crankshaft rear seal is leaking, replace it before reinstalling the flywheel/driveplate (see Section 17).

Installation

Refer to illustration 16.9

8 Remove any thread sealant from the crankshaft flywheel bolt holes and bolts.

9 For manual transaxle models, position the flywheel at the crankshaft. For automatic transaxle models, position the adapter, driveplate, and backing plate at the crankshaft **(see illustration)**. Be sure to align the marks made during removal. Before installing the bolts, apply thread-locking compound to the threads of the bolts.

10 Wedge a screwdriver in the ring gear teeth to keep the flywheel/driveplate from turning as you tighten the bolts to the torque listed in this Chapter's Specifications. Follow a criss-cross pattern and work up to the final torque in three or four steps.

11 The remainder of installation is the reverse of the removal procedure.

17 Rear main oil seal - replacement

Refer to illustrations 17.4 and 17.5

1 The transaxle must be removed from the vehicle for this procedure (see Chapter 7).

2 Remove the flywheel/driveplate (see Section 16).

3 Cut the rear main oil seal lip with a razor knife.

4 Pry out the old seal with a screwdriver taped or wrapped in a rag, or use a seal removal tool to pry the seal out **(see illustration)**.

17.5 . . . then lubricate the crankshaft journal and the lip of the new seal with engine oil and tap the new seal into place - the seal lip is stiff and can be easily damaged during installation if you are not careful

5 Apply engine oil to the crankshaft seal journal and to the lip of the new seal. Carefully push the new seal part way into place by hand. Carefully tap into place using a flat punch, large socket, or a suitable short pipe or tubing of the correct diameter until the oil seal is flush with the edge of the rear cover **(see illustration)**.

6 Reinstall the flywheel/driveplate (see Section 16).

7 The remaining steps are the reverse of removal.

18 Engine mounts - check and replacement

1 Engine mounts seldom require attention, but broken or deteriorated mounts should be replaced immediately or the added strain placed on the driveline components may cause damage or wear. **Warning:** *Do not remove any engine mounts or components of engine mounts if the engine is not properly supported as described. DO NOT place any part of your body directly under the engine when performing engine mount work and when the engine is supported only by a jack!*

Check

2 During the check, the engine must be raised slightly to remove the weight from the mounts. An overhead engine support cradle is the preferred method of safely raising the engine **(see illustration 16.1)**.

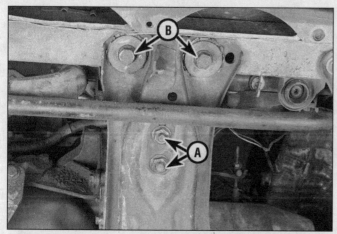

18.9 Front of transaxle crossmember (rear mounting similar). Remove the nuts (A) for the front transaxle mount (bolts at B are from removing the crossmember after both mounts are disconnected)

18.11 Remove the transaxle/engine mount through-bolt and nut (arrows)

3   Raise the vehicle and support it securely on jackstands, then position a jack with a block of wood under the engine oil pan or use an engine support fixture from above. Carefully raise the engine just enough to take the weight off the mounts. **Warning:** *DO NOT place any part of your body under the engine when it's supported only by a jack! Support the engine just enough to take the weight off the engine mounts but without lifting the weight of the car from the jackstands.*
4   Check the mounts to see if the rubber is cracked, hardened or separated from the metal portion. Occasionally, the rubber will split down the center.
5   Check for relative movement between the mounts and the engine or frame, using a large screwdriver or prybar to attempt to move the mounts. If movement is noted, lower the engine and tighten the mount fasteners.
6   Rubber preservative liquid, available from any automotive parts store, should be applied to the engine mounts (and other chassis rubber components) to help protect from deterioration.

Replacement

Refer to illustrations 18.9, 18.11 and 18.15
7   Disconnect the negative battery cable from the battery, then raise the vehicle and support it securely on jackstands (if not already done). Support the engine as described in Step 3.
8   The engine has three mounts. Two are on the transaxle crossmember, and one at the right side of the engine, connecting the block to the right fenderwell.
9   Detach the transaxle/engine mount nuts located on the bottom of the transaxle mount crossmember **(see illustration)**.
10 Remove the crossmember mounting bolts **(see illustrations 18.9)** and remove the crossmember for access. **Warning:** *Do not remove the crossmember if the engine is not supported as described above.*
11 Remove the transaxle/engine mount through-bolts/nuts to detach the rubber mount from the engine bracket **(see illustration)**.
12 Reinstall the transaxle/engine mounts by attaching the mounts to the brackets, installing the mount through-bolts, washers and nuts, and then installing the crossmember.
13 Tighten all bolts/nuts to the torque listed in this Chapter's Specifications.
14 Replace the engine mount No. 3 (located on the right side, near

18.15 Engine mount on the right side below the timing belt cover is removed by detaching the through-bolt (A), and the two mount-to-engine-bracket nuts (B)

the engine timing belt cover), with the bottom of the engine supported with a floorjack and block of wood under the oil pan.
15 Relieve the weight on the mount with the floorjack. Remove the through-bolt nut and washer and withdraw the through-bolt from the frame bracket **(see illustration)**.
16 Remove the two nuts securing the mount to the engine bracket **(see illustration 18.15)**.
17 Reinstall the right side engine mount by installing the mount onto the vehicle's frame bracket, inserting the through-bolt to connect the engine bracket, and using a new washer, install the washer and nut. Install the two nuts and washers to the engine bracket studs.
18 Tighten all bolts and nuts to the torque listed in this Chapter's Specifications.
19 If, after replacing the engine mounts, there is some noise or vibration that wasn't present before, you may need to "neutralize" the mounting bolts. Loosen the transaxle mount nuts/bolts (the ones to the crossmember) about three turns each.
20 Without driving, start the engine and shift the transmission into gear, then shift into Neutral and finally, shut the engine off. Now retighten all mounting bolts/nuts to this Chapter's Specifications.

Chapter 2 Part B
General engine overhaul procedures

Contents

	Section
Compression check	4
Crankshaft - inspection	19
Crankshaft - installation and main bearing oil clearance check	23
Crankshaft removal	14
Cylinder head - cleaning and inspection	10
Cylinder head - disassembly	9
Cylinder head - reassembly	12
Cylinder honing	17
Engine - removal and installation	6
Engine block - cleaning	15
Engine block - inspection	16
CHECK ENGINE light	See Chapter 6
Engine overhaul - disassembly sequence	8
Engine overhaul - reassembly sequence	21

	Section
Engine rebuilding alternatives	7
Engine removal - methods and precautions	5
General information	1
Initial start-up and break-in after overhaul	26
Main and connecting rod bearings - inspection and selection	20
Oil pressure check	2
Piston rings - installation	22
Pistons/connecting rods - inspection	18
Pistons/connecting rods - installation and rod bearing oil clearance check	25
Pistons/connecting rods - removal	13
Rear main oil seal installation	24
Vacuum gauge diagnostic check	3
Valves - servicing	11

Specifications

General
Displacement... 79.3 cubic inches (1.3 liters)
Cylinder compression pressure @ 300 rpm
  Standard.. 204 psi (1412 kPa)
  Minimum.. 149 psi (1030 kPa)
Oil pressure (engine warm)
  1000 rpm... 28 to 43 psi (196 to 294 kPa)
  3000 rpm... 43 to 57 psi (294 to 392 kPa)

Cylinder head
Warpage limit... 0.006 inch (0.15 mm)

Valves and related components
Valve margin thickness (minimum)
  Intake... 0.051 inch (1.30 mm)
  Exhaust.. 0.039 inch (1.0 mm)
Valve stem diameter
  Intake... 0.2744 to 0.2750 inch (6.970 to 6.985 mm)
  Exhaust.. 0.2742 to 0.2748 inch (6.965 to 6.980 mm)
Valve stem-to-guide clearance
  Intake... 0.0010 to 0.0024 inch (0.025 to 0.060 mm)
  Exhaust.. 0.0012 to 0.0026 inch (0.030 to 0.065 mm)
    Service limit.. 0.008 inch (0.2 mm)
Valve spring
  Out-of-square limit.................................. 0.059 inch (1.5 mm)
  Free length (minimum)............................ 1.717 inches (43.6 mm)

Crankshaft and connecting rods

Connecting rod journal	
Diameter	1.5724 to 1.5731 inches (39.940 to 39.956 mm)
Out-of-round limits	0.0020 inch (0.05 mm)
Main bearing journal	
Diameter	1.9661 to 1.9688 inches (49.938 to 49.956 mm)
Out-of-round	0.0020 inch (0.05 mm)
Runout limit	0.0016 inch (0.04 mm)
Main Bearing oil clearance	
Standard	0.0007 to 0.0014 inch (0.018 to 0.036 mm)
Service limit	0.0039 inch (0.10 mm)
Connecting rod oil clearance	
Standard	0.0009 to 0.0017 inch (0.024 to 0.042 mm)
Service limit	0.0039 inch (0.10 mm)
Connecting rod side clearance, service limit	0.012 inch (0.30 mm)
Crankshaft endplay	
Standard	0.0031 to 0.0111 inch (0.080 to 0.282 mm)
Service limit	0.012 inch (0.30 mm)

Engine block

Deck warpage limit	0.006 inch (0.15 mm)
Cylinder bore diameter (standard)	2.7953 to 2.7960 inches (71.000 to 71.019 mm)
Taper and out-of-round limits	0.0007 inch (0.019 mm)

Pistons and rings

Piston diameter (standard)	2.793 to 2.794 to inches (70.954 to 70.974 mm)
Piston-to-bore clearance, service limit	0.006 inch (0.15 mm)
Piston ring end gap	
Compression rings (both)	0.006 to 0.012 inch (0.15 to 0.30 mm)
Oil ring	
Standard	0.008 to 0.028 inches (0.20 to 0.70 mm)
Service limit	0.039 inch (1.0 mm)
Piston ring groove clearance, both compression rings	0.001 to 0.003 inch (0.03 to 0.065 mm)

Torque specifications*

	Ft-lbs (unless otherwise indicated)
Main bearing cap bolts	40 to 43
Connecting rod cap nuts/bolts	
Step 1	11 to 13
Step 2	22 to 25
Rear main seal retainer bolts	71 to 97 in-lbs

*Note: *Refer to Part A for additional torque specifications.*

1 General information

Included in this portion of Chapter 2 are the general overhaul procedures for the cylinder head and internal engine components.

The information ranges from advice concerning preparation for an overhaul and the purchase of replacement parts to detailed, step-by-step procedures covering removal and installation of internal engine components and the inspection of parts.

The following Sections have been written based on the assumption that the engine has been removed from the vehicle. For information concerning in-vehicle engine repair, as well as removal and installation of the external components necessary for the overhaul, see Chapter 2A and Section 8 of this Chapter.

The Specifications included in this Part are only those necessary for the inspection and overhaul procedures which follow. Refer to Chapter 2, Part A for additional Specifications.

It's not always easy to determine when, or if, an engine should be completely overhauled, as a number of factors must be considered.

High mileage is not necessarily an indication that an overhaul is needed, while low mileage doesn't preclude the need for an overhaul. Frequency of servicing is probably the most important consideration. An engine that's had regular and frequent oil and filter changes, as well as other required maintenance, will most likely give many thousands of miles of reliable service. Conversely, a neglected engine may require an overhaul very early in its life.

Excessive oil consumption is an indication that piston rings, valve seals and/or valve guides are in need of attention. Make sure that oil leaks aren't responsible before deciding that the rings and/or guides are bad. Perform a cylinder compression check to determine the extent of the work required (see Section 4). Also check the vacuum readings under various conditions (see Section 3).

Loss of power, rough running, knocking or metallic engine noises, excessive valve train noise and high fuel consumption rates may also point to the need for an overhaul, especially if they're all present at the same time. If a complete tune-up doesn't remedy the situation, major mechanical work is the only solution.

An engine overhaul involves restoring the internal parts to the specifications of a new engine. During an overhaul, the piston rings are replaced and the cylinder walls are reconditioned (re-bored and/or honed). If a re-bore is done by an automotive machine shop, new oversize pistons will also be installed. The main bearings, connecting rod bearings and camshaft bearings are generally replaced with new ones and, if necessary, the crankshaft may be reground to restore the journals. Generally, the valves are serviced as well, since they're usually in less-than-perfect condition at this point. While the engine is being

Chapter 2 Part B General engine overhaul procedures

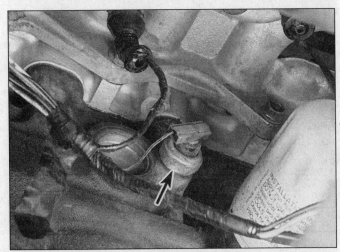

2.2 The oil pressure sending unit (arrow) is located on the block, under the intake manifold, near the oil filter - the oil pressure can be checked by removing the sending unit and connecting a pressure gauge in its place

3.4 The vacuum gauge is easily attached to an unused port on the intake manifold

overhauled, other components, such as the distributor, starter and alternator, can be rebuilt as well. The end result should be a like new engine that will give many trouble free miles. **Note:** *Critical cooling system components such as the hoses, drivebelts, thermostat and water pump should be replaced with new parts when an engine is overhauled. The radiator should be checked carefully to ensure that it isn't clogged or leaking* (see Chapter 3). If you purchase a rebuilt engine or short block, some rebuilders will not warranty their engines unless the radiator has been professionally flushed. Also, we don't recommend overhauling the oil pump - always install a new one when an engine is rebuilt.

Before beginning the engine overhaul, read through the entire procedure to familiarize yourself with the scope and requirements of the job. Overhauling an engine isn't difficult, but it is time-consuming. Plan on the vehicle being tied up for a minimum of two weeks, especially if parts must be taken to an automotive machine shop for repair or reconditioning. Check on availability of parts and make sure that any necessary special tools and equipment are obtained in advance. Most work can be done with typical hand tools, although a number of precision measuring tools are required for inspecting parts to determine if they must be replaced. Often an automotive machine shop will handle the inspection of parts and offer advice concerning reconditioning and replacement. **Note:** *Always wait until the engine has been completely disassembled and all components, especially the engine block, have been inspected before deciding what service and repair operations must be performed by an automotive machine shop.* Since the block's condition will be the major factor to consider when determining whether to overhaul the original engine or buy a rebuilt one, never purchase parts or have machine work done on other components until the block has been thoroughly inspected. As a general rule, time is the primary cost of an overhaul, so it doesn't pay to install worn or substandard parts.

As a final note, to ensure maximum life and minimum trouble from a rebuilt engine, everything must be assembled with care in a spotlessly-clean environment.

2 Oil pressure check

Refer to illustration 2.2

1 Low engine oil pressure can be a sign of an engine in need of rebuilding. A "low oil pressure" indicator (often called an "idiot light") is not a test of the oiling system. Such indicators only come on when the oil pressure is dangerously low. Even a factory oil pressure gauge in the instrument panel is only a relative indication, although much better for driver information than a warning light. A better test is with a mechanical (not electrical) oil pressure gauge. When used in conjunction with an accurate tachometer, an engine's oil pressure performance can be compared to manufacturer's Specifications for that year and model.

2 Locate the oil pressure indicator sending unit **(see illustration)**.

3 Remove the oil pressure sending unit and install a fitting which will allow you to directly connect your hand-held, mechanical oil pressure gauge. Use Teflon tape or sealant on the threads of the adapter and the fitting on the end of your gauge's hose.

4 Connect an accurate tachometer to the engine, according to the tachometer manufacturer's instructions.

5 Check the oil pressure with the engine running (full operating temperature) at the specified engine speed, and compare it to this Chapter's Specifications. If it's extremely low, the bearings and/or oil pump are probably worn out.

3 Vacuum gauge diagnostic check

Refer to illustration 3.4

A vacuum gauge provides valuable information about what is going on in the engine at a low-cost. You can check for worn rings or cylinder walls, leaking head or intake manifold gaskets, incorrect carburetor adjustments, restricted exhaust, stuck or burned valves, weak valve springs, improper ignition or valve timing and ignition problems.

Unfortunately, vacuum gauge readings are easy to misinterpret, so they should be used in conjunction with other tests to confirm the diagnosis.

Both the absolute readings and the rate of needle movement are important for accurate interpretation. Most gauges measure vacuum in inches of mercury (in-Hg). As vacuum increases (or atmospheric pressure decreases), the reading will increase. Also, for every 1,000 foot increase in elevation above sea level, the gauge readings will decrease about one inch of mercury.

Connect the vacuum gauge directly to intake manifold vacuum, not to ported (above the throttle plate) vacuum **(see illustration)**. Be sure no hoses are left disconnected during the test or false readings will result.

Before you begin the test, allow the engine to warm up completely. Block the wheels and set the parking brake. With the transmission in neutral (or Park, on automatics), start the engine and allow it to run at normal idle speed. **Warning:** *Carefully inspect the fan blades for cracks or damage before starting the engine. Keep your hands and the vacuum tester clear of the fan and do not stand in front of the vehicle or in line with the fan when the engine is running.*

Read the vacuum gauge; an average, healthy engine should normally produce between 17 and 20 inches of vacuum with a fairly steady needle.

Refer to the following vacuum gauge readings and what they indicate about the engines condition:

1 A low steady reading usually indicates a leaking gasket between the intake manifold and carburetor or throttle body, a leaky vacuum hose, late ignition timing or incorrect camshaft timing. Check ignition timing with a timing light and eliminate all other possible causes, utilizing the tests provided in this Chapter before you remove the timing belt cover to check the timing marks.

2 If the reading is three to eight inches below normal and it fluctuates at that low reading, suspect an intake manifold gasket leak at an intake port or a faulty injector.

3 If the needle has regular drops of about two to four inches at a steady rate the valves are probably leaking. Perform a compression or leak-down test to confirm this.

4 An irregular drop or down-flick of the needle can be caused by a sticking valve or an ignition misfire. Perform a compression or leak-down test and read the spark plugs.

5 A rapid vibration of about four in-Hg vibration at idle combined with exhaust smoke indicates worn valve guides. Perform a leak-down test to confirm this. If the rapid vibration occurs with an increase in engine speed, check for a leaking intake manifold gasket or head gasket, weak valve springs, burned valves or ignition misfire.

6 A slight fluctuation, say one inch up and down, may mean ignition problems. Check all the usual tune-up items and, if necessary, run the engine on an ignition analyzer.

7 If there is a large fluctuation, perform a compression or leak-down test to look for a weak or dead cylinder or a blown head gasket.

8 If the needle moves slowly through a wide range, check for a clogged PCV system, incorrect idle fuel mixture, throttle body or intake manifold gasket leaks.

9 Check for a slow return after revving the engine by quickly snapping the throttle open until the engine reaches about 2,500 rpm and let it shut. Normally the reading should drop to near zero, rise above normal idle reading (about 5 in.-Hg over) and then return to the previous idle reading. If the vacuum returns slowly and doesn't peak when the throttle is snapped shut, the rings may be worn. If there is a long delay, look for a restricted exhaust system (often the muffler or catalytic converter). An easy way to check this is to temporarily disconnect the exhaust ahead of the suspected part and redo the test.

4 Compression check

Refer to illustration 4.6

1 A compression check will tell you what mechanical condition the upper end (pistons, rings, valves, head gasket[s]) of your engine is in. Specifically, it can tell you if the compression is down due to leakage caused by worn piston rings, defective valves and seats or a blown head gasket. **Note:** *The engine must be at normal operating temperature and the battery must be fully charged for this check.*

2 Begin by cleaning the area around the spark plugs before you remove them (compressed air should be used, if available, otherwise a small brush or even a bicycle tire pump will work). The idea is to prevent dirt from getting into the cylinders as the compression check is being done.

3 Remove all of the spark plugs from the engine (see Chapter 1).

4 Block the throttle wide open.

5 Detach the coil wire from the center of the distributor cap and ground it on the engine block. Use a jumper wire with alligator clips on each end to ensure a good ground. It is also a good idea to pull the EFI fuse from the fuse panel to disable the fuel pump during the compression test.

6 Install the compression gauge in the spark plug hole **(see illustration)**.

7 Crank the engine over at least seven compression strokes and watch the gauge. The compression should build up quickly in a healthy engine. Low compression on the first stroke, followed by gradually increasing pressure on successive strokes, indicates worn piston rings. A low compression reading on the first stroke, which doesn't build up during successive strokes, indicates leaking valves or a blown

4.6 A compression gauge with a threaded fitting for the spark plug hole is preferred over the type that requires hand pressure to maintain the seal - be sure to block open the throttle valve as far as possible during the compression check!

head gasket (a cracked head could also be the cause). Deposits on the undersides of the valve heads can also cause low compression. Record the highest gauge reading obtained.

8 Repeat the procedure for the remaining cylinders and compare the results to this Chapter's Specifications.

9 Add some engine oil (about three squirts from a plunger-type oil can) to each cylinder, through the spark plug hole, and repeat the test.

10 If the compression increases after the oil is added, the piston rings are definitely worn. If the compression doesn't increase significantly, the leakage is occurring at the valves or head gasket. Leakage past the valves may be caused by burned valve seats and/or faces or warped, cracked or bent valves.

11 If two adjacent cylinders have equally low compression, there's a strong possibility that the head gasket between them is blown. The appearance of coolant in the combustion chambers or the crankcase would verify this condition.

12 If one cylinder is 20-percent lower than the others, and the engine has a slightly rough idle, a worn exhaust lobe on the camshaft could be the cause.

13 If the compression is unusually high, the combustion chambers are probably coated with carbon deposits. If that's the case, the cylinder head should be removed and decarbonized.

14 If compression is way down or varies greatly between cylinders, it would be a good idea to have a leak-down test performed by an automotive repair shop. This test will pinpoint exactly where the leakage is occurring and how severe it is.

5 Engine removal - methods and precautions

If you've decided that an engine must be removed for overhaul or major repair work, several preliminary steps should be taken.

Locating a suitable place to work is extremely important. Adequate work space, along with storage space for the vehicle, will be needed. If a shop or garage isn't available, at the very least a flat, level, clean work surface made of concrete or asphalt is required.

Cleaning the engine compartment and engine before beginning the removal procedure will help keep tools clean and organized.

An engine hoist or A-frame will also be necessary. Make sure the equipment is rated in excess of the combined weight of the engine and transaxle. Safety is of primary importance, considering the potential hazards involved in lifting the engine out of the vehicle.

If the engine is being removed by a novice, a helper should be available. Advice and aid from someone more experienced would also be helpful. There are many instances when one person cannot simultaneously perform all of the operations required when lifting the engine out of the vehicle.

Chapter 2 Part B General engine overhaul procedures

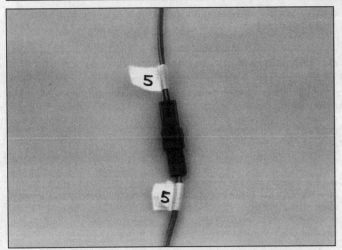

6.7 Label both ends of each wire and hose before disconnecting it

6.24 Attach the engine hoist to the engine lifting brackets (arrows) - lift the engine high enough to clear the vehicle, then move it away and lower the hoist - the engine can be removed with the transaxle still attached

Plan the operation ahead of time. Arrange for or obtain all of the tools and equipment you'll need prior to beginning the job. Some of the equipment necessary to perform engine removal and installation safely and with relative ease are (in addition to an engine hoist) a heavy duty floor jack, complete sets of wrenches and sockets as described in the front of this manual, wooden blocks and plenty of rags and cleaning solvent for mopping up spilled oil, coolant and gasoline. If the hoist must be rented, make sure that you arrange for it in advance and perform all of the operations possible without it beforehand. This will save you money and time.

Plan for the vehicle to be out of use for quite a while. A machine shop will be required to perform some of the work which the do-it-yourselfer can't accomplish without special equipment. These shops often have a busy schedule, so it would be a good idea to consult them before removing the engine in order to accurately estimate the amount of time required to rebuild or repair components that may need work.

Always be extremely careful when removing and installing the engine. Serious injury can result from careless actions. Plan ahead, take your time and a job of this nature, although major, can be accomplished successfully.

6 Engine - removal and installation

Refer to illustrations 6.7 and 6.24

Note: *Read through the entire Section before beginning this procedure. The manufacturer recommends removing the engine and transaxle from the top as a unit, then separating the engine from the transaxle on the shop floor.*

Removal

Warning: *Some models covered by this manual are equipped with Supplemental Restraint Systems (SRS), more commonly known as airbags. Always disconnect the negative battery cable, then the positive battery cable and wait two minutes before working in the vicinity of the impact sensors, steering column or instrument panel to avoid the possibility of accidental deployment of the airbag, which could cause personal injury (see Chapter 12). Do not use any electrical test equipment on any of the airbag system wires or tamper with them in any way.*

1 Relieve the fuel system pressure (see Chapter 4).
2 Disconnect the negative cable from the battery.
3 Place protective covers on the fenders and cowl and remove the hood (see Chapter 11).
4 Raise the vehicle and support it securely on jackstands.
5 Drain the cooling system, engine oil, transaxle oil and remove the drivebelts (see Chapter 1).
6 Remove the front wheels.
7 Clearly label, then disconnect all vacuum lines, coolant and emissions hoses, wiring harness connectors, ground straps and fuel lines. Masking tape and/or a touch up paint applicator work well for marking items **(see illustration)**. Take instant photos or sketch the locations of components and brackets.
8 Remove the air cleaner housing assembly (see Chapter 4) and the tube to the throttle body or carburetor.
9 Remove the engine undercover to provide access to the bottom of the engine (see Chapter 11).
10 Remove the battery bracket, the battery cover, the battery, the battery carrier and the battery duct.
11 Disconnect the accelerator cable.
12 Remove the windshield washer tank and coolant reservoir tank.
13 Disconnect the heater hoses.
14 Remove the cooling fan, the radiator hoses and the radiator (see Chapter 3).
15 Release the residual fuel pressure in the tank by removing the gas cap, then detach the fuel lines connecting the engine to the chassis (see Chapter 4). Plug or cap all open fittings.
16 On power steering-equipped vehicles, unbolt the power steering pump. If clearance allows, tie the pump aside without disconnecting the hoses. If necessary, remove the pump (see Chapter 10).
17 On air-conditioned models, unbolt the compressor and set it aside. Do not disconnect the refrigerant hoses. **Note:** *Wire the compressor out of the way with a coat hanger, don't let the compressor hang on the hoses.*
18 Disconnect the throttle linkage, transmission Throttle Valve (TV) linkage, speedometer cable, and speed control cable, if equipped, from the engine/transaxle (see Chapter 4).
19 Disconnect the clutch release cylinder, the shift control rod and cable and the extension bar (see Chapter 8). Drain the transaxle fluid.
20 Disconnect the front exhaust pipe, and refer to Chapter 8 for removal of the driveaxles.
21 Attach a lifting sling to the engine. Position a hoist and connect the sling to it. Take up the slack until there is slight tension on the hoist.
22 Recheck to be sure nothing except the mounts are still connecting the engine or transaxle to the vehicle. Disconnect and label anything still remaining.
23 Support the transaxle with a floor jack. Place a block of wood on the jack head to prevent damage to the transaxle. Refer to Part A of this Chapter and remove the transaxle crossmember, transaxle mounts, and the right-hand engine mount. **Warning:** *Do not place any part of your body under the engine/transaxle when it's supported only by a hoist or other lifting device.*
24 Slowly lift the engine and transaxle out of the vehicle **(see illustration)**. It may be necessary to pry the mounts away from the frame brackets.

25 Move the engine/transaxle away from the vehicle and carefully lower the hoist until the engine can be set on the floor.
26 On automatic transaxle-equipped models, remove the torque converter-to-driveplate fasteners (see Chapter 7B) and push the converter back slightly into the bellhousing.
27 Remove the engine-to-transaxle bolts and separate the engine from the transaxle (see Chapter 7A). The torque converter should remain in the transaxle. Remove the flywheel/driveplate and mount the engine on an engine stand. **Note:** *On automatic transaxle-equipped models, mark the front and rear spacer plates and keep them with the driveplate.*

Installation

28 Check the engine/transaxle mounts. If they're worn or damaged, replace them.
29 On manual transaxle-equipped models, inspect the clutch components (see Chapter 8) and on automatic models inspect the converter seal and bushing.
30 On automatic transaxle-equipped models, apply a dab of grease to the nose of the converter. Make sure the converter is completely seated on the transaxle input shaft and the front pump splines. To do this, push in on the converter and turn it, feeling for a "clunk" - it may even "clunk" more than once. If you feel nothing, the converter is already completely seated.
31 Carefully guide the transaxle into place, following the procedure outlined in Chapter 7. **Caution:** *Do not use the bolts to force the engine and transaxle into alignment. It may crack or damage major components.*
32 Install the engine-to-transaxle bolts and tighten them to the torque listed in the Chapter 7 Specifications.
33 Attach the hoist to the engine and carefully lower the engine/transaxle assembly into the engine compartment.
34 Install the mount bolts and tighten them securely.
35 Reinstall the remaining components and fasteners in the reverse order of removal.
36 Add coolant, oil, power steering and transmission fluids as needed (see Chapter 1).
37 Run the engine and check for proper operation and leaks. Shut off the engine and recheck the fluid levels.

7 Engine rebuilding alternatives

The do-it-yourselfer is faced with a number of options when performing an engine overhaul. The decision to replace the engine block, piston/connecting rod assemblies and crankshaft depends on a number of factors, with the number one consideration being the condition of the block. Other considerations are cost, access to machine shop facilities, parts availability, time required to complete the project and the extent of prior mechanical experience on the part of the do-it-yourselfer.

Some of the rebuilding alternatives include:

Individual parts - If the inspection procedures reveal that the engine block and most engine components are in reusable condition, purchasing individual parts may be the most economical alternative. The block, crankshaft and piston/connecting rod assemblies should all be inspected carefully. Even if the block shows little wear, the cylinder bores should be surface-honed.

Short block - A short block consists of an engine block with a crankshaft and piston/connecting rod assemblies already installed. All new bearings are incorporated and all clearances will be correct. The existing camshaft, valve train components, cylinder head and external parts can be bolted to the short block with little or no machine shop work necessary.

Long block - A long block consists of a short block plus an oil pump, oil pan, cylinder head, valve cover, camshaft and valve train components, timing sprockets, belt and timing covers. All components are installed with new bearings, seals and gaskets incorporated throughout. The installation of manifolds and external parts is all that's necessary.

Give careful thought to which alternative is best for you and discuss the situation with local automotive machine shops, auto parts dealers and experienced rebuilders before ordering or purchasing replacement parts.

8 Engine overhaul - disassembly sequence

1 It's much easier to disassemble and work on the engine if it's mounted on a portable engine stand. A stand can often be rented quite cheaply from an equipment-rental yard. Before the engine is mounted on a stand, the flywheel/driveplate and rear oil seal retainer should be removed from the engine.
2 If a stand isn't available, it's possible to disassemble the engine with it blocked up on the floor. Be extra careful not to tip or drop the engine when working without a stand.
3 If you're going to obtain a rebuilt engine, all external components must come off first, to be transferred to the replacement engine, just as they will if you're doing a complete engine overhaul yourself. These include:

Alternator and brackets
Emissions control components
Distributor, spark plug wires and spark plugs
Thermostat and housing cover
Water pump
EFI components
Intake/exhaust manifolds
Oil filter
Engine mounts
Clutch and flywheel/driveplate
Engine rear plate
Rear main seal retainer plate

Note: *When removing the external components from the engine, pay close attention to details that may be helpful or important during installation. Note the installed position of gaskets, seals, spacers, pins, brackets, washers, bolts and other small items.*

4 If you're obtaining a short block, which consists of the engine block, crankshaft, pistons and connecting rods all assembled, then the cylinder head, oil pan and oil pump will have to be removed as well from your engine so that your short block can be turned in to the rebuilder as a core. See *Engine rebuilding alternatives* for additional information regarding the different possibilities to be considered.
5 If you're planning a complete overhaul, the engine must be disassembled and the internal components removed. The engine cylinder head is first removed in Chapter 2 Part A, then the following engine block components are disassembled:

Timing belt, sprockets and tensioner assembly
Oil pan
Oil pump assembly
Rear cover
Piston/connecting rod assemblies
Crankshaft rear oil seal retainer
Crankshaft and main bearings

6 Before beginning the disassembly and overhaul procedures, make sure the following items are available. Also, refer to Section 21 for a list of tools and materials needed for engine reassembly.

Common hand tools
Small cardboard boxes or plastic bags for storing parts
Gasket scraper
Ridge reamer
Micrometers
Telescoping gauges
Dial indicator set
Valve spring compressor
Cylinder surfacing hone
Piston ring groove-cleaning tool
Electric drill motor
Tap and die set
Wire brushes
Oil gallery brushes
Cleaning solvent

9.2 A small plastic bag, with an appropriate label, can be used to store the valve train components so they can be kept together and reinstalled in the correct guide

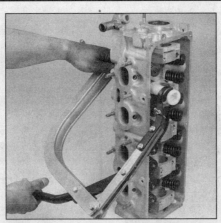

9.3a Compress the spring to expose the keepers - this type of compressor can only be used when the cylinder head is off the engine

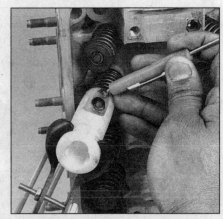

9.3b Remove the keepers with a small magnetic screwdriver or needle-nose pliers

9 Cylinder head - disassembly

Refer to illustrations 9.2, 9.3a and 9.3b

Note: *New and rebuilt cylinder heads are commonly available for most engines at dealerships and auto parts stores. Due to the fact that some specialized tools are necessary for the disassembly and inspection procedures, and replacement parts may not be readily available, it may be more practical and economical for the home mechanic to purchase a replacement head rather than taking the time to disassemble, inspect and recondition the original.*

1   Cylinder head disassembly involves removal of the intake and exhaust valves and related components. It's assumed that the rocker arm/shaft assembly and camshaft have already been removed (see Part A as needed).

2   Before the valves are removed, arrange to label and store them, along with their related components, so they can be kept separate and reinstalled in the same valve guides they are removed from **(see illustration)**.

3   Compress the springs on the first valve with a spring compressor and remove the keepers **(see illustrations)**. Carefully release the valve spring compressor and remove the retainer, the spring and the spring seat. **Caution:** *Be very careful not to nick or otherwise damage the lifter bores when compressing the valve springs.* **Note:** *If your spring compressor does not have an end (such as the one shown) with cutouts on the side, an adapter is available to use with a standard spring compressor.*

4   Pull the valve out of the head, then remove the oil seal from the guide. If the valve binds in the guide (won't pull through), push it back into the head and deburr the area around the keeper groove with a fine file or whetstone.

5   Repeat the procedure for the remaining valves. Remember to keep all the parts for each valve together so they can be reinstalled in the same locations.

6   Once the valves and related components have been removed and stored in an organized manner, the head should be thoroughly cleaned and inspected. If a complete engine overhaul is being done, finish the engine disassembly procedures before beginning the cylinder head cleaning and inspection process.

10 Cylinder head - cleaning and inspection

Refer to illustrations 10.12, 10.14, 10.16, 10.17 and 10.18

1   Thorough cleaning of the cylinder head(s) and related valve train components, followed by a detailed inspection, will enable you to decide how much valve service work must be done during the engine overhaul. **Note:** *If the engine was severely overheated, the cylinder head is probably warped* (see Step 12).

Cleaning

2   Scrape all traces of old gasket material and sealing compound off the head gasket, intake manifold and exhaust manifold sealing surfaces. Be very careful not to gouge the cylinder head. Special gasket-removal solvents that soften gaskets and make removal much easier are available at auto parts stores.

3   Remove all built up scale from the coolant passages.

4   Run a stiff wire brush through the various holes to remove deposits that may have formed in them. If there are heavy rust deposits in the water passages, the bare head should be professionally cleaned at a machine shop.

5   Run an appropriate-size tap into each of the threaded holes to remove corrosion and thread sealant that may be present. If compressed air is available, use it to clear the holes of debris produced by this operation. **Warning:** *Wear eye protection when using compressed air!*

6   Clean the exhaust and intake manifold stud threads with a wire brush.

7   Clean the cylinder head with solvent and dry it thoroughly. Compressed air will speed the drying process and ensure that all holes and recessed areas are clean. **Note:** *Decarbonizing chemicals are available and may prove very useful when cleaning cylinder heads and valve train components. They are very caustic and should be used with caution. Be sure to follow the instructions on the container.*

8   Clean the valvetrain components with solvent and dry them thoroughly. Compressed air will speed the drying process and can be used to clean out the oil passages. Don't mix them up during the cleaning process; keep them in a box with numbered compartments.

9   Clean all the valve springs, spring seats, keepers and retainers with solvent and dry them thoroughly. Work on the components from one valve at a time to avoid mixing up the parts.

10   Scrape off any heavy deposits that may have formed on the valves, then use a motorized wire brush to remove deposits from the valve heads and stems. Again, make sure the valves don't get mixed up.

Inspection

Note: *Be sure to perform all of the following inspection procedures before concluding that machine shop work is required. Make a list of the items that need attention. The inspection procedures for the rockers, shafts and camshaft can be found in Part A.*

Cylinder head

11   Inspect the head very carefully for cracks, evidence of coolant leakage and other damage. If cracks are found, check with an automotive machine shop concerning repair. If repair isn't possible, a new cylinder head should be obtained.

10.12 Check the cylinder head gasket surfaces for warpage by trying to slip a feeler gauge under the precision straightedge (see the Specifications for the maximum warpage allowed and use a feeler gauge of that thickness)

10.14 A dial indicator can be used to determine the valve stem-to-guide clearance

12 Using a straightedge and feeler gauge, check the head gasket mating surface for warpage **(see illustration)**. If the warpage exceeds the limit found in this Chapter's Specifications, it can be resurfaced at an automotive machine shop.

13 Examine the valve seats in each of the combustion chambers. If they're pitted, cracked or burned, the head will require valve service that's beyond the scope of the home mechanic.

14 Check the valve stem-to-guide clearance with a small hole gauge and micrometer, or a dial indicator. Also, check the valve stem deflection with a dial indicator attached securely to the head **(see illustration)**. The valve must be in the guide and approximately 1/16-inch off the seat. The total valve stem movement indicated by the gauge needle must be noted, then divided by two to obtain the actual clearance value. If it exceeds the stem-to-guide clearance limit found in this Chapter's Specifications, the valve guides should be replaced. After this is done, if there's still some doubt regarding the condition of the valve guides they should be checked by an automotive machine shop (the cost should be minimal). **Note:** *Most home mechanics will not have a precision small bore gauge, but your local machine shop can measure the guides for you.*

Valves

15 Carefully inspect each valve face for uneven wear, deformation, cracks, pits and burned areas. Check the valve stem for scuffing and galling and the neck for cracks. Rotate the valve and check for any obvious indication that it's bent. Look for pits and excessive wear on the end of the stem. The presence of any of these conditions indicates the need for valve service by an automotive machine shop.

16 Measure the margin width on each valve **(see illustration)**. Any valve with a margin narrower than that listed in this Chapter's Specifications will have to be replaced with a new one.

Valve components

17 Check each valve spring for wear (on the ends) and pits. Measure the free length and compare it to this Chapter's Specifications **(see illustration)**. Any springs that are shorter than specified have sagged and should not be re-used. The tension of all springs should be pressure checked with a special fixture before deciding that they're suitable for use in a rebuilt engine (take the springs to an automotive machine shop for this check).

18 Stand each spring on a flat surface and check it for squareness **(see illustration)**. If any of the springs are distorted or sagged, replace all of them with new parts.

19 Check the spring retainers and keepers for obvious wear and cracks. Any questionable parts should be replaced with new ones, as extensive damage will occur if they fail during engine operation.

20 Any damaged or excessively worn parts must be replaced with new ones.

21 If the inspection process indicates that the valve components are in generally poor condition and worn beyond the limits specified, which is usually the case in an engine that's being overhauled, reassemble the valves in the cylinder head and refer to Section 11 for valve servicing recommendations.

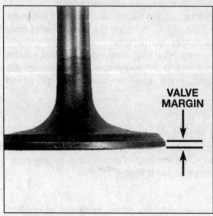

10.16 The margin width on each valve must be as specified (if no margin exists, the valve cannot be re-used)

10.17 Measure the free length of each valve spring with a dial or vernier caliper

10.18 Check each valve spring for squareness

11 Valves - servicing

1   Because of the complex nature of the job and the special tools and equipment needed, servicing of the valves, the valve seats and the valve guides, commonly known as a valve job, should be done by a professional.
2   The home mechanic can remove and disassemble the head(s), do the initial cleaning and inspection, then reassemble and deliver them to a dealer service department or an automotive machine shop for the actual service work. Doing the inspection will enable you to see what condition the head(s) and valvetrain components are in and will ensure that you know what work and new parts are required when dealing with an automotive machine shop.
3   The dealer service department, or automotive machine shop, will remove the valves and springs, recondition or replace the valves and valve seats, recondition the valve guides, check and replace the valve springs, spring retainers and keepers (as necessary), replace the valve seals with new ones, reassemble the valve components and make sure the installed spring height is correct. The cylinder head gasket surface will also be resurfaced if it's warped.
4   After the valve job has been performed by a professional, the head(s) will be in like new condition. When the heads are returned, be sure to clean them again before installation on the engine to remove any metal particles and abrasive grit that may still be present from the valve service or head resurfacing operations. Use compressed air, if available, to blow out all the oil holes and passages.

12 Cylinder head - reassembly

Refer to illustrations 12.3 and 12.6

1   Regardless of whether or not the head was sent to an automotive machine shop for valve servicing, make sure it's clean before beginning reassembly. Note that there are several small core plugs (also called freeze plugs or expansion plugs) in the head. These should be replaced whenever the engine is overhauled or the cylinder head is reconditioned (see Section 15 for the replacement procedure).
2   If the head was sent out for valve servicing, the valves and related components will already be in place. Begin the reassembly procedure with Step 8.
3   Install new seals on each of the valve guides. **Note:** *Intake and exhaust valves require different seals - DO NOT mix them up!* Gently tap each valve seal into place until it's seated on the guide **(see illustration)**. **Caution:** *Don't hammer on the valve seals once they're seated or you may damage them. Don't twist or cock the seals during installation or they won't seat properly on the valve stems.*
4   Beginning at one end of the head, lubricate and install the first valve. Apply moly-base grease or clean engine oil to the valve stem.
5   Drop the spring seat over the valve guide and set the valve spring

12.3 Gently tap the valve seals into place with a deep socket and hammer

and retainer in place.
6   Compress the springs with a valve spring compressor and carefully install the keepers in the upper groove, then slowly release the compressor and make sure the keepers seat properly. Apply a small dab of grease to each keeper to hold it in place if necessary **(see illustration)**.
7   Repeat the procedure for the remaining valves. Be sure to return the components to their original locations - don't mix them up!
8   Tap the end of the valve stem lightly two or three times with a plastic hammer to verify that the keepers are all fully seated.

13 Pistons/connecting rods - removal

Refer to illustrations 13.1, 13.3, 13.4 and 13.6
Note: *Prior to removing the piston/connecting rod assemblies, remove the cylinder head(s), the oil pan and the oil pump pick-up tube by referring to the appropriate Sections in Chapter 2A.*

1   Use your fingernail to feel if a ridge has formed at the upper limit of ring travel (about 1/4-inch down from the top of each cylinder). If carbon deposits or cylinder wear have produced ridges, they must be completely removed with a special tool **(see illustration)**. Follow the manufacturer's instructions provided with the tool. Failure to remove the ridges before attempting to remove the piston/connecting rod assemblies may result in piston damage.
2   After the cylinder ridges have been removed, turn the engine upside-down so the crankshaft is facing up.
3   Before the connecting rods are removed, check the endplay with feeler gauges. Slide them between the first connecting rod and the crankshaft throw until the play is removed **(see illustration)**. The

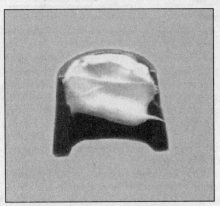

12.6 Apply a small dab of grease to each keeper before installation - it'll hold them in place on the valve stem as the spring is released

13.1 A ridge reamer is required to remove the ridge from the top of each cylinder - do this before removing the pistons!

13.3 Check the connecting rod side clearance with a feeler gauge as shown here

13.4 The connecting rods and caps should be marked to indicate which cylinder they're installed in - if they aren't, mark them with a center punch or scribe to avoid confusion during reassembly

13.6 To prevent damage to the crankshaft journals and cylinder walls, slip sections of hose over the rod bolts before removing the pistons

14.1 Checking crankshaft endplay with a dial indicator

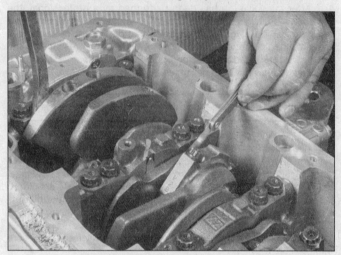

14.3 Checking crankshaft endplay with a feeler gauge

endplay is equal to the thickness of the feeler gauge(s). If the endplay exceeds the specified service limit, new connecting rods will be required. If new rods (or a new crankshaft) are installed, the endplay may fall under the service limit (if it does, the rods will have to be machined to restore it - consult an automotive machine shop for advice if necessary). Repeat the procedure for the remaining connecting rods.

4   Check the connecting rods and caps for identification marks. If they aren't plainly marked, use a small center punch to make the appropriate number of indentations on each rod and cap (1, 2, 3, etc., depending on the engine type and cylinder they're associated with) **(see illustration)**.

5   Loosen each of the connecting rod cap nuts 1/2-turn at a time until they can be removed by hand. Remove the number one connecting rod cap and bearing insert. Don't drop the bearing insert out of the cap.

6   Slip a short length of plastic or rubber hose over each connecting rod cap bolt to protect the crankshaft journal and cylinder wall as the piston is removed **(see illustration)**.

7   Remove the bearing insert and push the connecting rod/piston assembly out through the top of the engine. Use a wooden hammer handle to push on the upper bearing surface in the connecting rod. If resistance is felt, double-check to make sure that all of the ridge was removed from the cylinder.

8   Repeat the procedure for the remaining cylinders. **Note:** *Turn the crankshaft as needed to put the rod to be removed close to parallel with the cylinder bore, i.e. don't try to drive it out while at a large angle to the bore.*

9   After removal, reassemble the connecting rod caps and bearing inserts in their respective connecting rods and install the cap nuts/bolts finger tight. Leaving the old bearing inserts in place until reassembly will help prevent the connecting rod bearing surfaces from being accidentally nicked or gouged.

10   Don't separate the pistons from the connecting rods (see Section 18 for additional information).

14 Crankshaft removal

Refer to illustrations 14.1 and 14.3

Note: *The crankshaft can be removed only after the engine has been removed from the vehicle. It's assumed that the flywheel or driveplate, crankshaft sprocket, timing belt, oil pan, oil pick-up tube, oil pump and piston/connecting rod assemblies have already been removed. The rear main oil seal and retainer must be removed from the block before proceeding with crankshaft removal.*

1   Before the crankshaft is removed, check the endplay. Mount a dial indicator with the stem in line with the crankshaft and touching the end of the crankshaft **(see illustration)**.

2   Push the crankshaft all the way to the rear and zero the dial indicator. Next, pry the crankshaft to the front as far as possible and check the reading on the dial indicator. The distance that it moves is the endplay. If it's greater than specified, check the crankshaft thrust surfaces for wear. If no wear is evident, new thrust bearings should correct the endplay. **Note:** *Different thicknesses of thrust bearings are available to achieve the correct endplay.*

Chapter 2 Part B General engine overhaul procedures   2B-11

15.1a A hammer and a large punch can be used to knock the core plugs sideways in their bores

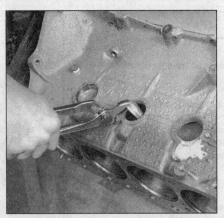

15.1b Pull the core plugs from the block with pliers

15.8 All bolt holes in the block - particularly the main bearing cap and head bolt holes - should be cleaned and restored with a tap (be sure to remove debris from the holes after this is done)

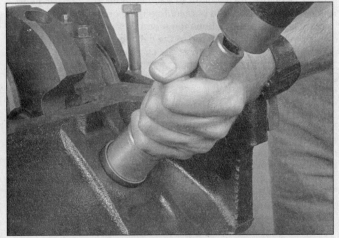

15.10 A large socket on an extension can be used to drive the new core plugs into the bores

3   If a dial indicator is not available, feeler gauges can be used. Gently pry or push the crankshaft all the way to the front of the engine. Slip feeler gauges between the crankshaft and the front face of the number 4 (thrust) main bearing to determine the clearance (see illustration).
4   Check the main bearing caps to see if they're marked to indicate their locations. They should be numbered consecutively from the front of the engine to the rear. If they aren't, mark them with number stamping dies or a center punch. Main bearing caps generally have a cast-in arrow, which points to the front of the engine. Loosen the main bearing cap bolts 1/4-turn at a time each, in the reverse order of the recommended tightening sequence (see illustration 23.12), until they can be removed by hand.
5   Gently tap the caps with a soft-face hammer, then separate them from the engine block. If necessary, use the bolts as levers to remove the caps. Try not to drop the bearing inserts if they come out with the caps.
6   Carefully lift the crankshaft out of the engine. It may be a good idea to have an assistant available, since the crankshaft is quite heavy. With the bearing inserts in place in the engine block and main bearing caps or cap assembly, return the caps to their respective locations on the engine block and tighten the bolts finger tight.

15 Engine block - cleaning

Refer to illustrations 15.1a, 15.1b, 15.8 and 15.10
Caution: *The core plugs (also known as freeze or soft plugs) may be difficult or impossible to retrieve if they're driven completely into the block coolant passages.*

1   Using the blunt end of a punch, tap in on the outer edge of the core plug to turn the plug sideways in the bore. Then using pliers, pull the core plug from the engine block (see illustrations).
2   Using a gasket scraper, remove all traces of gasket material from the engine block. Be very careful not to nick or gouge the gasket sealing surfaces.
3   Remove the main bearing caps or cap assembly and separate the bearing inserts from the caps and the engine block. Tag the bearings, indicating which cylinder they were removed from and whether they were in the cap or the block, then set them aside.
4   Remove all of the threaded oil gallery plugs from the block. The plugs are usually very tight - they may have to be drilled out and the holes retapped. Use new plugs when the engine is reassembled.
5   If the engine is extremely dirty, it should be taken to an automotive machine shop to be steam cleaned or hot tanked.
6   After the block is returned, clean all oil holes and oil galleries one more time. Brushes specifically designed for this purpose are available at most auto parts stores. Flush the passages with warm water until the water runs clear, dry the block thoroughly and wipe all machined surfaces with a light, rust preventive oil. If you have access to compressed air, use it to speed the drying process and to blow out all the oil holes and galleries. **Warning:** *Wear eye protection when using compressed air!*
7   If the block is not extremely dirty or sludged up, you can do an adequate cleaning job with hot soapy water and a stiff brush. Take plenty of time and do a thorough job. Regardless of the cleaning method used, be sure to clean all oil holes and galleries very thoroughly, dry the block completely and coat all machined surfaces with light oil.
8   The threaded holes in the block must be clean to ensure accurate torque readings during reassembly. Run the proper size tap into each of the holes to remove rust, corrosion, thread sealant or sludge and restore damaged threads (see illustration). If possible, use compressed air to clear the holes of debris produced by this operation. Now is a good time to clean the threads on the head bolts and the main bearing cap bolts as well.
9   Reinstall the main bearing caps and tighten the bolts finger tight.
10 After coating the sealing surfaces of the new core plugs with Permatex no. 2 sealant, install them in the engine block (see illustration). Make sure they are driven in straight and seated properly or leakage could result. Special tools are available for this purpose, but a large socket, with an outside diameter that will just slip into the core plug, a 1/2-inch drive extension and a hammer will work just as well.
11 Apply non-hardening sealant (such as Permatex no. 2 or Teflon pipe sealant) to the new oil gallery plugs and thread them into the holes in the block. Make sure they are tightened securely.
12 If the engine isn't going to be reassembled right away, cover it with a large plastic trash bag to keep it clean.

16 Engine block - inspection

Refer to illustrations 16.4a, 16.4b, 16.4c and 16.9

1   Before the block is inspected, it should be cleaned as described in Section 15.
2   Visually check the block for cracks, rust and corrosion. Look for stripped threads in the threaded holes. It's also a good idea to have the block checked for hidden cracks by an automotive machine shop that has the special equipment to do this type of work, especially if the vehicle had a history of overheating or using coolant. If defects are found, have the block repaired, if possible, or replaced.
3   Check the cylinder bores for scuffing and scoring.
4   Check the cylinders for taper and out-of-round conditions as follows **(see illustrations)**:
5   Measure the diameter of each cylinder at the top (just under the ridge area), center and bottom of the cylinder bore, parallel to the crankshaft axis.
6   Next, measure each cylinder's diameter at the same three locations perpendicular to the crankshaft axis.
7   The taper of each cylinder is the difference between the bore diameter at the top of the cylinder and the diameter at the bottom. The out-of-round specification of the cylinder bore is the difference between the parallel and perpendicular readings. Compare your results to this Chapter's Specifications.
8   If the cylinder walls are badly scuffed or scored, or if they're out-of-round or tapered beyond the limits given in this Chapter's Specifications, have the engine block rebored and honed at an automotive machine shop. If a rebore is done, oversize pistons and rings will be required.
9   Using a precision straightedge and feeler gauge, check the block deck (the surface that mates with the cylinder head) for distortion **(see illustration)**. If it's distorted beyond the specified limit, it can be resurfaced by an automotive machine shop.
10 If the cylinders are in reasonably good condition and not worn to the outside of the limits, and if the piston-to-cylinder clearances can be maintained properly, then they don't have to be rebored. Honing is all that's necessary (Section 17).

17 Cylinder honing

Refer to illustrations 17.3a and 17.3b

1   Prior to engine reassembly, the cylinder bores must be honed so the new piston rings will seat correctly and provide the best possible combustion chamber seal. **Note:** *If you don't have the tools or don't want to tackle the honing operation, most automotive machine shops will do it for a reasonable fee.*

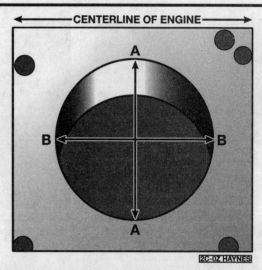

16.4a Measure the diameter of each cylinder at a right angle to engine centerline (A), and parallel to engine centerline (B) - out-of-round is the difference between A and B; taper is the difference between A and B at the top of the cylinder and A and B at the bottom of the cylinder

2   Before honing the cylinders, install the main bearing caps or cap assembly (without bearing inserts) and tighten the bolts to the specified torque.
3   Two types of cylinder hones are commonly available - the flex hone or "bottle brush" type and the more traditional surfacing hone with spring-loaded stones. Both will do the job, but for the less-experienced mechanic the "bottle brush" hone will probably be easier to use. You'll also need some kerosene or honing oil, rags and an electric drill motor. The drill motor should be operated at a steady, slow speed. Use a large 1/2-inch drill or a 3/8-inch variable-speed drill. Proceed as follows:

a) *Mount the hone in the drill motor, compress the stones and slip it into the first cylinder* **(see illustration)**. **Warning:** *Be sure to wear safety goggles or a face shield!*
b) *Lubricate the cylinder with plenty of honing oil, turn on the drill and move the hone up-and-down in the cylinder at a pace that will produce a fine crosshatch pattern on the cylinder walls. Ideally, the crosshatch lines should intersect at approximately a 60-degree angle* **(see illustration)**. *Be sure to use plenty of lubricant and don't take off any more material than is absolutely necessary to produce the desired finish.* **Note:** *Piston ring manufacturers may specify a smaller crosshatch angle than the traditional 60-degrees - read and follow any instructions included with the new rings.*

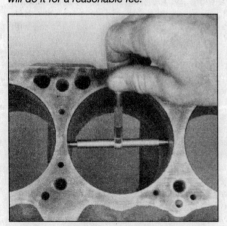

16.4b The ability to "feel" when the telescoping gauge is at the correct point will be developed over time, so work slowly and repeat the check until you're satisfied that the bore measurement is accurate

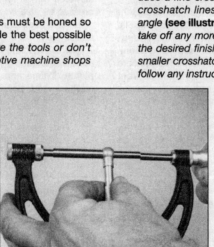

16.4c The gauge is then measured with a micrometer to determine the bore size

16.9 Check the block deck for distortion with a precision straightedge and feeler gauges

Chapter 2 Part B General engine overhaul procedures

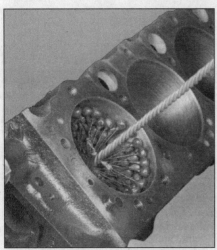

17.3a A "bottle brush" hone will produce better results if you have never done cylinder honing before

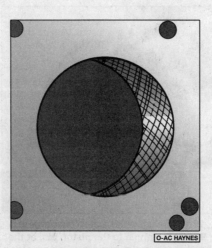

17.3b The cylinder hone should leave a smooth, crosshatch pattern with the lines intersecting at approximately a 60-degree angle

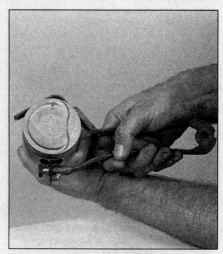

18.4a The piston ring grooves can be cleaned with a special tool, as shown here . . .

c) Don't withdraw the hone from the cylinder while it's running. Instead, shut off the drill and continue moving the hone up-and-down in the cylinder until it comes to a complete stop, then compress the stones and withdraw the hone. If you're using a "bottle brush" type hone, stop the drill motor, then turn the chuck in the normal direction of rotation while withdrawing the hone from the cylinder.

d) Wipe the oil out of the cylinder and repeat the procedure for the remaining cylinders.

4 After the honing job is complete, chamfer the top edges of the cylinder bores with a small file so the rings won't catch when the pistons are installed. Be very careful not to nick the cylinder walls with the end of the file.

5 The entire engine block must be washed again very thoroughly with warm, soapy water to remove all traces of the abrasive grit produced during the honing operation. **Note:** *The bores can be considered clean when a lint-free white cloth - dampened with clean engine oil - used to wipe them out doesn't pick up any more honing residue, which will show up as gray areas on the cloth. Be sure to run a brush through all oil holes and galleries and flush them with running water.*

6 After rinsing, dry the block and apply a coat of light rust preventive oil to all machined surfaces. Wrap the block in a plastic trash bag to keep it clean and set it aside until reassembly.

18 Pistons/connecting rods - inspection

Refer to illustrations 18.4a, 18.4b, 18.10 and 18.11

1 Before the inspection process can be carried out, the piston/connecting rod assemblies must be cleaned and the original piston rings removed from the pistons. **Note:** *Always use new piston rings when the engine is reassembled.*

2 Using a piston ring installation tool, carefully remove the rings from the pistons. Be careful not to nick or gouge the pistons in the process.

3 Scrape all traces of carbon from the top of the piston. A handheld wire brush or a piece of fine emery cloth can be used once the majority of the deposits have been scraped away. Do not, under any circumstances, use a wire brush mounted in a drill motor to remove deposits from the pistons. The piston material is soft and may be eroded away by the wire brush.

4 Use a piston ring groove-cleaning tool to remove carbon deposits from the ring grooves. If a tool isn't available, a piece broken off the old ring will do the job. Be very careful to remove only the carbon deposits - don't remove any metal and do not nick or scratch the sides of the ring grooves **(see illustrations)**.

5 Once the deposits have been removed, clean the piston/rod assemblies with solvent and dry them with compressed air (if available). Make sure the oil return holes in the back sides of the ring grooves and the oil hole in the lower end of each rod are clear.

6 If the pistons and cylinder walls aren't damaged or worn excessively, and if the engine block is not rebored, new pistons won't be necessary. Normal piston wear appears as even vertical wear on the piston thrust surfaces and slight looseness of the top ring in its groove. New piston rings, however, should always be used when an engine is rebuilt.

7 Carefully inspect each piston for cracks around the skirt, at the pin bosses and at the ring lands.

8 Look for scoring and scuffing on the thrust faces of the skirt, holes in the piston crown and burned areas at the edge of the crown. If the skirt is scored or scuffed, the engine may have been suffering from overheating and/or abnormal combustion, which caused excessively high operating temperatures. The cooling and lubrication systems should be checked thoroughly. A hole in the piston crown is an indication that abnormal combustion (preignition) was occurring. Burned areas at the edge of the piston crown are usually evidence of spark knock (detonation). If any of the above problems exist, the causes must be corrected or the damage will occur again. The causes may include intake air leaks, incorrect air/fuel mixture, incorrect ignition timing and EGR system malfunctions.

18.4b . . . or a section of a broken ring

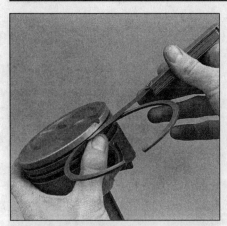

18.10 Check the ring groove clearance with a feeler gauge at several points around the groove

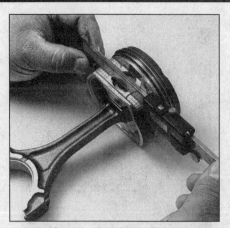

18.11 Measure the piston diameter at a 90-degree angle to the piston pin, at the bottom of the piston pin area - a precision caliper may be used if a micrometer isn't available

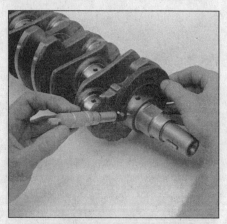

19.6 Measure the diameter of each crankshaft journal at several points to detect taper and out-of-round conditions

9   Corrosion of the piston, in the form of small pits, indicates that coolant is leaking into the combustion chamber and/or the crankcase. Again, the cause must be corrected or the problem may persist in the rebuilt engine.

10   Measure the piston ring groove clearance by laying a new piston ring in each ring groove and slipping a feeler gauge in beside it **(see illustration)**. Check the clearance at three or four locations around each groove. Be sure to use the correct ring for each groove - they are different. If the clearance is greater than that listed in this Chapter's Specifications, new pistons will have to be used.

11   Check the piston-to-bore clearance by measuring the bore (see Section 16) and the piston diameter. Make sure the pistons and bores are correctly matched. Measure the piston across the skirt, at a 90-degree angle to the piston pin **(see illustration)**. Subtract the piston diameter from the bore diameter to obtain the clearance. If it's greater than specified, the block will have to be rebored and new pistons and rings installed.

12   Check the piston-to-rod clearance by twisting the piston and rod in opposite directions. Any noticeable play indicates excessive wear, which must be corrected.

13   If the pistons must be removed from the connecting rods for any reason, the rods should be taken to an automotive machine shop, to be checked for bend and twist, since automotive machine shops have special equipment for this purpose.

14   Check the connecting rods for cracks and other damage. Temporarily remove the rod caps, lift out the old bearing inserts, wipe the rod and cap bearing surfaces clean and inspect them for nicks, gouges and scratches. After checking the rods, replace the old bearings, slip the caps into place and tighten the nuts finger tight. **Note:** *If the engine is being rebuilt because of a connecting rod knock, be sure to install new rods.*

19   Crankshaft - inspection

Refer to illustration 19.6

1   Clean the crankshaft with solvent and dry it with compressed air (if available).

2   Check the main and connecting rod bearing journals for uneven wear, scoring, pits and cracks.

3   Remove all burrs from the crankshaft oil holes with a stone, file or scraper.

4   Clean the oil holes with a stiff brush and flush them with solvent.

5   Check the rest of the crankshaft for cracks and other damage. It should be magnafluxed to reveal hidden cracks - an automotive machine shop will handle the procedure.

6   Using a micrometer, measure the diameter of the main and connecting rod journals and compare the results to this Chapter's Specifications **(see illustration)**. By measuring the diameter at a number of points around each journal's circumference, you'll be able to determine whether or not the journal is out-of-round. Take the measurement at each end of the journal, near the crank throws, to determine if the journal is tapered. Crankshaft runout should be checked also, but large V-blocks and a dial indicator are needed to do it correctly. If you don't have the equipment, have a machine shop check the runout.

7   If the crankshaft journals are damaged, tapered, out-of-round or worn beyond the limits given in the Specifications, have the crankshaft reground by an automotive machine shop. Be sure to use the correct size bearing inserts if the crankshaft is reconditioned.

8   Check the oil seal journals at each end of the crankshaft for wear and damage. If the seal has worn a groove in the journal, or if it's nicked or scratched, the new seal may leak when the engine is reassembled. In some cases, an automotive machine shop may be able to repair the journal by pressing on a thin sleeve. If repair isn't feasible, a new or different crankshaft should be installed.

9   Refer to Section 20 and examine the main and rod bearing inserts.

20   Main and connecting rod bearings - inspection and selection

Inspection

Refer to illustration 20.1

1   Even though the main and connecting rod bearings should be replaced with new ones during the engine overhaul, the old bearings should be retained for close examination, as they may reveal valuable information about the condition of the engine **(see illustration)**.

2   Bearing failure occurs because of lack of lubrication, the presence of dirt or other foreign particles, overloading the engine and corrosion. Regardless of the cause of bearing failure, it must be corrected before the engine is reassembled to prevent it from happening again.

3   When examining the bearings, remove them from the engine block, the main bearing caps, the connecting rods and the rod caps and lay them out on a clean surface in the same general position as their location in the engine. This will enable you to match any bearing problems with the corresponding crankshaft journal.

4   Dirt and other foreign particles get into the engine in a variety of ways. It may be left in the engine during assembly, or it may pass through filters or the PCV system. It may get into the oil, and from there into the bearings. Metal chips from machining operations and normal

Chapter 2 Part B General engine overhaul procedures

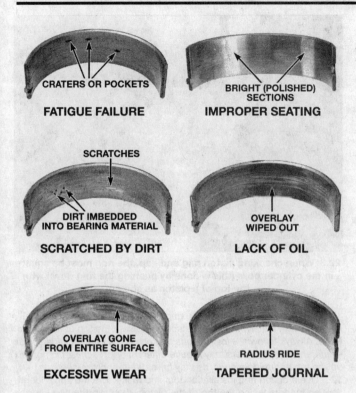

20.1 When inspecting the main and connecting rod bearings, look for these problems

engine wear are often present. Abrasives are sometimes left in engine components after reconditioning, especially when parts are not thoroughly cleaned using the proper cleaning methods. Whatever the source, these foreign objects often end up embedded in the soft bearing material and are easily recognized. Large particles will not embed in the bearing and will score or gouge the bearing and journal. The best prevention for this cause of bearing failure is to clean all parts thoroughly and keep everything spotlessly clean during engine assembly. Frequent and regular engine oil and filter changes are also recommended.

5 Lack of lubrication (or lubrication breakdown) has a number of interrelated causes. Excessive heat (which thins the oil), overloading (which squeezes the oil from the bearing face) and oil leakage or throw off (from excessive bearing clearances, worn oil pump or high engine speeds) all contribute to lubrication breakdown. Blocked oil passages, which usually are the result of misaligned oil holes in a bearing shell, will also oil starve a bearing and destroy it. When lack of lubrication is the cause of bearing failure, the bearing material is wiped or extruded from the steel backing of the bearing. Temperatures may increase to the point where the steel backing turns blue from overheating.

6 Driving habits can have a definite effect on bearing life. Low speed operation in too high a gear (lugging the engine) puts very high loads on bearings, which tends to squeeze out the oil film. These loads cause the bearings to flex, which produces fine cracks in the bearing face (fatigue failure). Eventually the bearing material will loosen in pieces and tear away from the steel backing. Short trip driving leads to corrosion of bearings because insufficient engine heat is produced to drive off the condensed water and corrosive gases. These products collect in the engine oil, forming acid and sludge. As the oil is carried to the engine bearings, the acid attacks and corrodes the bearing material.

7 Incorrect bearing installation during engine assembly will lead to bearing failure as well. Tight-fitting bearings leave insufficient bearing oil clearance and will result in oil starvation. Dirt or foreign particles trapped behind a bearing insert result in high spots on the bearing which lead to failure.

Selection

8 If the original bearings are worn or damaged, or if the oil clearances are incorrect (see Section 23 or 25), the following procedures should be used to select the correct new bearings for engine reassembly. However, if the crankshaft has been reground, new undersize bearings must be installed - the following procedure should not be used if undersize bearings are required! The automotive machine shop that reconditions the crankshaft will provide or help you select the correct size bearings. Regardless of how the bearing sizes are determined, use the oil clearance, measured with Plastigage, as a guide to ensure the bearings are the right size.

Main bearings

9 If you need to use a STANDARD size main bearing, install one that is the same size as the original bearing listed in the specifications in the front of this Chapter. If the diameter is less than the minimum, have a qualified engine machine shop grind the journals to match standard undersize bearings.

Connecting rod bearings

10 If you need to use a STANDARD size rod bearing, install one that is the same size as the original bearing listed in the specifications in the from of this Chapter. If the diameter is less than the minimum, grind the journals to match standard undersize bearings.

All bearings

11 Remember, the oil clearance is the final judge when selecting new bearing sizes. If you have any questions or are unsure which bearings to use, get help from a dealer parts or service department.

21 Engine overhaul - reassembly sequence

1 Before beginning engine reassembly, make sure you have all the necessary new parts, gaskets and seals as well as the following items on hand:

Common hand tools
A 1/2-inch drive torque wrench
Piston ring installation tool
Piston ring compressor
Short lengths of rubber or plastic hose
to fit over connecting rod bolts
Plastigage
Feeler gauges
A fine-tooth file
New engine oil
Engine assembly lube or moly-base grease
Camshaft installation lube
Gasket sealant
Thread locking compound

2 In order to save time and avoid problems, engine reassembly must be done in the following general order:

Piston rings (Part B)
Crankshaft and main bearings (Part B)
Piston/connecting rod assemblies (Part B)
Rear main (crankshaft) oil seal (Part B)
Oil pump (Part A)
Oil pick-up (Part A)
Oil pan (Part A)
Cylinder head (Part A)
Camshaft (Part A)
Timing belt and sprockets (Part A)
Water pump (Chapter 3)
Timing belt covers (Part A)
Intake and exhaust manifolds (Part A)
Valve cover (Part A)
Flywheel/driveplate (Part A)

22 Piston rings - installation

Refer to illustrations 22.3, 22.4, 22.9a, 22.9b and 22.12

1   Before installing the new piston rings, the ring end gaps must be checked. It's assumed that the piston ring groove clearance has been checked and verified correct (see Section 18).
2   Lay out the piston/connecting rod assemblies and the new ring sets so the ring sets will be matched with the same piston and cylinder during the end gap measurement and engine assembly.
3   Insert the top (number one) ring into the first cylinder and square it up with the cylinder walls by pushing it in with the top of the piston **(see illustration)**. The ring should be near the bottom of the cylinder, at the lower limit of ring travel.
4   To measure the end gap, slip feeler gauges between the ends of the ring until a gauge equal to the gap width is found **(see illustration)**. The feeler gauge should slide between the ring ends with a slight amount of drag. Compare the measurement to that found in this Chapter's Specifications. If the gap is larger or smaller than specified, double-check to make sure you have the correct rings before proceeding.
5   If the gap is too small, it must be enlarged or the ring ends may come in contact with each other during engine operation, which can cause serious damage to the engine. The end gap can be increased by filing the ring ends very carefully with a fine file. Mount the file in a vise equipped with soft jaws, slip the ring over the file with the ends contacting the file face and slowly move the ring to remove material from the ends. **Caution:** *When performing this operation, file only from the outside in. After the proper gap is achieved, deburr the sharp edges of the filed ends with a whetstone.*
6   Excess end gap is not critical unless it's greater than the service limit listed in this Chapter's Specifications. Again, double-check to make sure you have the correct rings for your engine.
7   Repeat the procedure for each ring that will be installed in the first cylinder and for each ring in the remaining cylinders. Remember to keep rings, pistons and cylinders matched up.
8   Once the ring end gaps have been checked/corrected, the rings can be installed on the pistons.
9   The oil control ring (lowest one on the piston) is usually installed first. It's composed of three separate components. Slip the spacer/expander into the groove **(see illustration)**. If an anti-rotation tang is used, make sure it's inserted into the drilled hole in the ring groove. Next, install the lower side rail. Don't use a piston ring installation tool on the oil ring side rails, as they may be damaged. Instead, place one end of the side rail into the groove between the spacer/expander and the ring land, hold it firmly in place and slide a finger around the piston while pushing the rail into the groove **(see illustration)**. Next, install the upper side rail in the same manner.
10   After the three oil ring components have been installed, check to make sure that both the upper and lower side rails can be turned smoothly in the ring groove.

22.3 When checking piston ring end gap, the ring must be square in the cylinder bore (this is done by pushing the ring down with the top of a piston as shown)

11   The number two (middle) ring is installed next. It's usually stamped with a mark which must face up, toward the top of the piston. **Note:** *Always follow the instructions printed on the ring package or box - different manufacturers may require different approaches. Do not mix up the top and middle rings, as they have different cross-sections.*
12   Use a piston ring installation tool and make sure the ring's identification mark is facing the top of the piston, then slip the ring into the middle groove on the piston **(see illustration)**. Don't expand the ring any more than necessary to slide it over the piston.
13   Install the number one (top) ring in the same manner. Make sure the mark is facing up. Be careful not to confuse the number one and number two rings.
14   Repeat the procedure for the remaining pistons and rings.

23 Crankshaft - installation and main bearing oil clearance check

Oil clearance check

Refer to illustrations 23.10, 23.12, 23.14, 23.19a and 23.19b

1   Crankshaft installation is the first major step in engine reassembly. It's assumed at this point that the engine block and crankshaft have been cleaned, inspected and repaired or reconditioned.
2   Position the engine with the bottom facing up.
3   Remove the main bearing cap bolts and lift out the caps. Lay the caps out in the proper order.

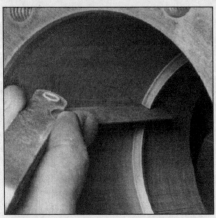

22.4 With the ring square in the cylinder, measure the end gap with a feeler gauge

22.9a Install the spacer/expander in the oil control ring groove

22.9b DO NOT use a piston ring installation tool when installing the oil ring side rails

Chapter 2 Part B General engine overhaul procedures

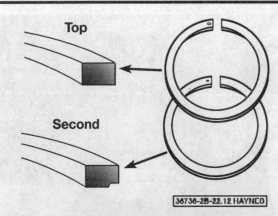

22.12 Install the compression rings with a ring expander - the mark must face up

23.10 Lay the Plastigage strips (arrow) on the main bearing journals, parallel to the crankshaft centerline

4   If they're still in place, remove the old bearing inserts from the block and the main bearing caps. Wipe the main bearing surfaces of the block and caps with a clean, lint free cloth. They must be kept spotlessly clean!

Main bearing oil clearance check

5   Clean the back sides of the new main bearing inserts and lay the bearing half with the oil groove in each main bearing saddle in the block. Lay the other bearing half from each bearing set in the corresponding main bearing cap. Make sure the tab on each bearing insert fits into the recess in the block or cap. Also, the oil holes in the block must line up with the oil holes in the bearing insert. **Caution:** *Do not hammer the bearings into place and don't nick or gouge the bearing faces. No lubrication should be used at this time.*

6   The thrust bearings (washers) must be installed in the number four cap and saddle.

7   Clean the faces of the bearings in the block and the crankshaft main bearing journals with a clean, lint free cloth. Check or clean the oil holes in the crankshaft, as any dirt here can go only one way - straight through the new bearings.

8   Once you're certain the crankshaft is clean, carefully lay it in position in the main bearings.

9   Before the crankshaft can be permanently installed, the main bearing oil clearance must be checked.

10 Trim several pieces of the appropriate size Plastigage (they must be slightly shorter than the width of the main bearings) and place one piece on each crankshaft main bearing journal, parallel with the journal axis **(see illustration)**.

11 Clean the faces of the bearings in the caps and install the caps in their respective positions (don't mix them up) with the arrows pointing toward the front of the engine. Don't disturb the Plastigage. Apply a light coat of oil to the bolt threads and the under- sides of the bolt heads, then install them.

12 Following the recommended sequence **(see illustration)**, tighten the main bearing cap bolts, in three steps, to the torque listed in this Chapter's Specifications. Do not rotate the crankshaft at any time during this operation!

13 Remove the bolts and carefully lift off the main bearing caps. Keep them in order. Don't disturb the Plastigage or rotate the crankshaft. If any of the main bearing caps are difficult to remove, tap them gently from side-to-side with a soft-face hammer to loosen them.

14 Compare the width of the crushed Plastigage on each journal to the scale printed on the Plastigage envelope to obtain the main bearing oil clearance **(see illustration)**. Check the Specifications to make sure it's correct.

15 If the clearance is not as specified, the bearing inserts may be the wrong size (which means different ones will be required - see Section 20). Before deciding that different inserts are needed, make sure that no dirt or oil was between the bearing inserts and the caps or block when the clearance was measured. If the Plastigage is noticeably wider at one end than the other, the journal may be tapered (see Section 19).

16 Carefully scrape all traces of the Plastigage material off the main bearing journals and/or the bearing faces. Don't nick or scratch the bearing faces.

23.12 Main bearing cap bolt tightening sequence

23.14 Compare the width of the crushed Plastigage to the scale on the envelope to determine the main bearing oil clearance (always take the measurement at the widest point of the Plastigage) - be sure to use the correct scale; standard and metric scales are included

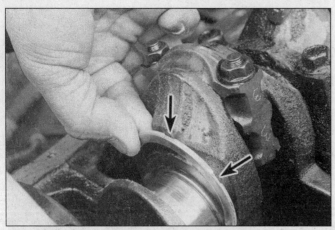

23.19a Insert the thrust washer into position with the oil grooves facing OUT (arrow)

23.19b Rotate the thrust washer into position

Final crankshaft installation

17 Carefully lift the crankshaft out of the engine. Clean the bearing faces in the block, then apply a thin, uniform layer of clean moly-base grease or engine assembly lube to each of the bearing surfaces. Coat the thrust washers as well.

18 Lubricate the crankshaft surfaces that contact the oil seals with multi-purpose grease, engine assembly lube or clean engine oil.

19 Make sure the crankshaft journals are clean, then lay the crankshaft back in place in the block. Clean the faces of the bearings in the caps, then apply lubricant to them. Install the caps in their respective positions with the arrows pointing toward the front of the engine. **Note:** *Be sure to install the thrust washers with the number 4 main journal.* The upper (block side) thrust washers can be rotated into position around the crank with the crank in the block, with the thrust washer grooves facing OUT. The tanged lower thrust washers should be placed on the caps with their grooves OUT and the tangs fitting into the cap slots **(see illustrations)**.

20 Apply a light coat of oil to the bolt threads and the undersides of the bolt heads, then install them. Tighten all main bearing cap bolts to the torque listed in this Chapter's Specifications, following the recommended sequence.

21 Rotate the crankshaft a number of times by hand to check for any obvious binding.

22 Check the crankshaft endplay with a feeler gauge or a dial indicator as described in Section 14. The endplay should be correct if the crankshaft thrust faces aren't worn or damaged and new thrust washers have been installed.

23 Install a new rear main oil seal, then bolt the retainer to the block (see Section 24).

24 Rear main oil seal - installation

Refer to illustrations 24.3 and 24.5

1 The crankshaft must be installed first and the main bearing caps bolted in place, then the new seal should be installed in the retainer and the retainer bolted to the block.

2 Check the seal contact surface on the crankshaft very carefully for scratches and nicks that could damage the new seal lip and cause oil leaks. If the crankshaft is damaged, the only alternative is a new or different crankshaft.

3 The old seal can be removed from the retainer by driving it out from the back side with a hammer and punch **(see illustration)**. Be sure to note how far it's recessed into the bore before removing it; the new seal will have to be recessed an equal amount. Be very careful not to scratch or otherwise damage to the bore in the retainer or oil leaks could develop.

4 Make sure the retainer is clean, then apply a thin coat of engine oil to the outer edge of the new seal. The seal must be pressed squarely into the bore, so hammering it into place isn't recommended. If you

24.3 After removing the retainer from the block, support it on a couple of wood blocks and drive out the old seal with a punch or screwdriver and hammer

don't have access to a press, sandwich the housing and seal between two smooth pieces of wood and press the seal into place with the jaws of a large vise. The pieces of wood must be thick enough to distribute the force evenly around the entire circumference of the seal. Work slowly and make sure the seal enters the bore squarely.

5 As a last resort, the seal can be tapped into the retainer with a hammer. Use a block of wood to distribute the force evenly and make sure the seal is driven in squarely **(see illustration)**.

24.5 Drive the new seal into the retainer with a wood block or a section of pipe, if you have one large enough - make sure you don't cock the seal in the retainer bore

Chapter 2 Part B General engine overhaul procedures

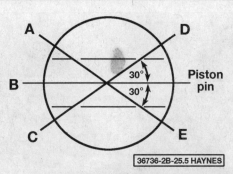

25.5 Ring end gap positions - stagger the ring end gaps around the piston, as shown, before installing the pistons

25.9 Check to assure the "F" mark on the piston (arrow) and the oil groove on the big end of the connecting rod are aligned towards the front of the engine

6 The seal lips must be lubricated with clean engine oil or multi-purpose grease before the seal/retainer is slipped over the crankshaft and bolted to the block, using a new gasket.

7 Tighten the bolts a little at a time to the torque listed in the Chapter 2A Specifications.

25 Pistons/connecting rods - installation and rod bearing oil clearance check

Refer to illustrations 25.5, 25.9, 25.11, 25.13 and 25.17

1 Before installing the piston/connecting rod assemblies, the cylinder walls must be perfectly clean, the top edge of each cylinder must be chamfered, and the crankshaft must be in place.

2 Remove the cap from the end of the number one connecting rod (refer to the marks made during removal). Remove the original bearing inserts and wipe the bearing surfaces of the connecting rod and cap with a clean, lint-free cloth. They must be kept spotlessly clean.

Connecting rod bearing oil clearance check

3 Clean the back side of the new upper bearing insert, then lay it in place in the connecting rod. Make sure the tab on the bearing fits into the recess in the rod so the oil holes line up. Don't hammer the bearing insert into place and be very careful not to nick or gouge the bearing face. Don't lubricate the bearing at this time.

4 Clean the back side of the other bearing insert and install it in the rod cap. Again, make sure the tab on the bearing fits into the recess in the cap, and don't apply any lubricant. It's critically important that the mating surfaces of the bearing and connecting rod are perfectly clean and oil free when they're assembled.

5 Position the piston ring gaps at staggered intervals around the piston **(see illustration)**.

6 Slip a section of plastic or rubber hose over each connecting rod cap bolt.

7 Lubricate the piston and rings with clean engine oil and attach a piston ring compressor to the piston. Leave the skirt protruding about 1/4-inch to guide the piston into the cylinder. The rings must be compressed until they're flush with the piston.

8 Rotate the crankshaft until the number one connecting rod journal is at BDC (bottom dead center) and apply a coat of engine oil to the cylinder wall.

9 With the "F" mark on the side of the piston **(see illustration)** facing the front of the engine, gently insert the piston/connecting rod assembly into the number one cylinder bore and rest the bottom edge of the ring compressor on the engine block.

10 Tap the top edge of the ring compressor to make sure it's contacting the block around its entire circumference.

11 Gently tap on the top of the piston with the end of a wooden hammer handle **(see illustration)** while guiding the end of the connecting rod into place on the crankshaft journal. The piston rings may try to pop out of the ring compressor just before entering the cylinder bore, so keep some downward pressure on the ring compressor. Work slowly, and if any resistance is felt as the piston enters the cylinder, stop immediately. Find out what's hanging up and fix it before proceeding. **Caution:** *Do not, for any reason, force the piston into the cylinder - you might break a ring and/or the piston.*

12 Once the piston/connecting rod assembly is installed, the connecting rod bearing oil clearance must be checked before the rod cap is permanently bolted in place.

13 Cut a piece of the appropriate size Plastigage slightly shorter than the width of the connecting rod bearing and lay it in place on the number one connecting rod journal, parallel with the journal axis **(see illustration)**.

25.11 The piston can be driven (gently) into the cylinder bore with the end of a wooden or plastic hammer handle

25.13 Lay the Plastigage strips on each rod bearing journal, parallel to the crankshaft centerline

14 Clean the connecting rod cap bearing face, remove the protective hoses from the connecting rod bolts and install the rod cap. Make sure the mating mark on the cap is on the same side as the mark on the connecting rod. Check the cap to make sure the front mark is facing the timing belt end of the engine.
15 Apply a light coat of oil to the undersides of the nuts, then install and tighten them to the torque listed in this Chapter's Specifications, working up to it in three steps. Use a thin-wall socket to avoid erroneous torque readings that can result if the socket is wedged between the rod cap and nut. If the socket tends to wedge itself between the nut and the cap, lift up on it slightly until it no longer contacts the cap. Do not rotate the crankshaft at any time during this operation.
16 Remove the nuts and detach the rod cap, being very careful not to disturb the Plastigage.
17 Compare the width of the crushed Plastigage to the scale printed on the Plastigage envelope to obtain the oil clearance (see illustration). Compare it to this Chapter's Specifications to make sure the clearance is correct.
18 If the clearance is not as specified, the bearing inserts may be the wrong size (which means different ones will be required). Before deciding that different inserts are needed, make sure that no dirt or oil was between the bearing inserts and the connecting rod or cap when the clearance was measured. Also, recheck the journal diameter. If the Plastigage was wider at one end than the other, the journal may be tapered (refer to Section 19).

Final connecting rod installation

19 Carefully scrape all traces of the Plastigage material off the rod journal and/or bearing face. Be very careful not to scratch the bearing, use your fingernail or the edge of a credit card to remove the Plastigage.
20 Make sure the bearing faces are perfectly clean, then apply a uniform layer of clean moly-base grease or engine assembly lube to both of them. You'll have to push the piston higher into the cylinder to expose the face of the bearing insert in the connecting rod, be sure to slip the protective hoses over the rod bolts first.
21 Slide the connecting rod back into place on the journal, remove the protective hoses from the rod cap bolts, install the rod cap and tighten the nuts to the torque listed in this Chapter's Specifications. Again, work up to the torque in three steps.
22 Repeat the entire procedure for the remaining pistons/connecting rods.
23 The important points to remember are:
   a) Keep the back sides of the bearing inserts and the insides of the connecting rods and caps perfectly clean when assembling them.
   b) Make sure you have the correct piston/rod assembly for each cylinder.
   c) The "F" mark on the piston must face the front of the engine.
   d) Lubricate the cylinder walls with clean oil.
   e) Lubricate the bearing faces when installing the rod caps after the oil clearance has been checked.
24 After all the piston/connecting rod assemblies have been properly installed, rotate the crankshaft a number of times by hand to check for any obvious binding.
25 As a final step, the connecting rod endplay must be checked. Refer to Section 13 for this procedure.
26 Compare the measured endplay to this Chapter's Specifications to make sure it's correct. If it was correct before disassembly and the original crankshaft and rods were reinstalled, it should still be right. If

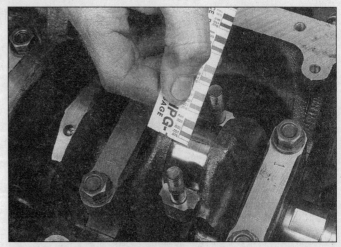

25.17 Measure the width of the crushed Plastigage to determine the rod bearing oil clearance (be sure to use the correct scale - standard and metric scales are included)

new rods or a new crankshaft were installed, the endplay may be inadequate. If so, the rods will have to be removed and taken to an automotive machine shop for resizing.

26 Initial start-up and break-in after overhaul

Warning: *Have a fire extinguisher handy when starting the engine for the first time.*

1 Once the engine has been installed in the vehicle, double-check the engine oil and coolant levels.
2 With the spark plugs out of the engine and the ignition system and fuel pump disabled (see Section 4), crank the engine until oil pressure registers on the gauge or the light goes out.
3 Install the spark plugs, hook up the plug wires and restore the ignition system and fuel pump functions (see Section 4).
4 Start the engine. It may take a few moments for the fuel system to build up pressure, but the engine should start without a great deal of effort.
5 After the engine starts, it should be allowed to warm up to normal operating temperature. While the engine is warming up, make a thorough check for fuel, oil and coolant leaks.
6 Shut the engine off and recheck the engine oil and coolant levels.
7 Drive the vehicle to an area with minimum traffic, accelerate from 30 to 50 mph, then allow the vehicle to slow to 30 mph with the throttle closed. Repeat the procedure 10 or 12 times. This will load the piston rings and cause them to seat properly against the cylinder walls. Check again for oil and coolant leaks.
8 Drive the vehicle gently for the first 500 miles (no sustained high speeds) and keep a constant check on the oil level. It is not unusual for an engine to use oil during the break-in period.
9 At approximately 500 to 600 miles, change the oil and filter.
10 For the next few hundred miles, drive the vehicle normally. Do not pamper it or abuse it.
11 After 2000 miles, change the oil and filter again and consider the engine broken in.

Chapter 3
Cooling, heating and air conditioning systems

Contents

Section		Section	
Air conditioning and heating system - check and maintenance	12	Drivebelt check, adjustment and replacement	See Chapter 1
Air conditioning compressor - removal and installation	14	Engine cooling fan and circuit - check and component replacement	4
Air conditioning condenser - removal and installation	15	General information	1
Air conditioning evaporator and expansion valve - removal and installation	16	Heater and air conditioning control assembly - removal, installation, check and adjustment	11
Air conditioning receiver/drier - removal and installation	13	Heater core - replacement	10
Antifreeze - general information	2	Radiator and coolant reservoir - removal and installation	5
Blower motor and circuit check - check and component replacement	9	Thermostat - check and replacement	3
Coolant level check	See Chapter 1	Underhood hose check and replacement	See Chapter 1
Coolant temperature gauge sending unit - check and replacement	8	Water pump - check	6
Cooling system check	See Chapter 1	Water pump - removal and installation	7
Cooling system servicing (draining, flushing and refilling)	See Chapter 1		

Specifications

General

Radiator cap pressure rating	11 to 15 psi (74 to 103 kPa)
Thermostat rating (opening)	188 to 193 degrees F (86.5 to 89.5 degrees C)
Refrigerant type	
1993 and earlier	R-12
1994 and later	R-134a
Refrigerant capacity	
1993 and earlier	25 ounces (739.3 ml)
1994 and later	24.9 ounces (736.4 ml)

Torque specifications

Ft-lbs (unless otherwise indicated)

Thermostat housing bolts	14 to 19
Water pump-to-block bolts	14 to 19
Water pump inlet bolts	14 to 220
Radiator drain plug (1995 and later)	6 to 10 in-lbs
Compressor mounting bolts	30 to 40

1 General information

Engine cooling system

All vehicles covered by this manual employ a pressurized engine cooling system with thermostatically-controlled coolant circulation. An impeller type water pump mounted on the front of the block pumps coolant through the engine. The coolant flows around each cylinder and toward the rear of the engine. Cast-in coolant passages direct coolant around the intake and exhaust ports, near the spark plug areas and in proximity to the exhaust valve guides.

A wax-pellet type thermostat is located in the thermostat housing at the transaxle end of the engine. During warm up, the closed thermostat prevents coolant from circulating through the radiator. When the engine reaches normal operating temperature, the thermostat opens and allows hot coolant to travel through the radiator, where it is cooled before returning to the engine.

The cooling system is sealed by a pressure-type radiator cap. This raises the boiling point of the coolant, and the higher boiling point of the coolant increases the cooling efficiency of the radiator. If the system pressure exceeds the cap pressure-relief value, the excess pressure in the system forces the spring-loaded valve inside the cap off its seat and allows the coolant to escape through the overflow tube into a coolant reservoir. When the system cools, the excess coolant is automatically drawn from the reservoir back into the radiator.

The coolant reservoir does double duty as both the point at which fresh coolant is added to the cooling system to maintain the proper fluid level and as a holding tank for overheated coolant.

This type of cooling system is known as a closed design because coolant that escapes past the pressure cap is saved and reused.

Heating system

The heating system consists of a blower fan and heater core located within the heater box under the passenger end of the dashboard, the inlet and outlet hoses connecting the heater core to the engine cooling system and the heater/air conditioning control head on the dashboard. Hot engine coolant is circulated through the heater core. When the heater mode is activated, a flap door opens to expose the heater box to the passenger compartment. A fan switch on the control head activates the blower motor, which forces air through the core, heating the air.

Air conditioning system

The air conditioning system consists of a condenser mounted in front of the radiator, an evaporator mounted adjacent to the heater core, a compressor mounted on the engine, a filter-drier which contains a high pressure relief valve and the plumbing connecting all of the above.

A blower fan forces the warmer air of the passenger compartment through the evaporator core (similar to a radiator in reverse), transferring the heat from the air to the refrigerant. The liquid refrigerant boils off into low pressure vapor, taking the heat with it when it leaves the evaporator. The compressor keeps refrigerant circulating through the system, pumping the warmed coolant through the condenser where it is cooled and then circulated back to the evaporator.

2 Antifreeze - general information

Refer to illustration 2.4

Warning: *Do not allow antifreeze to come in contact with your skin or painted surfaces of the vehicle. Rinse off spills immediately with plenty of water. Antifreeze is highly toxic if ingested. Never leave antifreeze lying around in an open container or in puddles on the floor; children and pets are attracted by it's sweet smell and may drink it. Check with local authorities about disposing of used antifreeze. Many communities have collection centers which will see that antifreeze is disposed of safely. Never dump used antifreeze on the ground or into drains.*

2.4 An inexpensive hydrometer can be used to test the condition of your coolant - also available are antifreeze test strips

Note: *Non-toxic antifreeze is now manufactured and available at local auto parts stores, but even these types should be disposed of properly.*

1 The cooling system should be filled with a water/ethylene-glycol based antifreeze solution, which will prevent freezing down to at least -20 degrees F, or lower if local climate requires it. It also provides protection against corrosion and increases the coolant boiling point.

2 The cooling system should be drained, flushed and refilled every 30,000 miles or every two years (see Chapter 1). The use of antifreeze solutions for periods of longer than two years is likely to cause damage and encourage the formation of rust and scale in the system. If your tap water is "hard," i.e. contains a lot of dissolved minerals, use distilled water with the antifreeze.

3 Before adding antifreeze to the system, check all hose connections, because antifreeze tends to search out and leak through very minute openings. Engines do not normally consume coolant. Therefore, if the level goes down, find the cause and correct it.

4 The exact mixture of antifreeze-to-water you should use depends on the relative weather conditions. The mixture should contain at least 50 percent antifreeze, but should never contain more than 70 percent antifreeze. Consult the mixture ratio chart on the antifreeze container before adding coolant. Hydrometers are available at most auto parts stores to test the ratio of antifreeze to water **(see illustration)** or antifreeze test strips are available instead of the hydrometer gauge. Use antifreeze which meets the vehicle manufacturer's specifications.

3 Thermostat - check and replacement

Warning: *Wait until the engine is completely cool before beginning this procedure. Do not allow antifreeze to come in contact with your skin or painted surfaces of the vehicle. Rinse off spills immediately with plenty of water. Antifreeze is highly toxic if ingested. Never leave antifreeze lying around in an open container or in puddles on the floor; children and pets are attracted by it's sweet smell and may drink it. Check with local authorities about disposing of used antifreeze. Many communities have collection centers which will see that antifreeze is disposed of safely.*

Check

1 Before assuming the thermostat is responsible for a cooling system problem, check the coolant level (Chapter 1), drivebelt tension (Chapter 1) and temperature gauge (or light) operation.

2 If the engine takes a long time to warm up (as indicated by the temperature gauge or heater operation), the thermostat is probably stuck open. Replace the thermostat with a new one.

Chapter 3 Cooling, heating and air conditioning systems

3.7a Disconnect the electrical connector (A) from the thermoswitch located in the thermostat housing - arrow (B) indicates one of the thermostat cover bolts, the other is behind the distributor (1988 model shown)

3.7b On later models (1997 shown), There are two hoses to disconnect (A) - arrow (B) indicates one of the thermostat housing bolts, the other is below the housing

3.10 The thermostat is installed with the spring end into the cylinder head and the bypass hole (arrow) up (or subvalve on later models)

3   If the engine runs hot, use your hand to check the temperature of the lower radiator hose. If the hose is not hot, but the engine is, the thermostat is probably stuck in the closed position, preventing the coolant inside the engine from traveling through the radiator. Replace the thermostat. **Caution:** *Do not drive the vehicle without a thermostat. The computer may stay in open loop and emissions and fuel economy will suffer.*

4   If the lower radiator hose is hot, it means that the coolant is flowing and the thermostat is open. Consult the Troubleshooting Section at the front of this manual for further diagnosis.

Replacement

Refer to illustrations 3.7a, 3.7b and 3.10

5   Disconnect the negative cable from the battery.
6   Drain the coolant from the radiator (see Chapter 1).
7   Disconnect the thermoswitch electrical connector from the thermostat cover located at the left end of the cylinder head **(see illustrations)**.
8   Loosen the radiator hose clamp and remove the radiator hose from the thermostat cover. **Note:** *The radiator hose can be left attached to the thermostat cover, unless the thermostat cover itself is to be replaced.*
9   Detach the thermostat cover from the engine **(see illustrations 3.7a and 3.7b)**. Be prepared for some coolant to spill as the gasket seal is broken.
10 Remove the thermostat, noting the direction in which it was installed in the block **(see illustration)**.
11 Remove the gasket and thoroughly clean the sealing surfaces.
12 Install the thermostat with the spring end toward the engine, and with the small bypass hole in the thermostat at the high point (top).
13 Coat a new gasket with RTV sealant and install it, aligning the gasket with the bolt holes in the block. On 1995 and later models, install the gasket with the print side facing the cylinder head.
14 Installation is the reverse of removal. Tighten the thermostat cover fasteners to the torque listed in this Chapter's Specifications.
15 Refill the cooling system, run the engine and check for leaks and proper operation.

4   Engine cooling fan and circuit - check and component replacement

Warning: *Some models covered by this manual are equipped with Supplemental Restraint Systems (SRS), more commonly known as airbags. Always disconnect the negative battery cable, then the positive battery cable and wait two minutes before working in the vicinity of the impact sensors, steering column or instrument panel to avoid the possibility of accidental deployment of the airbag, which could cause personal injury (see Chapter 12). Do not use any electrical test equipment on any of the airbag system wires or tamper with them in any way.*

Check

Refer to illustration 4.2

1   In normal operation the fan comes on whenever the key is ON and the engine is warmed up (97 degrees F or more). On 1993 and earlier models, it will also operate anytime the key is ON and the thermostatic switch **(see illustration 3.7a)** on the thermostat housing is grounded. If the vehicle is equipped with air conditioning, the fan will come on any time the air conditioning is engaged, regardless of the engine temperature. To test for normal fan operation, disconnect the thermoswitch and turn the ignition key ON. If the fan does not operate, check the fan system components including the fan relay, fan thermoswitch, and fan motor itself.

2   To test an inoperative fan motor (one that doesn't come on when the engine gets hot or when the air conditioner is on), first check the fuses and/or fusible links (see Chapter 12). Check both the cooling fan fuse and the engine fuse. Then disconnect the electrical connector at the fan motor and use fused jumper wires to connect the fan directly to the battery **(see illustration)**. If the fan still does not work, replace the

4.2 Disconnect the radiator fan electrical connector (arrow) and connect fused jumper wires - the yellow wire is connected to battery positive, and the yellow/red wire or black wire to a chassis ground

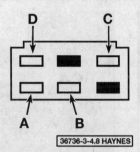

4.8 Cooling fan relay terminal guide for continuity test (models through 1993 shown)

fan motor. A further check is to test the resistance between the two terminals on the fan side of the connector. Resistance should be less than 50 ohms. **Warning:** *Do not allow the test clips to contact each other or any metallic part of the vehicle.*

3   If the fuses are good and the fan motor tested OK in the previous test but is still inoperative, then the fault lies in the relay, thermoswitch (1993 and earlier), PCM (1994 and later) or wiring. Begin by checking for power at the harness side of the fan connector, with the thermoswitch connected and the key ON, there should be at least 10 volts in the yellow wire (with the relay energized on 1994 and later models). The relays and thermoswitch can be tested as described below. **Note:** *On 1994 and later models, the cooling fan relay is controlled by the PCM with information supplied by the engine coolant temperature sensor. See Chapter 6 for testing the coolant temperature sensor. If the fan motor, relay and related circuits all test good and the cooling fan fails to operate normally, the fault may lie with the PCM. Have the PCM diagnosed by a dealer service department or other qualified repair facility.*

4   On all models through 1993 models, the fan thermoswitch can be tested for continuity with an ohmmeter. Disconnect the wiring connector and attach one lead of the ohmmeter to the electrical connector prong on the thermoswitch, and the other lead to the body of the thermoswitch. When the engine is cold (below 194-degrees F) there should be NO continuity. When the engine is hot (above 207-degrees F), there should be continuity. Replace the thermoswitch if necessary.

5   To replace the thermoswitch, first drain the cooling system (see Chapter 1). Disconnect the electrical connector and unscrew the switch from the thermostat housing **(see illustration 3.7b)**. Install the new switch with a new O-ring and tighten it securely. Connect the electrical connector and refill the cooling system.

6   On 1994 and later models, the engine cooling fan is controlled by the computer with a signal from the main coolant temperature sensor, not a separate fan thermoswitch. To test the main coolant temperature sensor, see Chapter 6.

4.15 Upper bolts (arrows) holding the fan shroud to the radiator

Relay check

Refer to illustration 4.8

7   On all models, locate the fan relay, behind the left headlight and just ahead of the battery (see Chapter 12).

8   Remove the cooling fan relay. On models through 1993, attach a jumper wire to the terminal D and attach it to battery power **(see illustration)**. Measure the resistance between terminals B and C with an ohmmeter. Resistance should read more than 10,000 ohms. Now connect a jumper wire from terminal A to a chassis ground and again check the resistance between B and C. It should be less than 5 ohms. If not replace the relay.

9   On 1994 and later models, connect jumper wires from both the white/black and the black/white wire terminals of the relay to battery power. Measure the voltage at the yellow wire terminal. It should be less than 1 volt. Now connect another jumper wire from the brown wire terminal to chassis ground and check the voltage at the yellow wire terminal again. If it isn't at least 10 volts, replace the relay.

10 If the fan relay, thermoswitch and fan motor all test good, take the vehicle to a dealer service department or other qualified repair facility for further diagnosis, due to the complexity and variety of the circuits involved.

Fan replacement

Refer to illustrations 4.15, 4.16 and 4.17

11 Disconnect the negative battery cable.
12 Drain the cooling system (see Chapter 1) below the level of the

4.16 Hold the cooling fan blades and remove the fan retaining nut (arrow)

4.17 Remove the screws (arrows, one screw is hidden behind the motor in this view) retaining the motor to the shroud

Chapter 3 Cooling, heating and air conditioning systems

5.7 Remove the two radiator upper mounting brackets (arrows)

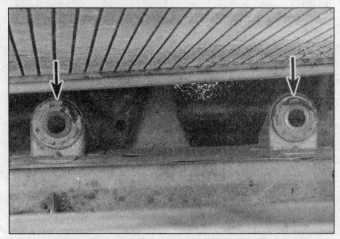

5.10 The lower mounts (arrows) must be in place on the bottom radiator support when the radiator is reinstalled

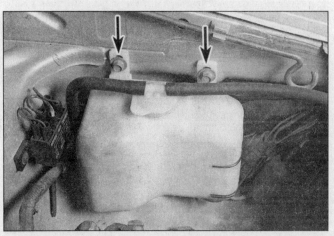

5.14 Remove the reservoir mounting bolts (arrows) and lift it out - on early models (shown here with battery removed for clarity) it is on the fenderwell to the left of the battery, on later models in front of the battery

upper radiator hose.
13   Disconnect the wiring connector at the fan motor and pull the harness out of the clips on the fan shroud.
14   Loosen the radiator upper hose clamps and remove the upper hose.
15   Loosen the two bolts holding the bottom of the fan shroud, remove the two upper bolts attaching the fan shroud to the radiator, then lift the fan/shroud from the engine compartment **(see illustration)**.
16   While holding the fan blades, remove the fan retaining nut **(see illustration)**. Remove the fan from the fan motor.
17   Remove the fan motor from the fan shroud **(see illustration)**.
18   Installation is the reverse of removal.

5   Radiator and coolant reservoir - removal and installation

Warning 1: *Some models covered by this manual are equipped with Supplemental Restraint Systems (SRS), more commonly known as airbags. Always disconnect the negative battery cable, then the positive battery cable and wait two minutes before working in the vicinity of the impact sensors, steering column or instrument panel to avoid the possibility of accidental deployment of the airbag, which could cause personal injury (see Chapter 12). Do not use any electrical test equipment on any of the airbag system wires or tamper with them in any way.*
Warning 2: *Wait until the engine is completely cool before beginning this procedure. Do not allow antifreeze to come in contact with your skin or painted surfaces of the vehicle. Rinse off spills immediately with plenty of water. Antifreeze is highly toxic if ingested. Never leave antifreeze lying around in an open container or in puddles on the floor; children and pets are attracted by it's sweet smell and may drink it. Check with local authorities about disposing of used antifreeze. Many communities have collection centers which will see that antifreeze is disposed of safely.*

Radiator

Refer to illustrations 5.7 and 5.10

1   Disconnect the negative battery cable.
2   Drain the engine coolant into a container (see Chapter 1).
3   Refer to Section 4 and remove the engine cooling fan and shroud.
4   Disconnect the coolant reservoir hose from the radiator filler neck.
5   Loosen the radiator hose clamps and remove both the upper and lower hoses.
6   If equipped with an automatic transaxle, disconnect the oil cooler hoses from the radiator. Place a drip pan to catch the transmission fluid and cap the fittings.
7   Remove the bolts holding the upper radiator mounting brackets to the upper radiator support and lift the radiator from the vehicle **(see illustration)**.
8   With the radiator removed, it can be inspected for leaks, damage and internal blockage. If in need of repairs, have a professional radiator shop or dealer service department perform the work as special techniques are required.
9   Bugs and dirt can be cleaned from the radiator with compressed air and a soft brush. Don't bend the cooling fins as this is done. **Warning:** *Wear eye protection when using compressed air.*
10   Installation is the reverse of the removal procedure. Be sure the bottom of the radiator is located properly, with the bottom projections fitting into the rubber mounts on the lower radiator support **(see illustration)**.
11   After installation, fill the cooling system with the proper mixture of antifreeze and water. Refer to Chapter 1 if necessary.
12   Start the engine and check for leaks. Allow the engine to reach normal operating temperature, indicated by both radiator hoses becoming hot. Recheck the coolant level and add more if required.
13   On automatic transaxle equipped models, check and add automatic transmission fluid as needed.

Coolant reservoir

Refer to illustration 5.14

14   Unbolt the coolant reservoir and lift it out of engine compartment **(see illustration)**.

Chapter 3 Cooling, heating and air conditioning systems

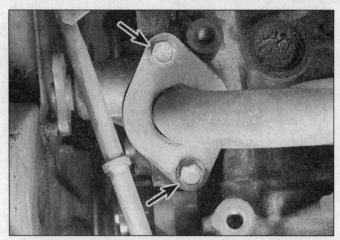

7.4 Remove the bolts (arrows) holding the water pump inlet pipe to the left side of the water pump

7.6 After removing the front covers and timing belt, remove the water pump mounting bolts (arrows) and remove the water pump - arrow A indicates the weep hole

15   Pour the coolant into a container. Wash out and inspect the reservoir for cracks and chafing. Replace it if damaged.
16   Installation is the reverse of removal.

6   Water pump - check

1   A failure in the water pump can cause serious engine damage due to overheating. If the pump is defective, it should be replaced with a new or rebuilt unit
2   Remove the timing belt cover(s) (see Chapter 2A).
3   Water pumps are equipped with weep or vent holes **(see illustration 7.6)**. If a failure occurs in the pump seal, coolant will leak from the hole.
4   Check the water pump shaft bearing for wear by grasping the pump hub and gently rocking the hub and shaft from side to side. If any looseness is apparent, excessive water pump shaft/bearing wear is possible.
5   If the water pump shaft bearings fail there may be a howling sound at the drivebelt end of the engine while it's running. Don't mistake drivebelt slippage, which causes a squealing sound, for water pump bearing failure. If a squealing sound is heard, check belt condition and belt tension.

7   Water pump - removal and installation

Warning: *Wait until the engine is completely cool before beginning this procedure. Do not allow antifreeze to come in contact with your skin or painted surfaces of the vehicle. Rinse off spills immediately with plenty of water. Antifreeze is highly toxic if ingested. Never leave antifreeze lying around in an open container or in puddles on the floor; children and pets are attracted by it's sweet smell and may drink it. Check with local authorities about disposing of used antifreeze. Many communities have collection centers which will see that antifreeze is disposed of safely.*

Removal

Refer to illustrations 7.4 and 7.6
1   Disconnect the negative battery cable from the battery.
2   Drain the cooling system (see Chapter 1). If the coolant is relatively new or in good condition, save it and reuse it.
3   Remove the water pump drivebelt, pulley, timing belt covers and timing belt (see Chapter 2A).
4   Remove the water inlet pipe and gasket from the water pump **(see illustration)**.
5   On later models, remove the water bypass pipe and O-ring,

located on the water inlet pipe.
6   Remove the water pump mounting bolts **(see illustration)** and detach the water pump from the engine. If the water pump is stuck, gently tap it with a soft-faced hammer to break the seal.

Installation

7   Clean the bolt threads and the threaded holes in the engine to remove corrosion and sealant.
8   Remove all traces of old gasket material from the sealing surfaces.
9   Compare the new water pump to the old one to make sure they are identical.
10   Apply a thin film of silicone sealant to the new gasket and install it on the water pump.
11   Carefully mate the water pump to the engine.
12   Install the water pump mounting bolts. Tighten them to the torque listed in this Chapter's Specifications. Don't over-tighten them or the pump may be damaged.
13   Reinstall all parts removed for access to the water pump. Clean the water pump inlet mounting flange and install it with a new gasket.
Note: *Make sure you install a new O-ring where the bypass pipe connects to the water pump inlet.*
14   Refill the cooling system (see Chapter 1) and check the timing belt tension (see Chapter 2A). Run the engine and check for leaks.

8   Coolant temperature gauge sending unit - check and replacement

Refer to illustration 8.3
Warning: *Wait until the engine is completely cool before beginning this procedure. Do not allow antifreeze to come in contact with your skin or painted surfaces of the vehicle. Rinse off spills immediately with plenty of water. Antifreeze is highly toxic if ingested. Never leave antifreeze lying around in an open container or in puddles on the floor; children and pets are attracted by it's sweet smell and may drink it. Check with local authorities about disposing of used antifreeze. Many communities have collection centers which will see that antifreeze is disposed of safely.*

Check

1   If the coolant temperature gauge is inoperative, check the fuses first (see Chapter 12).
2   If the temperature gauge indicates excessive temperature after running awhile, see the Troubleshooting Section in the front of the manual.

Chapter 3 Cooling, heating and air conditioning systems

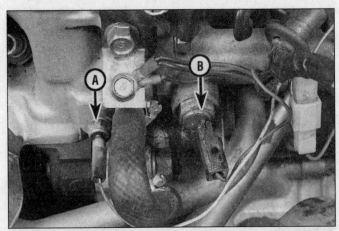

8.3 Location of the temperature gauge sending unit (A) - (B) is the thermoswitch for the cooling fan

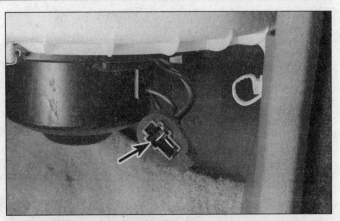

9.2 With the blower motor electrical connector (arrow indicates blower side of connector) disconnected, check for shorts between the electrical connector (harness side) and any convenient chassis ground - If no shorts are found, reconnect the blower electrical connector to the blower terminal for additional tests

9.5 Remove the blower motor resistor assembly mounting screws (arrows)

3 If the temperature gauge indicates HOT as soon as the engine is started cold, disconnect the electrical connector at the coolant gauge sending unit, located on the cylinder head, near the thermostat housing (see illustration). If the gauge reading drops, replace the sending unit. If the reading remains high, the wire to the gauge may be shorted to ground or the gauge is faulty.

4 If the coolant temperature gauge fails to show any indication after the engine has been warmed up, (approximately 10 minutes) and the fuses are good, shut off the engine. Disconnect the electrical connector at the sending unit and, using a jumper wire, connect the wire to a clean ground on the engine. Briefly turn on the ignition without starting the engine. If the gauge now indicates HOT, replace the sending unit.

5 Additionally, the sending unit may be checked for resistance using an ohmmeter connected to the terminal and the body of the sending unit. With the engine coolant warm (90 to 155 degrees F), resistance should be 190 to 260 ohms. With the engine fully HOT (above 206 degrees F) the resistance should drop to 24 to 28 ohms.

6 If the sending unit is OK, but gauge fails to respond, the circuit may be open or the temperature gauge may be faulty. Check the wire from the sender to the gauge.

Replacement

7 Drain the coolant (see Chapter 1).
8 Disconnect the wiring connector from the sending unit.
9 Using a deep socket or a wrench, remove the sending unit.
10 Install the new sending unit, and tighten it securely. Do not use thread sealer as it may electrically insulate the sending unit. Connect the electrical connector.
11 Refill the cooling system and check for coolant leakage and proper gauge operation.

9 Blower motor and circuit - check and component replacement

Warning: *Some models covered by this manual are equipped with Supplemental Restraint Systems (SRS), more commonly known as airbags. Always disconnect the negative battery cable, then the positive battery cable and wait two minutes before working in the vicinity of the impact sensors, steering column or instrument panel to avoid the possibility of accidental deployment of the airbag, which could cause personal injury (see Chapter 12). Do not use any electrical test equipment on any of the airbag system wires or tamper with them in any way.*

Check

Refer to illustrations 9.2, 9.5, 9.6a and 9.6b

1 On 1990 through 1993 models, if heater blower fan is inoperative, first check the 15A heater fuse, located in the fuse panel in the dashboard, next to the steering wheel (see Chapter 12). On 1994 and later models, check the 30A circuit breaker located in the dash fuse panel (see Chapter 12), and also check the 20A wiper fuse.

2 If the fuse is blown on early models, or if the reset button is OUT on circuit-breaker-equipped models, first check for a short circuit in the blower wiring by disconnecting the electrical connector at the blower (see illustration). Check for any shorts (continuity) between the blower electrical connector and any convenient body ground. Replace or repair the harness if necessary.

3 On models so equipped, depress the reset button to reset the heater circuit breaker.

4 Measure the voltage at the heater blower motor as follows:
   a) Remove the panel below the glove box for access to the blower (see Chapter 11).
   b) Turn the ignition switch on, and turn the blower switch to the Position 4 (high speed).
   c) Using a voltmeter, test the blower motor terminal wires.
   d) The voltage should measure 12-volts. If 12-volts is present and the blower doesn't run, attach a jumper lead from the ground side of the blower connector to a chassis ground. If the blower runs now, check the ground circuit to the blower. If the blower still doesn't run, replace the blower motor.

5 With the voltmeter still connected to the blower electrical connector, turn the blower switch down to the medium and low settings, checking that the voltage is lower at each lower position of the blower switch. Replace the resistor assembly if necessary (see illustration).

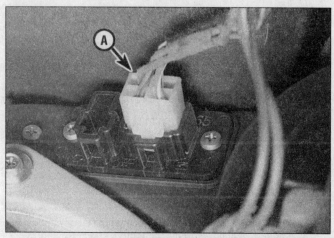

9.6a On models before 1994, test for power at the blue/red wire (A) - test ground the other wires one at a time to check for change in blower speed

9.6b On later models, check for resistance between terminals 1 and 2 (should be 2.1 ohms) and terminals 1 and 3 (should be 0.8 ohms)

Caution: *On models through 1993, remove only the two screws holding the resistor to the case. There is another screw close to the resistor which holds a mounting plate inside the case. Removal of this screw could cause the plate to fall down inside the case, requiring case removal and disassembly to retrieve it.* **Note:** *If the blower motor resistor is OK in all tests, refer to Section 11 for a check of the blower motor speed switch in the dashboard climate control panel.*

6 Further checks of the resistors can be made by removing the resistor assembly and checking terminal continuity as follows **(see illustration):**

a) On 1990 through 1993 models, leave the connector in place on the resistor and with the key On and the blower switch OFF, check for power at the blue/red wire. There should be battery voltage. Now ground the other three wires, one at a time. The blower should change speed as each different wire is grounded. If not, replace the resistor.

b) On 1994 and later models (three-pin connector), disconnect the harness connector from the resistor and test the pins for proper resistance **(see illustration)**. If the resistance isn't as specified, replace the resistor assembly.

Blower motor - removal and installation

Refer to illustrations 9.9, 9.10a and 9.10b

7 Remove the trim panel underneath the glove box. **Note:** *On models through 1993, the blower motor is in the heater housing, under the*

9.9 Remove the three screws (arrows) and pull the blower motor down and out of the housing

center of the dash, while on later models it is in a separate blower housing under the passenger side of the dashboard.

8 Disconnect the electrical connector from the blower motor.

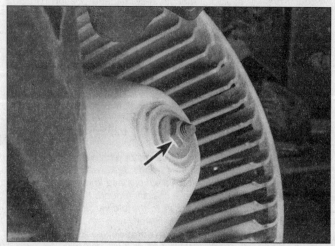

9.10a Depending on model, the blower fan can be removed by taking off a nut (arrow) or ...

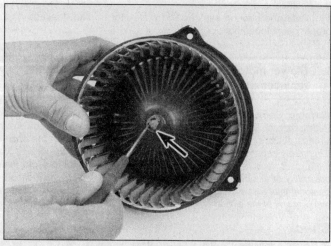

9.10b ... prying off a clip (arrow)

Chapter 3 Cooling, heating and air conditioning systems   3-9

10.4a On models through 1993, disconnect the heater core hoses (arrows) at the firewall, just below the brake line distribution block

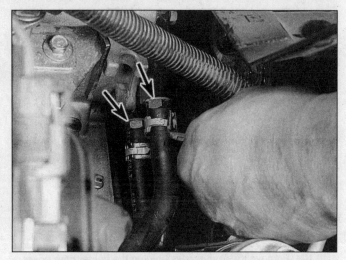

10.4b On 1994 and later models, the heater hose connections (arrows) are below the air-conditioning receiver-drier

9 Remove the three screws retaining the blower motor to the housing **(see illustration)**. Withdraw the blower motor straight down and out of the housing.
10 To remove the blower fan, remove the fastener from the shaft and withdraw the fan from the motor **(see illustrations)**.
11 Installation is the reverse of removal.

10 Heater core - replacement

Refer to illustrations 10.4a, 10.4b, 10.7, 10.9, 10.10a and 10.10b
Warning: *Some models covered by this manual are equipped with Supplemental Restraint Systems (SRS), more commonly known as airbags. Always disconnect the negative battery cable, then the positive battery cable and wait two minutes before working in the vicinity of the impact sensors, steering column or instrument panel to avoid the possibility of accidental deployment of the airbag, which could cause personal injury (see Chapter 12). Do not use any electrical test equipment on any of the airbag system wires or tamper with them in any way.*
1 Disconnect the cable from the negative battery terminal.

2 Drain the engine coolant (see Chapter 1).
3 Remove the instrument panel (see Chapter 11).
4 Disconnect the heater hoses from the heater core by unlocking the hose connections at the firewall **(see illustrations)**. Keep plenty of towels or rags on the carpeting to catch any coolant that may drip.
5 Disconnect the electrical connectors from the blower motor and blower resistor.
6 Disconnect or move the wiring harness and antenna cable from in front of the heater distribution box.
7 The heater core is contained within a housing bolted to the firewall at the center of the dash. Remove the bolt and nut retaining the clamp connecting the right side of the heater housing to the evaporator housing **(see illustration)**.
8 Disconnect the defroster ducts that are attached to the heater housing. On models through 1993, disconnect the blower motor connector and the blower resistor connector.
9 Remove the three nuts from the studs retaining the heater unit to the dash. Remove the heater unit from the dash **(see illustration)**.
Note: *The cables attached to the heater housing may be left in place while the heater core is removed from the housing, or may be discon-*

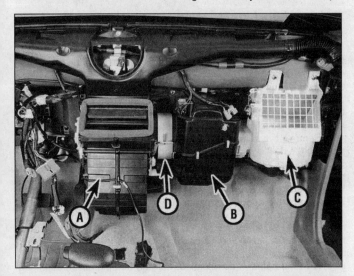

10.7 Components of the heating/air conditioning system include the heater case (A), the evaporator case (B) and the blower motor case (C) - remove the clamp (D) connecting the heater case to the evaporator case

10.9 Remove the nuts from the studs at the firewall and pull the heater housing straight back

3-10 Chapter 3 Cooling, heating and air conditioning systems

10.10a Remove the four screws holding the heater core cover plate (arrow) and remove the cover (1997 model shown)

10.10b Pull the heater core (1997 model shown) from the case - save the foam sealing material (arrow) for use with the new core

11.4a On models through 1993, remove the four screws (arrows) on the control panel

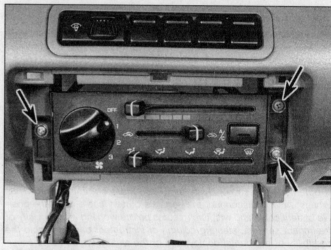

11.4b On later models, there are three screws (arrows) holding the control panel to the dashboard

nected (see Section 11) to remove the housing completely from the vehicle.

10   Remove the heater core from the heater unit **(see illustrations)**. **Note:** *On models through 1993, the heater housing must be separated into two halves to remove the core inside. Remove the case screws and clips to separate the halves. On later models, a simple plate (fours screws) can be removed to allow withdrawal of the core.*

11   Reassembly is the reverse of removal. When installing the heater hoses on the heater core, make sure the hose connectors are securely locked. **Note:** *Transfer the foam sealing material from the old core to the new core before inserting it in the case.*

11 Heater and air conditioning control assembly - removal, installation, check and adjustment

Warning: *Some models covered by this manual are equipped with Supplemental Restraint Systems (SRS), more commonly known as airbags. Always disconnect the negative battery cable, then the positive battery cable and wait two minutes before working in the vicinity of the impact sensors, steering column or instrument panel to avoid the possibility of accidental deployment of the airbag, which could cause personal injury (see Chapter 12). Do not use any electrical test equipment on any of the airbag system wires or tamper with them in any way.*

Removal and installation

Refer to illustrations 11.4a, 11.4b, 11.5a, 11.5b and 11.6

1   Disconnect the negative battery cable.
2   Remove the glove box and the underdash panel on the passenger side (see Chapter 11).
3   Remove the center heating/air conditioning and radio control panel bezel (see Chapter 12). On models through 1993, remove the radio (see Chapter 12). On later models, the shift console must be removed before removing the bezel around the controls (see Chap-ter 12).
4   Remove the screws holding the heating/air conditioning control panel to the dashboard **(see illustrations)**.
5   There are three levers on the heating/air conditioning control panel, each connected to a cable that operates the "doors" in the heating/ventilating system. The operating ends of the cables are retained by clips near the door they operate. Depress the clips and remove the cable end from each door-operating lever **(see illustrations)**. On connects to the Recirc/Fresh air door, one to a lever on the right side of the below-dash heating/ventilating box, and one to a lever on the left side of the box.
6   With the cable ends released, pull the heating/air-conditioning control panel out from the dashboard. It will only pull out a few inches, but once you disconnect the electrical connector at the back of the blower motor switch, it can be pulled completely out of the dashboard **(see illustration)**. **Note:** *On some models, there is also an electrical*

Chapter 3 Cooling, heating and air conditioning systems

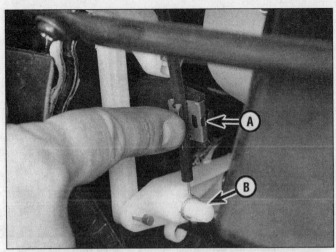

11.5a To release the cable end from the function-control lever (right side of heating box), open the spring clip (A) to release, then pull the cable end (B) from the post on the lever

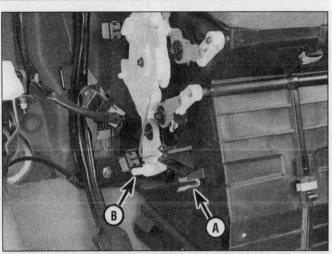

11.5b On the left side of the heating box, open the temperature control cable retaining clip (A) and pull the cable from the post (B) - (shown with cable removed)

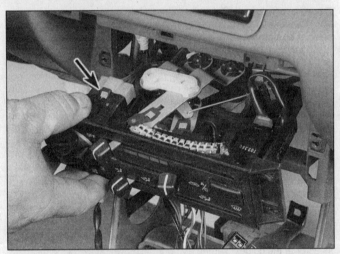

11.6 To release the electrical connector from the blower speed switch, depress the tab (arrow) and pull the connector off

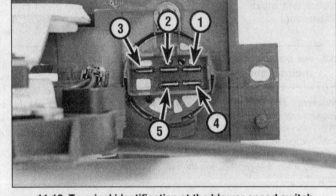

11.12 Terminal identification at the blower speed switch (1997 model shown)

connector just below and behind the air-conditioning switch that must be disconnected.

7 Installation is the reverse of the removal procedure. When installation of the heater/air conditioning control assembly is complete, connect and adjust the heater control wires as follows.

Cable adjustments

8 To connect and adjust the REC-FRESH wire, set the REC-FRESH lever to the FRESH position, connect the REC-FRESH wire to the REC-FRESH door, set the door to FRESH position and clamp the wire in place. Check that the lever moves its full stroke.

9 To connect and adjust the MIX wire, set the MIX lever to the COLD position, connect the MIX wire to the MIX door, set the door to the COLD position and clamp the wire in place. Check that the lever moves its full stroke.

10 To connect and adjust the MODE wire, set the MODE lever to the DEFROST position, push the door lever post forward (see illustration 11.5b), attach the cable end and snap the cable into the clamp.

Electrical checks

Refer to illustrations 11.12 and 11.15

11 To test the blower motor switch, disconnect the electrical connector from the back of the switch, and check the resistance between terminals on the switch. In all the test positions, the resistance between the specified terminals should be less than 5 ohms. On models through 1993, put the switch in position 1 and measure resistance between the black and blue/white wire terminals. In position 2, measure resistance between black and blue/yellow wire terminals, and between blue/yellow and blue/red. In position 3, check between black and blue/black, and between blue/black and blue/red. If the switch doesn't pass the tests, replace the switch.

12 On 1994 and later models, disconnect the electrical connector from the back of the switch (see illustration), and test for resistance as follows:

a) With the switch in the OFF position, there should be resistance between all terminals of more than 10,000 ohms.
b) In position 1, there should be less than 5 ohms between terminals 1 and 4, but greater than 10,000 between all others.
c) In position 2, there should be less than 5 ohms between terminals 2 and 4 and terminals 4 and 5, but greater than 10,000 between all others.
d) In position 3, there should be less than 5 ohms between terminals 3 and 4 and terminals 4 and 5, but greater than 10,000 between all others.

13 The air-conditioning switch can also be tested. Disconnect the electrical connector on the back of the switch, and using an ohmmeter, test for resistance between the terminals. On models through 1993, there should be more than 10,000 ohms between the blue wire termi-

11.15 depress the tabs on the AC switch and withdraw it from the control panel

12.1 The fins of the condenser can be cleaned with a fin comb - use the side of the tool with the correct fins-per-inch spacing for your core

nal and the blue/white wire terminal when the switch is not engaged. When the switch is pushed in, resistance between those terminals should be less than 5 ohms.

14 On 1994 and later models, perform the same test as above, but between the blue/yellow and green/black wire terminals.

15 To replace the AC switch, depress the two tabs on the back of the switch and withdraw it from the heating/air conditioning panel **(see illustration)**.

12 Air conditioning and heating system - check and maintenance

Warning: *The air conditioning system is under high pressure. Do not loosen any hose fittings or remove any components until the system has been discharged. Air conditioning refrigerant should be properly discharged into an EPA-approved recovery/recycling unit by a dealer service department or an automotive air conditioning repair facility. Always wear eye protection when disconnecting air conditioning system fittings.*

Air conditioning system

Refer to illustration 12.1

1 The following maintenance checks should be performed on a regular basis to ensure that the air conditioning system continues to operate at peak efficiency:

a) Inspect the condition of the compressor drivebelt. If it is worn or deteriorated, replace it (see Chapter 1).
b) Check the drivebelt tension and, if necessary, adjust it (see Chapter 1).
c) Inspect the system hoses. Look for cracks, bubbles, hardening and deterioration. Inspect the hoses and all fittings for oil bubbles or seepage. If there is any evidence of wear, damage or leakage, replace the hose(s).
d) Inspect the condenser fins for leaves, bugs and any other foreign material that may have embedded itself in the fins. Use a "fin comb" or compressed air to remove debris from the condenser **(see illustration)**.
e) Make sure the system has the correct refrigerant charge.

2 It's a good idea to operate the system for about ten minutes at least once a month. This is particularly important during the winter months because long term non-use can cause hardening, and subsequent failure, of the seals.

3 Leaks in the air conditioning system are best spotted when the system is brought up to operating temperature and pressure, by running the engine with the air conditioning ON for five minutes. Shut the engine off and inspect the air conditioning hoses and connections. Traces of oil usually indicate refrigerant leaks.

4 Because of the complexity of the air conditioning system and the special equipment required to effectively work on it, accurate troubleshooting of the system should be left to a professional technician.

5 If the air conditioning system doesn't operate at all, check the fuse panel and the air conditioning relay, located in the fuse/relay box in the engine compartment. Refer to Section 11 for electrical checks of heating/air conditioning system control components.

6 The most common cause of poor cooling is simply a low system refrigerant charge. If a noticeable drop in cool air output occurs, the following quick check will help you determine if the refrigerant level is low. For more complete information on the air conditioning system, refer to the Haynes Automotive Heating and Air Conditioning Manual.

Checking the refrigerant charge

Refer to illustration 12.10

7 Warm the engine up to normal operating temperature.

8 With the engine at fast idle, place the air conditioning temperature selector at the coldest setting and put the blower at the highest setting. Open the doors (to make sure the air conditioning system doesn't cycle off as soon as it cools the passenger compartment).

9 With the compressor engaged, the clutch will make an audible click and the center of the clutch will rotate. After the system reaches operating temperature, feel the two pipes connected to the evaporator at the firewall.

10 The pipe leading from the condenser outlet to the evaporator (small tubing) should be cold, and the evaporator outlet line (the larger tubing that leads back to the compressor) should be slightly colder (3 to 10 degrees F). If the evaporator outlet is considerably warmer than the inlet, the system needs a charge. Insert a thermometer in the center air distribution duct while operating the air conditioning system **(see illustration)** - the temperature of the output air should be 35 to 40 degrees F below the ambient air temperature (down to approximately 40 degrees F). If the ambient (outside) air temperature is very high, say 110 degrees F, the duct air temperature may be as high as 60 degrees F, but generally the air conditioning is 35 to 40 degrees F cooler than the ambient air. If the air isn't as cold as it used to be, the system probably needs a charge. Further inspection or testing of the system is beyond the scope of the home mechanic and should be left to a professional.

11 Inspect the sight glass in the top of the receiver-drier. If the refrigerant looks foamy when running, it's low **(see illustration 13.3)**. When ambient temperatures are very hot, bubbles may show in the sight glass even with the proper amount of refrigerant. With the proper amount of refrigerant, when the air conditioning is turned off, the sight glass should show refrigerant that foams, then clears.

Chapter 3 Cooling, heating and air conditioning systems

12.10 Check the temperature of the output air in the center register with a thermometer - it should be 35 to 40 degrees below the ambient air temperature

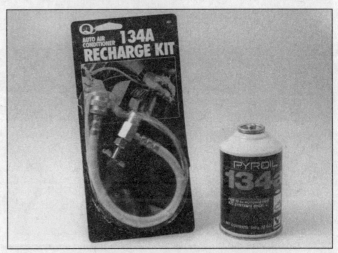

12.13a Refrigerant kit for R-134 recharging

12.13b Add new 134-A refrigerant only at the low-side service port (arrow)

Adding refrigerant

Caution: *Refrigerant has changed from the use of R-12 through 1993 models, to the "environmentally friendly" R-134a used in 1994 and later models. The two refrigerants are NOT compatible. Even after purging and evacuating an R-12 system, there is enough residual oil and refrigerant in the hoses and components that simply filling the system with R-134a cannot be done. Special fittings and manifold gauge sets are used on the different refrigerant types so that an accidental hook-up of the two systems cannot be made. When replacing entire components, additional refrigerant oil should be added equal to the amount that is removed with the component being replaced. Refrigerant oils, just like refrigerant R-12 vs. R-134a, are not compatible. Be sure to read the can before adding any oil to the system, to make sure it is compatible with the type of system being repaired.*

1993 and earlier models

12 Because of environmental regulations, R-12 refrigerant is not available for home-mechanic use. Model years 1990 through 1993 use R-12 refrigerant. Have the system discharged, evacuated, charged and leak tested by a qualified shop. Use only refrigerant oil compatible with your system. Refrigerant oils for use with refrigerant R-12 are not compatible with other oils.

1994 and later models

Refer to illustration 12.13a and 12.13b

13 Buy an automotive charging kit at an auto parts store **(see illustration)**. A charging kit includes a 14-ounce can of R-134a refrigerant, a tap valve and a short section of hose that can be attached between the tap valve and the system low side service valve . The system low side service valve (charging port) is located on the larger air conditioning tubing line going to the firewall **(see illustration)**. Because one can of refrigerant may not be sufficient to bring the system charge up to the proper level, it's a good idea to buy a couple of additional cans. Try to find at least one can that contains red refrigerant dye. If the system is leaking, the red dye will leak out with the refrigerant and help you pinpoint the location of the leak.

14 Connect the charging kit by following the manufacturer's instructions. Back off the valve handle on the charging kit and screw the kit onto the refrigerant can, making sure first that the O-ring or rubber seal inside the threaded portion of the kit is in place. **Warning:** *Wear protective eye wear when dealing with pressurized refrigerant cans.*

15 Remove the dust cap from the low-side charging port and attach the quick-connect fitting on the kit hose. **Warning:** *DO NOT hook the charging kit hose to the system high side! The fittings on the charging kit are designed to fit only on the low side of the system.*

16 Warm the engine to normal operating temperature and turn on the air conditioner. Keep the charging kit hose away from the fan and other moving parts.

17 Turn the valve handle on the kit until the stem pierces the can, then back the handle out to release the refrigerant. You should be able to hear the rush of gas. Add refrigerant to the low side of the system until both the outlet and the evaporator inlet pipe feel about the same temperature. Allow stabilization time between each addition. **Warning:** *Never add more than two cans of refrigerant to the system.* The can may tend to frost up, slowing the procedure. Wrap a shop towel wet with hot water around the bottom of the can to keep it from frosting.

18 If you have an accurate thermometer, you can place it in the center air conditioning duct inside the vehicle to monitor the air temperature. A charged system that is working properly, should output air down to approximately 40 degrees F.

19 When the can is empty, turn the valve handle to the closed position and release the connection from the low-side port. Replace the dust cap.

20 Remove the charging kit from the can and store the kit for future use with the piercing valve in the UP position, to prevent inadvertently piercing the can on the next use.

Heating systems

Refer to illustration 12.25

21 If the air coming out of the heater vents isn't hot, the problem could stem from any of the following causes:

Chapter 3 Cooling, heating and air conditioning systems

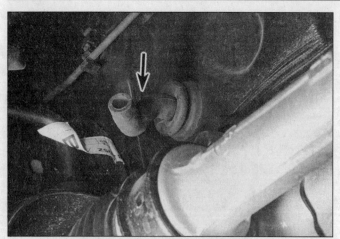

12.25 Check that the drain hose (arrow) from the evaporator case is clear - view is from below, looking up at the passenger side of the firewall

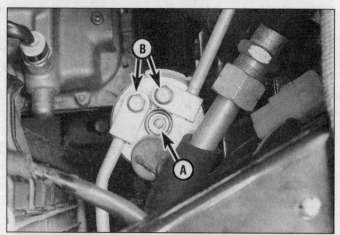

13.3 On 1994 and later models, the receiver-drier is mounted to the chassis to the left of the condenser - A is the sight glass, B are the bolts holding the refrigerant lines to the receiver-drier

13.5 Remove the receiver/drier clamp bolt and remove the receiver/drier

a) *The thermostat is stuck open, preventing the engine coolant from warming up enough to carry heat to the heater core. Replace the thermostat (see Section 3).*
b) *A heater hose is blocked, preventing the flow of coolant through the heater core. Feel both heater hoses at the firewall. They should be hot. If one of them is cold, there is an obstruction in one of the hoses or in the heater core, or the heater control valve is shut. Detach the hoses and back flush the heater core with a water hose. If the heater core is clear but circulation is impeded, remove the two hoses and flush them out with a water hose.*
c) *If flushing fails to remove the blockage from the heater core, the core must be replaced. (see Section 10).*

22 If the blower motor speed does not correspond to the setting selected on the blower switch, the problem could be a bad fuse, circuit, switch, blower motor resistor or motor (see Section 9).
23 If there isn't any air coming out of the vents:
a) *Turn the ignition ON and activate the fan control. Place your ear at the heating/air conditioning register (vent) and listen. Most motors are audible. Can you hear the motor running?*
b) *If you can't (and have already verified that the blower switch and the blower motor resistor are good), the blower motor itself is probably bad (see Section 9).*

24 If the carpet under the heater core is damp, or if antifreeze vapor or steam is coming through the vents, the heater core is leaking. Remove it (see Section 10) and install a new unit (most radiator shops will not repair a leaking heater core).
25 Inspect the drain hose from the heater/air conditioning assembly located under the vehicle; make sure it isn't clogged **(see illustration)**. If there is a humid mist coming from the system ducts, this hose may be plugged with leaves or road debris.

Eliminating air-conditioner odors

26 Unpleasant odors that often develop in air-conditioning systems are caused by the growth of a fungus, usually on the surface of the evaporator core. The warm, humid environment there is a perfect breeding ground for mildew to develop.
27 The evaporator core on most vehicles is difficult to access, and factory dealerships have a lengthy, expensive process for eliminating the fungus by opening up the evaporator case and using a powerful disinfectant and rinse on the core until the fungus is gone. You can service your own system at home, but it takes something much stronger than basic household germ-killers or deodorizers.
28 Aerosol disinfectants for automotive air-conditioning systems are available in most auto parts stores, but remember when shopping for them that the most effective treatments are also the most expensive. The basic procedure for using these sprays is to start by running the system in the RECIRC mode for ten minutes with the blower on its highest speed. Use the highest heat mode to dry out the system and keep the compressor from engaging by disconnecting the wiring plug at the compressor (see Section 14).
29 The disinfectant can usually comes with a long spray hose. Remove the blower motor resistor (see Section 9), point the nozzle inside the hole and spray towards the evaporator core, according to the manufacturer's recommendations. Try to cover the whole surface of the evaporator core, by aiming the spray up, down and sideways. Follow the manufacturer's recommendations for the length of spray and waiting time between applications.
30 Once the evaporator has been cleaned, the best way to prevent the mildew from coming back again is to always run your air-conditioning system on the exterior-air source, don't use the MAX/AC setting. The drier outside air keeps the fungus from starting. Also make sure your evaporator housing drain tube is clear **(see illustration 12.25)**.

13 Air conditioning receiver/drier - removal and installation

Refer to illustration 13.3
Warning: *The air conditioning system is under high pressure. Do not loosen any hose fittings or remove any components until the system has been discharged. Air conditioning refrigerant should be properly discharged into an EPA-approved recovery/recycling unit by a dealer*

Chapter 3 Cooling, heating and air conditioning systems   3-15

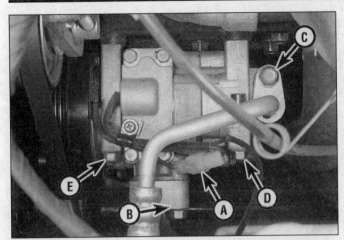

14.5 Disconnect the wiring harness connector (A), the two refrigerant lines (B and C), then loosen the compressor mounting bolts (E and D indicate the two upper bolts on a 1997 model)

15.4 To remove the condenser fan, disconnect the two electrical connectors (A), the remove the three mounting bolts (B)

service department or an automotive air conditioning repair facility. Always wear eye protection when disconnecting air conditioning system fittings.
1 Have the refrigerant discharged and recovered by a qualified repair facility.
2 On models through 1993, refer to Chapter 11 and remove the radiator grille - the receiver-drier is attached to the side of the condenser on these models. On 1994 and later models, remove the battery and the coolant recovery tank (see Section 5).
3 Remove the refrigerant line clamp bolt and disconnect the refrigerant lines from the receiver/drier **(see illustration)**. Cap the open fittings to prevent entry of moisture. On 1994 and later models, remove the receiver/drier pressure switch electrical connector.
4 On models through 1993, refer to Section 15 and remove the condenser with the receiver-drier still attached, then unbolt the receiver-drier from the condenser.
5 Remove the receiver/drier clamp bolt and remove the receiver/drier **(see illustration)**.
6 Installation is the reverse of removal. Replace any O-rings with new ones specifically for the type of refrigerant in your system and lubricate them with refrigerant oil prior to installation. **Warning:** *Do not apply compressor oil to the fitting nuts. Tighten the receiver/drier inlet and outlet fittings securely.*
7 Have the system evacuated, charged and leak tested by the shop that discharged it. If the receiver was replaced, have them add about 0.34 ounces (10 ml) of new refrigerant oil to the high pressure side of the compressor. Use only refrigerant oil compatible with your system.

14 Air conditioning compressor - removal and installation

Refer to illustration 14.5
Warning: *The air conditioning system is under high pressure. Do not loosen any hose fittings or remove any components until the system has been discharged. Air conditioning refrigerant should be properly discharged into an EPA-approved recovery/recycling unit by a dealer service department or an automotive air conditioning repair facility. Always wear eye protection when disconnecting air conditioning system fittings.*
1 Have the refrigerant discharged and recovered by a qualified repair facility.
2 Disconnect the negative cable from the battery.
3 Remove the undercover and splash shield on the passenger side from under the vehicle.
4 Remove the drivebelt from the compressor (see Chapter 1).

5 Disconnect the refrigerant lines and compressor electrical connector **(see illustration)**.
6 Unbolt the compressor and lift it from the vehicle. On 1994 and later models, remove three compressor mounting bolts, but the top-right bolt should just be loosened enough to come clear. Remove that bolt with the compressor, in order to clear the radiator and fan assembly coming out.
7 If a new or rebuilt compressor is being installed, drain the compressor of oil and then reinstall 5.9 ounces (175 ml) of refrigerant oil. **Caution:** *Make sure the refrigerant oil is compatible with the refrigerant type (R-12 or R-134a) used in your vehicle.*
8 Installation is the reverse of removal. Tighten the compressor mounting bolts securely. Replace any O-rings with new ones specifically for the type of refrigerant in your system and lubricate them with refrigerant oil prior to installation. **Warning:** *Do not apply compressor oil to the fitting nuts. Tighten the refrigerant line bolts securely.*
9 Have the system evacuated, recharged and leak tested by the shop that discharged it.

15 Air conditioning condenser - removal and installation

Refer to illustrations 15.4, 15.5a, 15.5b and 15.6
Warning 1: *Some models covered by this manual are equipped with Supplemental Restraint Systems (SRS), more commonly known as airbags. Always disconnect the negative battery cable, then the positive battery cable and wait two minutes before working in the vicinity of the impact sensors, steering column or instrument panel to avoid the possibility of accidental deployment of the airbag, which could cause personal injury (see Chapter 12). Do not use any electrical test equipment on any of the airbag system wires or tamper with them in any way.*
Warning 2: *The air conditioning system is under high pressure. Do not loosen any hose fittings or remove any components until the system has been discharged. Air conditioning refrigerant should be properly discharged into an EPA-approved recovery/recycling unit by a dealer service department or an automotive air conditioning repair facility. Always wear eye protection when disconnecting air conditioning system fittings.*
1 Have the refrigerant discharged and recovered by a qualified repair facility.
2 Refer to Chapter 11 and remove the radiator grille (all except SE models). Disconnect both negative and positive battery cables and wait two minutes.
3 On models equipped with airbags, unbolt and set aside the airbag sensor on the top radiator support.
4 Where equipped, remove the condenser cooling fan (in front of the condenser), by disconnecting the electrical connector and removing the mounting bolts **(see illustration)**.

Chapter 3 Cooling, heating and air conditioning systems

15.5a Disconnect the condenser inlet pipe (arrow), using two wrenches

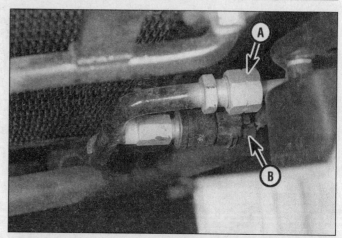

15.5b At the left side of the condenser, disconnect the condenser outlet pipe (A) and the electrical connector at the condenser fan switch (B)

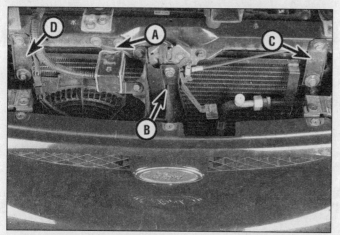

15.6 To access the condenser for removal, remove the airbag sensor (A), the center hood latch brace (B), and the two condenser mounting brackets (C and D)

16.2 Use a back-up wrench when disconnecting the air conditioning lines at the firewall

5   Disconnect the condenser inlet and outlet fittings (see illustrations). Immediately cap the open fittings to keep moisture and contamination out of the system.
6   Remove the center hood latch brace from the upper radiator support, then remove the condenser mounting brackets and lift out the condenser, being careful not to damage the fins (see illustration).
7   Check the condenser for cracks, damage, refrigerant leakage, bent fins, and distorted or damaged condenser inlet and outlet. Repair or replace the condenser as necessary.
8   Installation is the reverse of removal. Replace any O-rings with new ones specifically for the type of refrigerant in your system and lubricate them with refrigerant oil prior to installation. **Warning:** *Do not apply compressor oil to the fitting nuts.* Tighten the condenser inlet and outlet fittings securely.
9   Have the system evacuated, recharged and leak tested by the shop that discharged it. If the condenser was replaced, have the shop add one ounce (30 ml) of new refrigerant oil to the condenser. Use only refrigerant oil compatible with your system.

16 Air conditioning evaporator and expansion valve - removal and installation

Refer to illustrations 16.2, 16.6, 16.7, 16.9, 16.10 and 16.11
Warning: *Some models covered by this manual are equipped with Supplemental Restraint Systems (SRS), more commonly known as airbags. Always disconnect the negative battery cable, then the positive battery cable and wait two minutes before working in the vicinity of the impact sensors, steering column or instrument panel to avoid the possibility of accidental deployment of the airbag, which could cause personal injury (see Chapter 12). Do not use any electrical test equipment on any of the airbag system wires or tamper with them in any way.*
Warning 2: *The air conditioning system is under high pressure. Do not loosen any hose fittings or remove any components until the system has been discharged. Air conditioning refrigerant should be properly discharged into an EPA-approved recovery/recycling unit by a dealer service department or an automotive air conditioning repair facility. Always wear eye protection when disconnecting air conditioning system fittings.*

1   Have the refrigerant discharged and recovered by a qualified repair facility.
2   Disconnect the air conditioning lines at the firewall; using a back-up wrench to prevent damage the fittings (see illustration). Cap the open fittings after disassembly to prevent the entry of air or dirt.
3   Remove the glove box (see Chapter 11). Remove the glove box cover inside the dash.
4   Remove the side panel from the dash on the passenger side.
5   Remove the underdash panel on the passenger side.
6   Disconnect the two wires from the evaporator thermostat (see illustration).
7   Remove the bolts and open the clamps on the left and right sides of the evaporator housing (see illustration).

Chapter 3 Cooling, heating and air conditioning systems

16.6 Disconnect the wires (arrows) to the evaporator thermostat on the top left of the evaporator housing

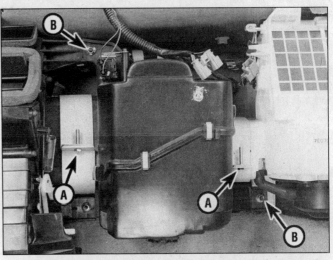

16.7 Remove the two clamp bolts (A), then remove the two nuts (B) mounting the evaporator case to the firewall studs

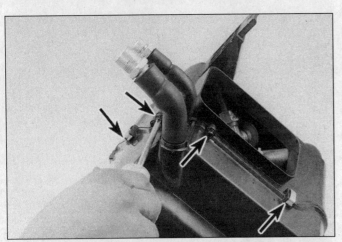

16.9 Remove the screws and clips (arrows indicate four shown here) to separate the housing halves

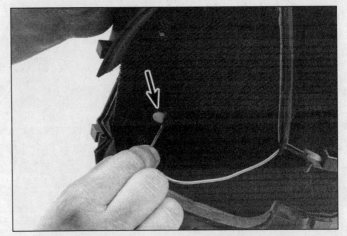

16.10 Pull the thermistor probe (arrow) from the core

8   Remove the evaporator cooling unit nuts from the mounting studs at the firewall, and remove the evaporator housing **(see illustration 16.7)**. Note: *The upper left mounting nut is shared with the heater core housing, and the lower right nut is shared by the blower motor housing.*
9   Remove the clips and screws to separate the lower and upper halves of the evaporator housing **(see illustration)**.
10 Carefully pull the thermistor probe from the evaporator core **(see illustration)**. Remove the evaporator.
11 Disconnect the expansion valve fittings and remove the expansion valve from the evaporator **(see illustration)**. Immediately cap the open fittings to keep moisture and contamination out of the system.
12 Check the evaporator core and fittings for cracks or any other damage. Replace the evaporator if necessary.
13 Reinstall the expansion valve, replacing the gaskets on the expansion valve. Tighten the expansion valve inlet and outlet fittings securely.
14 Evaporator cooling unit installation is the reverse of removal. Replace any O-rings with new ones specifically for the type of refrigerant in your system and lubricate them with refrigerant oil prior to installation. **Warning:** *Do not apply compressor oil to the fitting nuts.* Tighten the evaporator cooling unit inlet and outlet fittings securely. Make sure you reinstall any foam or rubber sealing materials with the new core.
15 Have the system evacuated, charged and leak tested by the shop

16.11 Evaporator core removed from the case - A is the expansion valve, B are insulation/sealing pieces

that discharged it. If the evaporator is replaced with a new unit, add 1.7 ounces (50 ml) of new refrigerant oil the high pressure side of the compressor. Use only refrigerant oil compatible with your system.

Chapter 3 Cooling, heating and air conditioning systems

Notes

Chapter 4
Fuel and exhaust systems

Contents

Section		Section	
Air cleaner assembly - removal and installation	9	Fuel level sending unit - check and replacement	8
Accelerator cable - removal, installation and adjustment	10	Fuel lines and fittings - inspection and replacement	4
Carburetor - removal and installation	12	Fuel pressure relief (fuel injected models only)	2
Carburetor - diagnosis, overhaul and adjustment	11	Fuel pump/fuel pressure - check	3
Catalytic converter	See Chapter 6	Fuel pump - removal and installation	7
Electronic Fuel Injection (EFI) system - check	14	Fuel system check	See Chapter 1
Electronic Fuel Injection (EFI) system - component check and replacement	15	Fuel tank cap gasket replacement	See Chapter 1
		Fuel tank cleaning and repair - general information	6
Electronic Fuel Injection (EFI) system - general information	13	Fuel tank - removal and installation	5
Exhaust manifold - removal and installation	See Chapter 2A	General information	1
Exhaust system check	See Chapter 1	Intake manifold - removal and installation	See Chapter 2A
Exhaust system servicing - general information	16	Underhood hose check and replacement	See Chapter 1
Fuel filter replacement	See Chapter 1		

Specifications

Fuel pressure
Feedback carbureted systems	2 to 5 psi (14 to 34 kPa)
Electronic fuel injection systems	
Fuel pump test connector jumped (engine not running)	38 to 44 psi (265 to 320 kPa)
Vacuum sensing hose attached (at idle)	30 to 38 psi (210 to 265 kPa)
Vacuum sensing hose detached (at idle)	36 to 44 psi (252 to 320 kPa)

Carburetor adjustments
Choke unloader	1/16 to 5/64 inch (1.55 to 2.06 mm)
Choke breaker	1/16 to 3/32 inch (1.73 to 2.24 mm)
Throttle opening	0.009 to 0.014 inch (0.25 to 0.36 mm)

Fuel injector resistance
11 to 18 ohms

Idle Speed
Automatic transmission	See Chapter 1
Manual transmission	See Chapter 1

Fast Idle Speed (carbureted engines)
1,650 to 2,150 rpm

Torque specifications
	Ft-lbs
Throttle body mounting bolts	14
Air intake plenum-to-manifold bolts	14 to 20
Fuel rail mounting bolts	14

1 General information

1988 and 1989 models are equipped with feedback carburetors. These systems use a mechanical fuel pump along with a carburetor that incorporates a series of independent systems (oxygen sensor system, deceleration system, fuel control system etc.) that all contribute to the control of the fuel mixture at various operating conditions. Later models are equipped with an Electronic Fuel Injection system.

Feedback carburetor systems

Feedback carburetors rely on an electronic signal which is generated by the oxygen sensor. Although the air/fuel mixture is controlled mechanically, many of the operating parameters are monitored by electronically controlled devices. For more information on the feedback carburetor systems, refer to Chapter 6.

The carburetor is a 2-barrel downdraft design with a primary and secondary venturi. All use an electrically controlled automatic choke that incorporates a bimetallic spring, an electric heater and a choke

2.2 Connect the fuel pressure gauge to the test port and bleed the residual fuel into a suitable container

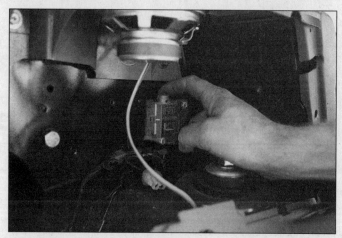

2.7 Unplug the electrical connector for the inertia switch - after repair have been made, plug in the connector and, if the switch "popped," push the reset button on top of the inertia switch to activate the fuel pump system (early system shown)

breaker diaphragm. A stepper motor and valve are used to modify the fuel flow through the primary and secondary main metering air bleed passages. The stepper motor operates an Air Bleed Control (ABCV) valve which varies fuel mixture by controlling airflow in idle and primary main metering air bleed passages. The secondary throttle valve is opened by a secondary vacuum diaphragm. This secondary throttle valve opens when the primary throttle valve is open between 47 and 53 degrees.

Multi Point Fuel Injection (MPFI) system

Multi point fuel injection uses timed impulses to inject the fuel directly into the intake port of each cylinder. The injectors are controlled by the Electronic Control Module (ECM). The ECM monitors various engine parameters and delivers the exact amount of fuel into the intake ports, by controlling the injector "on" time, or duration. The throttle body serves only to control the amount of air passing into the system. Because each cylinder is equipped with an injector mounted directly behind the intake valve, much better control of the fuel/air mixture ratio is possible.

Fuel pump and lines

On fuel-injected models, fuel is circulated from the fuel tank to the fuel injection system, and back to the fuel tank, through a pair of metal lines running along the underside of the vehicle. An electric fuel pump is attached to the fuel level sending unit inside the fuel tank. A vapor return system routes all vapors and hot fuel back to the fuel tank through a separate return line. **Note:** *Carbureted engines are equipped with a mechanical fuel pump mounted on the left rear corner of the cylinder head.*

The electric fuel pump will operate as long as the engine is cranking or running and the ECM is receiving ignition reference pulses from the electronic ignition system (see Chapter 5). If there are no reference pulses, the fuel pump will shut off after two or three seconds.

Exhaust system

The exhaust system includes an exhaust manifold fitted with an exhaust oxygen sensor, a catalytic converter, an exhaust pipe, and a muffler.

The catalytic converter is an emission control device added to the exhaust system to reduce pollutants. A single-bed converter is used in conjunction with a three-way (reduction) catalyst. Refer to Chapter 6 for more information regarding the catalytic converter.

2 Fuel pressure relief (fuel-injected models only)

Warning 1: *Gasoline is extremely flammable, so take extra precautions when you work on any part of the fuel system. Don't smoke or allow open flames or bare light bulbs near the work area, and don't work in a garage where a natural gas-type appliance (such as a water heater or a clothes dryer) with a pilot light is present. Since gasoline is carcinogenic, wear latex gloves when there's a possibility of being exposed to fuel, and, if you spill any fuel on your skin, rinse it off immediately with soap and water. Mop up any spills immediately and do not store fuel-soaked rags where they could ignite. The fuel system on fuel-injected models is under constant pressure, so, if any fuel lines are to be disconnected, the fuel pressure in the system must be relieved first. When you perform any kind of work on the fuel system, wear safety glasses and have a Class B type fire extinguisher on hand.*

Warning 2: *After the fuel pressure has been relieved, wrap shop towels around any fuel connection you'll be disconnecting. They'll absorb the residual fuel that may leak out, reducing the risk of fire and preventing contact with your skin.*

1 There are two methods for relieving the fuel system pressure; the easiest and most accessible is using a special fuel pressure gauge with a bleedoff valve. This special tool can be purchased most auto parts stores. In the event the tool is not available, locate the inertia switch and disable the fuel pump.

Fuel pressure gauge bleeding method

Refer to illustration 2.2

2 Locate the fuel pressure test port **(see illustration)** on the fuel rail and connect the fuel pressure gauge to the Schrader valve.
3 Direct the bleedoff hose into a metal cup or suitable container for gasoline storage.
4 Turn the valve and allow the excess fuel to bleed into the container.
5 Close the valve, remove the fuel pressure gauge and cap the test port.

Inertia switch method

Refer to illustrations 2.7

6 The fuel pump switch - sometimes called the "inertia switch" - which shuts off fuel to the engine in the event of a collision, affords a simple and convenient means by which fuel pressure can be relieved before servicing fuel injection components. The switch is located behind the driver's side trim panel in the rear compartment of the vehicle.
7 Unplug the inertia switch electrical connector **(see illustration)**.
Warning: *Some models covered by this manual are equipped with a Supplemental Restraint system (SRS), more commonly known as airbags. Always disconnect the negative battery cable, then the positive battery cable and wait two minutes before working in the vicinity of the impact sensors, steering column or instrument panel to avoid the possibility of accidental employment of the airbag, which could cause personal injury (see Chapter 12). Do not use electrical test equipment*

Chapter 4 Fuel and exhaust systems

3.4 Detach the outlet line from the fuel pump and install a fuel pressure gauge

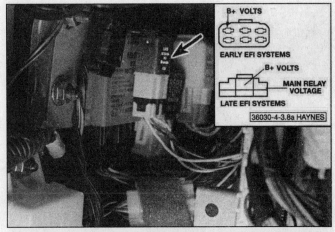

3.8a Check for battery voltage to the fuel pump relay with the ignition key ON (engine not running)

3.8b If the fuel pump isn't receiving voltage, first make sure the inertia switch hasn't "popped." Remove this small access panel . . .

3.8c . . . and depress the button on top of the switch to reset it

on any of the airbag system components or wiring, or tamper with them in any way.

8   Start the engine and allow it to run until it stops. This should take only a few seconds.

9   The fuel system pressure is now relieved. When you're finished working on the fuel system, simply plug the electrical connector back into the switch. If the inertia switch was "popped" (activated) during this procedure, push the reset button on the top of the switch.

3   Fuel pump/fuel pressure - check

Warning: *Gasoline is extremely flammable, so take extra precautions when you work on any part of the fuel system. See the Warning at the beginning of Section 2.*

1   Check that there is adequate fuel in the fuel tank. If you doubt the reading on the gauge, insert a long wooden dowel into the fuel filler opening; it will serve as a dipstick.

Mechanical pump (carbureted engines)

Refer to illustration 3.4

2   Remove the air cleaner assembly (see Section 9). Examine all fuel lines between the fuel tank and fuel pump for leaks, loose connections, kinks or distortion of the rubber hoses. Air leaks upstream of the fuel pump can seriously affect the pump's output.

3   Start the engine and check the body of the pump for leaks. Shut off the engine.

4   Remove the fuel filler cap to relieve the fuel tank pressure. Dis-

connect the fuel line at the carburetor. Disconnect the primary (low voltage) electrical connectors to the ignition coil so the engine can be cranked without it firing. Place an approved gasoline container at the end of the detached fuel line and have an assistant crank the engine for several seconds. There should be a strong spurt of gasoline from the line on every second revolution. If necessary, install a fuel pressure gauge and observe the pressure reading at idle **(see illustration)**. Refer to the Specifications listed in this Chapter.

5   If little or no gasoline emerges from the line during engine cranking, either the fuel line is clogged or the fuel pump is not working properly. Disconnect the fuel feed line from the pump and blow air through it to be sure that the line is clear. If the line is not clogged, the pump should be replaced with a new one.

Electric pump (fuel-injected engines)

Preliminary inspection

Refer to illustrations 3.8a, 3.8b and 3.8c

6   Should the fuel system fail to deliver the proper amount of fuel, or any fuel at all, inspect it as follows. Remove the fuel filler cap. Have an assistant turn the ignition key to the On position (engine not running) while you listen at the fuel filler opening. You should hear a whirring sound that lasts for a couple of seconds.

7   If you don't hear anything, check the fuel pump fuse (see Chapter 12). If the fuse is blown, replace it and see if it blows again. If it does, trace the fuel pump circuit for a short. Refer to the wiring diagrams at the end of Chapter 12.

8   Without removing the relay, backprobe the connector and check for battery voltage at the indicated terminal of the fuel pump relay connector **(see illustration)**. If there is no battery voltage present, have the relay(s) tested at a dealer service department or other qualified auto-

3.12 Connect a fuel pressure gauge to the Schrader valve on the fuel rail (late model EFI system shown)

3.14a Bridge the designated terminals of the Fuel Pump Test Connector using a jumper wire or paper clip. The check connector is located near the left shock tower (1993 and later model shown)

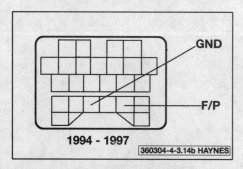

3.14b Fuel pump test terminals of the check connector - 1994 and later models

3.18 Detach the vacuum line from the fuel pressure regulator (arrow) and connect a hand-held vacuum pump. Observe the fuel pressure, first, with vacuum applied and then without vacuum applied. Fuel pressure should increase as vacuum decreases

motive repair shop. **Warning:** *Some models covered by this manual are equipped with the Supplemental Restraint system (SRS), more commonly known as airbags. Do not use electrical test equipment on any of the airbag system wiring or components or tamper with them in any way.* **Note 1:** *The inertia switch is an electrical device wired into the fuel pump circuit that will shut down power to the fuel pump in an accident. Be sure to check that the inertia switch is activated and in working order if the fuel pump is not receiving voltage* **(see illustrations)**. **Note 2:** *The fuel pump relay provides the circuit ground to operate the fuel pump. On 1993 and earlier models, after the engine is started, the fuel pump circuit ground is provided by the fuel pump switch located inside the airflow meter. If the engine starts and then suddenly shuts down, check the airflow meter. Refer to the wiring schematics at the end of Chapter 12 for additional information on the wiring color designations for the fuel pump relay.* **Note 3:** *The fuel pump relay is located under the driver's side dash panel and the power relay is located in the engine compartment front corner on the driver's side (left). On 1996 and 1997 OBD II models, the power relay (main relay) is located under the dash near the fuel pump relay.*

9   If the relay is good and there is no voltage present in Step 8, check the fuse(s) and the wiring circuit for the fuel pump relay and/or power relay (see Chapter 12). If voltage is present, check for battery voltage at the fuel pump harness connector located near the fuel tank. If voltage is reaching the fuel pump, remove the fuel pump and have it checked by a dealer service department or other qualified automotive repair facility.

Pressure check

Refer to illustrations 3.12, 3.14a, 3.14b and 3.18

10   Relieve the fuel system pressure (see Section 2).
11   Detach the cable from the negative battery terminal.
12   Remove the cap from the fuel pressure test port and attach a fuel pressure gauge **(see illustrations)**. If you don't have the correct adapter for the test port, remove the Schrader valve and connect the gauge hose to the fitting, using a hose clamp.
13   Attach the cable to the negative battery terminal.
14   First, check the fuel pressure by activating the Fuel Pump Test Connector **(see illustrations)**. This will activate the fuel pump and give a static (engine not running) fuel pressure reading. This also checks the integrity of the fuel pump circuit. **Note:** *The fuel pump test connector is located in the corner of the engine compartment in the driver's (left) side.*
15   Check the fuel pressure at idle. Compare your readings with the values listed in this Chapter's Specifications. Disconnect the vacuum hose from the fuel pressure regulator and watch the fuel pressure gauge - the fuel pressure should jump up considerably as soon as the hose is disconnected. If it doesn't, check for a vacuum signal to the fuel pressure regulator (see Step 20).
16   If the fuel pressure is low, pinch the fuel return line shut and watch the gauge. If the pressure doesn't rise, the fuel pump is defective or there is a restriction in the fuel feed line. If the pressure rises sharply, replace the pressure regulator. **Note:** *If the vehicle is equipped with a*

nylon fuel return line (or fuel lines made up of steel or other rigid material), it will be necessary to install a special fuel testing harness between the fuel rail and the return line. This can be made up from compatible fuel line connectors (available at dealer parts departments and some auto parts stores), fuel hose and hose clamps.

17 If the fuel pressure is too high, turn the engine off. Relieve the fuel pressure, disconnect the fuel return line and blow through it to check for a blockage. If there is no blockage, replace the fuel pressure regulator.

18 To check for smooth operation of the pressure regulator, hook up a hand-held vacuum pump to the vacuum hose port on the fuel pressure regulator **(see illustration)**.

19 Read the fuel pressure gauge with vacuum applied to the fuel pressure regulator and also with no vacuum applied. The fuel pressure should decrease as vacuum increases (and increase as vacuum decreases).

20 Connect a vacuum gauge to the pressure regulator vacuum hose. Start the engine and check for vacuum. If there isn't vacuum present, check for a clogged hose or vacuum port. If the amount of vacuum is adequate but the fuel pressure didn't increase when the vacuum hose was detached, replace the fuel pressure regulator.

21 Turn the ignition switch to OFF, and recheck the pressure on the gauge. If the pressure immediately begins to drop to zero:
  a) The fuel lines may be leaking.
  b) The fuel pressure regulator may be allowing the fuel pressure to bleed through to the return line
  c) A fuel injector (or injectors) may be leaking.
  d) The fuel pump may be defective.

4 Fuel lines and fittings - inspection and replacement

Warning: *Gasoline is extremely flammable, so take extra precautions when you work on any part of the fuel system. See the **Warning** at the beginning of Section 2.*

Inspection

1 Once in a while, you will have to raise the vehicle to service or replace some component (an exhaust pipe hanger, for example). Whenever you work under the vehicle, always inspect fuel lines and all fittings and connections for damage or deterioration.

2 Check all hoses and pipes for cracks, kinks, deformation or obstructions.

3 Make sure all hoses and pipe clips attach their associated hoses or pipes securely to the underside of the vehicle.

4 Verify all hose clamps attaching rubber hoses to metal fuel lines or pipes are snug enough to ensure a tight fit between the hoses and pipes.

Replacement

5 If you must replace any damaged sections, use original equipment replacement hoses or pipes constructed from exactly the same material as the section you are replacing. Do not install substitutes constructed from inferior or inappropriate material or you could cause a fuel leak or a fire.

6 Always, before detaching or disassembling any part of the fuel system, note the routing of all hoses and pipes and the orientation of all clamps and clips to ensure that replacement sections are installed in exactly the same manner. When attaching hoses to metal lines, overlap them and secure them properly.

7 Before detaching any part of the fuel system, be sure to relieve the fuel pressure (see Section 2) and disconnect the battery. Cover the fitting being disconnected with a rag to absorb any fuel that may spray out.

8 While you're under the vehicle, it's a good idea to check the condition of the fuel filter - make sure that it's not damaged (see Chapter 1).

5 Fuel tank - removal and installation

Refer to illustration 5.13

Warning: *Gasoline is extremely flammable, so take extra precautions when you work on any part of the fuel system. See the **Warning** at the beginning of Section 2.*

1 This procedure is much easier to perform if the fuel tank is empty. If the fuel tank is not equipped with a drain plug, postpone the job until the tank is empty, or siphon the fuel into an approved container using a siphoning kit (available at most auto parts stores). **Warning:** *Do not start the siphoning action by mouth!*

2 Remove the fuel filler cap to relieve fuel tank pressure. If you're working on a fuel-injected model, relieve the fuel system pressure (see Section 2).

3 Detach the cable from the negative terminal of the battery.

4 Raise the vehicle and place it securely on jackstands.

5 Remove the rear seat and cushion (see Chapter 11).

6 Remove the quarter trim panel from the rear of the vehicle. Also, remove the luggage compartment floor cover.

7 Remove the sending unit access plate and disconnect the fuel pump/fuel level sending unit electrical connector. Also, disconnect the fuel lines from the top of the fuel pump assembly.

8 Remove the inner floor filler cover bolts and separate the cover from the body.

9 Loosen the filler pipe hose clamp and overflow hose clamp and remove the hoses from the fuel tank.

10 Disconnect the fuel lines, the vapor return line and the fuel filler pipe. **Note:** *The fuel feed and return lines and the vapor return line are different diameters, so reattachment is simplified. If you have any doubts, however, clearly label the three lines and their respective inlet or outlet pipes. Be sure to plug the hoses to prevent leakage and contamination of the fuel system.*

11 Support the fuel tank with a floor jack. Place a sturdy plank between the jack head and the fuel tank to protect the tank.

12 Detach the fuel line bracket and remove the fuel line or tank protectors.

13 On two-door models, remove the mounting bolts **(see illustration)** from the fuel tank and angle the tank to the left to remove it from the vehicle. On four-door models, remove the mounting bolts from the tank straps, lower the straps and allow the tank to drop slowly. **Note:** *On four-door models, push the brake hose clip through the bracket on the fuel tank and position the front parking brake cable and conduit out of the way.*

14 Lower the tank enough to disconnect the wires and ground strap from the fuel pump/fuel gauge sending unit, if you have not already done so.

15 Installation is the reverse of removal.

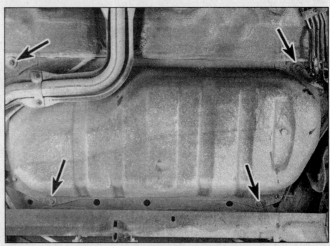

5.13 Remove the fuel tank mounting bolts (arrows) (two-door model shown)

7.4 Location of the mechanical fuel pump on carbureted models

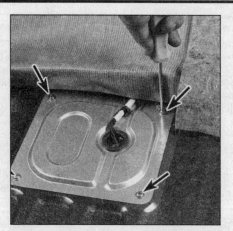

7.11 Remove the mounting screws (arrows) from the fuel pump access cover

7.12 Loosen the clamps and detach the fuel lines

6 Fuel tank cleaning and repair - general information

1 Any repairs to the fuel tank or filler neck should be carried out by a professional who has experience in this critical and potentially dangerous work. Even after cleaning and flushing of the fuel system, explosive fumes can remain and ignite during repair of the tank.
2 If the fuel tank is removed from the vehicle, it should not be placed in an area where sparks or open flames could ignite the fumes coming out of the tank. Be especially careful inside garages where a natural gas-type appliance is located, because the pilot light could cause an explosion.

7 Fuel pump - removal and installation

Warning: *Gasoline is extremely flammable, so take extra precautions when you work on any part of the fuel system. See the **Warning** at the beginning of Section 2.*
1 Relieve the fuel system pressure (see Section 2).
2 Disconnect the cable from the negative terminal of the battery.

Mechanical pump (carbureted models)

Refer to illustration 7.4
3 Relieve the fuel tank pressure by removing the fuel filler cap.
4 Locate the fuel pump mounted on the left rear corner of the cylinder head **(see illustration)**. Place rags underneath the pump to catch any spilled fuel.
5 Loosen hose clamps and slide them down the hoses, past the fittings. Disconnect the hoses from the pump, using a twisting motion as you pull them from the fittings. Immediately plug the hoses to prevent leakage of fuel and the entry of dirt.
6 Unscrew the fasteners that retain the pump to the cylinder head, then detach the pump from the head. Inspect the fuel pump arm for wear. Coat it with clean engine oil before installing it.
7 Using a gasket scraper or putty knife, remove all traces of old gasket material from the mating surfaces on the cylinder head (and the fuel pump, if the same one will be reinstalled). While scraping, be careful not to gouge the soft aluminum surfaces.
8 Installation is the reverse of the removal procedure, but be sure to use a new gasket and tighten the mounting fasteners securely.

Electric pump (fuel-injected models)

Refer to illustrations 7.11, 7.12, 7.13, 7.14, 7.15, 7.16 and 7.17
9 Remove the fuel tank cap.
10 Disconnect the cable from the negative terminal of the battery.
11 Remove the rear seat and fold the luggage compartment floor cover towards the front of the vehicle to reveal the fuel pump/sending unit access cover. Remove the access cover from inside the vehicle **(see illustration)**.
12 Disconnect the fuel lines **(see illustration)** and the electrical connector from the fuel pump/sending unit assembly.
13 Unscrew the fuel pump/sending unit assembly screws **(see illustration)** from the top of the fuel tank.
14 Carefully withdraw the fuel pump assembly from the fuel tank **(see illustration)**. Be careful not to bend the float arm of the fuel level sending unit.
15 Disconnect the fuel pump electrical connector **(see illustration)**.

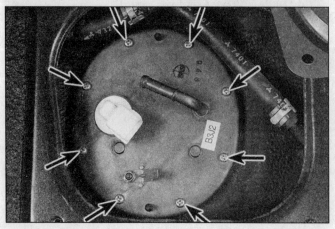

7.13 Remove the mounting screws from the fuel pump assembly (arrows)

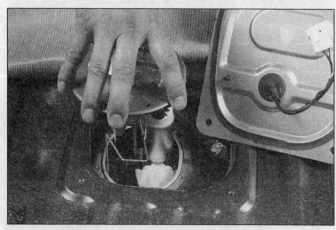

7.14 Simultaneously, lift and angle the assembly out of the fuel tank (late model EFI system shown)

Chapter 4 Fuel and exhaust systems

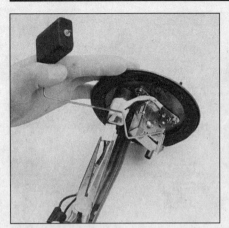

7.15 Disconnect the fuel pump connector

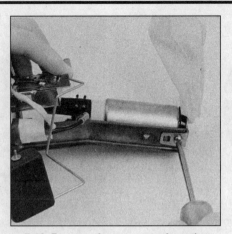

7.16 Remove the set screw from the fuel pump

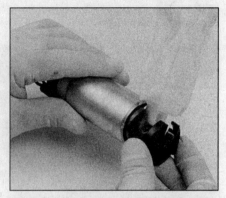

7.17 Pull the lower end of the fuel pump loose from the bracket and remove the rubber cushion that insulates the bottom of the pump

16 Remove the set screw from the bottom of the fuel pump bracket **(see illustration)**.
17 Remove the rubber cushion from the lower end of the fuel pump **(see illustration)**.
18 Remove the clip securing the filter to the pump.
19 Pull out the filter and inspect it for contamination. If it is dirty, replace it.
20 If you are only replacing the fuel pump filter, install the new filter, the clip and the rubber cushion, push the lower end of the pump back into the bracket, install the set screw and install the pump/bracket assembly in the fuel tank.
21 If you are replacing the fuel pump, loosen the hose clamp at the upper end of the pump and disconnect the pump from the hose.
22 Disconnect the wires from the pump terminals and remove the pump.
23 Installation is the reverse of removal.

8 Fuel level sending unit - check and replacement

Warning: *Gasoline is extremely flammable, so take extra precautions when you work on any part of the fuel system. See the* **Warning** *at the beginning of Section 2.*

Check

Refer to illustration 8.3

1 Before performing any tests on the fuel level sending unit, try to determine the actual fuel level in the tank. If this is not possible, either fill the tank with fuel or drain out any fuel in the tank into an approved gasoline container (see Section 5).
2 Raise the vehicle and secure it with jackstands.
3 Disconnect the fuel level sending unit electrical connector **(see illustration)** located at the front of the fuel tank.
4 Connect the probes of an ohmmeter to the electrical connector terminals and check for resistance. Use the 200 scale on the ohmmeter.
5 With the fuel tank completely full, the resistance should be about 6.0 ohms.
6 With the fuel tank empty, the resistance of the sending unit should be about 95 ohms.
7 If the readings are incorrect, replace the sending unit.
8 A more accurate check of the sending unit can be made by removing it from the fuel tank and checking its resistance while manually operating the float arm.

Replacement

Refer to illustrations 8.13 and 8.14

9 Remove the rear seat (see Chapter 11).
10 Remove the fuel level sending unit cover bolts the fuel tank.
11 Remove the screws that retain the cover to the fuel level sending unit.
12 Carefully angle the sending unit out of the opening without damaging the fuel level float located at the bottom of the assembly.
13 Disconnect the electrical connector from the sending unit **(see illustration)**.

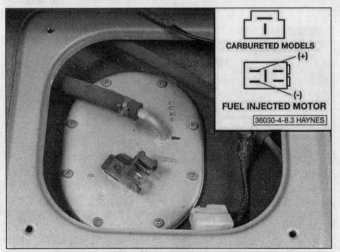

8.3 Fuel level sending unit terminals

8.13 Disconnect the fuel level sending unit electrical connector

4-8   Chapter 4   Fuel and exhaust systems

8.14 Remove the sending unit mounting nuts

14   Remove the sending unit mounting fasteners and remove it **(see illustration)**.
15   Installation is the reverse of removal.

9   Air cleaner assembly - removal and installation

Refer to illustrations 9.5 and 9.7

Carbureted engines

1   Remove the air filter from the air cleaner housing (see Chapter 1).

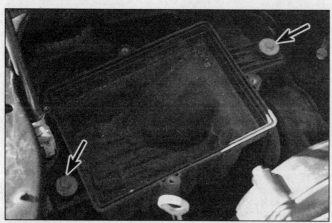

9.7 Remove the air cleaner housing bolts (arrows) using a socket and extension

10.2 Loosen the locknut on the threaded portion of the accelerator cable

9.5 Detach the intake air duct from the airflow meter assembly

2   Disconnect and mark any hoses, vacuum lines, electrical connectors, etc. from the air cleaner assembly.
3   Lift the air cleaner assembly from the engine.
4   Installation is the reverse of removal.

Fuel-injected engines

5   Disconnect the air intake hose from the airflow meter **(see illustration)**.
6   Detach the clips and remove the air filter cover/airflow meter and the filter element (see Chapter 1).
7   Remove the three bolts and remove the air cleaner housing from the engine compartment **(see illustration)**.
8   Installation is the reverse of removal.

10   Accelerator cable - removal, installation and adjustment

Removal

Refer to illustrations 10.2 and 10.3

1   Detach the cable from the negative terminal of the battery.
2   Loosen the locknut on the threaded portion of the accelerator cable at the throttle body **(see illustration)**.
3   Grasp the throttle lever arm and rotate it to put some slack in the cable, then slip the cable end out of its slot in the arm **(see illustration)**.

10.3 Rotate the throttle lever and pass the cable through the slot

4 Detach the accelerator cable from the accelerator pedal.
5 From inside the vehicle, pull the cable through the cable casing.

Installation and adjustment

6 Installation is the reverse of removal.
7 To adjust the cable, fully depress the accelerator pedal and check that the throttle is fully opened.
8 If not fully opened, loosen the locknuts, depress accelerator pedal and adjust the cable.
9 Tighten the locknuts and recheck the adjustment. Make sure the throttle closes fully when the pedal isn't depressed.

11 Carburetor - diagnosis, overhaul and adjustment

Warning: *Gasoline is extremely flammable, so take extra precautions when you work on any part of the fuel system. See the* **Warning** *at the beginning of Section 2.*

General information

1 The electronically controlled carburetor emission system, more commonly called the electronic feedback carburetor system, relies on an electronic signal generated by an exhaust gas oxygen sensor to control a variety of devices and keep emissions within limits. The system works in conjunction with a three-way catalyst to control the levels of carbon monoxide, hydrocarbons and oxides of nitrogen. The feedback carburetor system also works in conjunction with the computer. The two systems share certain sensors and output actuators; therefore, diagnosing the feedback carburetor system will require a thorough check of all the feedback carburetor components (refer to Chapter 6 for additional information).
2 The system operates in two modes: open loop and closed loop. When the engine is cold, the air/fuel mixture is controlled by the computer in accordance with a program designed in at the time of production. The air/fuel mixture during open loop mode will be richer to allow for proper engine warm-up. When the engine is at operating temperature, the system operates in closed loop and the air/fuel mixture is varied depending on the information supplied by the exhaust gas oxygen sensor.
3 Here is a list of the various sensors and output actuators involved with these feedback carburetor systems:

Air bleed control valve
Atmospheric pressure switch
Bowl vent solenoid valve
Clutch switch
Slow fuel cut solenoid
Positive Temperature Coefficient (PTC) Heater or Electronic Fuel Evaporator (EFE)
Engine Coolant Temperature sensor
Engine Coolant Temperature switch
EGR valve position sensor (California models only)
Idle switch
Manifold Absolute Pressure sensor
Vacuum solenoid valve
Vacuum switch

4 Refer to Chapter 6 for the diagnostic checks for the feedback carburetor system components.

General diagnosis

5 A thorough road test and check of carburetor adjustments should be done before any major carburetor work. Specifications for some adjustments are listed on the Vehicle Emissions Control Information (VECI) label found in the engine compartment.
6 Carburetor problems usually show up as flooding, hard starting, stalling, severe backfiring and poor acceleration. A carburetor that's leaking fuel and/or covered with wet looking deposits definitely needs attention.
7 Some performance complaints directed at the carburetor are actually a result of loose, out-of-adjustment or malfunctioning engine or electrical components. Others develop when vacuum hoses leak, are disconnected or are incorrectly routed. The proper approach to analyzing carburetor problems should include the following items:

a) *Inspect all vacuum hoses and actuators for leaks and correct installation* (see Chapters 1 and 6).
b) *Tighten the intake manifold and carburetor mounting nuts/bolts evenly and securely.*
c) *Perform a cylinder compression test* (see Chapter 2).
d) *Clean or replace the spark plugs as necessary* (see Chapter 1).
e) *Check the spark plug wires* (see Chapter 1).
f) *Inspect the ignition primary wires.*
g) *Check the ignition timing (follow the instructions printed on the Emissions Control Information label).*
h) *Check the fuel pump* (see Section 3).
i) *Check the heat control valve in the air cleaner for proper operation* (see Chapter 6).
j) *Check/replace the air filter element* (see Chapter 1).
k) *Check the PCV system* (see Chapters 1 and 6).
l) *Check/replace the fuel filter* (see Chapter 1). *Also, the strainer in the tank could be restricted.*
m) *Check for a plugged exhaust system.*
n) *Check EGR valve operation* (see Chapter 6).
o) *Check the choke - it should be completely open at normal engine operating temperature* (see Chapter 1).
p) *Check for fuel leaks and kinked or dented fuel lines* (see Chapters 1 and 4).
q) *Check accelerator pump operation with the engine off (remove the air cleaner cover and operate the throttle as you look into the carburetor throat - you should see a stream of gasoline enter the carburetor).*
r) *Check for incorrect fuel or bad gasoline.*
s) *Check the valve clearances (if applicable) and camshaft lobe lift* (see Chapters 1 and 2).
t) *Have a dealer service department or repair shop check the electronic engine and carburetor controls.*

8 Diagnosing carburetor problems may require that the engine be started and run with the air cleaner off. While running the engine without the air cleaner, backfires are possible. This situation is likely to occur if the carburetor is malfunctioning, but just the removal of the air cleaner can lean the fuel/air mixture enough to produce an engine backfire. **Warning:** *Don't position any part of your body, especially your face, directly over the carburetor during inspection and servicing procedures. Wear eye protection!*

Overhaul

9 Once it's determined that the carburetor needs an overhaul, several options are available. If you're going to attempt to overhaul the carburetor yourself, first obtain a good quality carburetor rebuild kit (which will include all necessary gaskets, internal parts, instructions and a parts list). You'll also need some special solvent and a means of blowing out the internal passages of the carburetor with air.
10 An alternative is to obtain a new or rebuilt carburetor. They're readily available from dealers and auto parts stores. Make absolutely sure the exchange carburetor is identical to the original. A tag is usually attached to the top of the carburetor or a number is stamped on the float bowl. It'll help determine the exact type of carburetor you have. When obtaining a rebuilt carburetor or a rebuild kit, make sure the kit or carburetor matches your application exactly. Seemingly insignificant differences can make a large difference in engine performance.
11 If you choose to overhaul your own carburetor, allow enough time to disassemble it carefully, soak the necessary parts in the cleaning solvent (usually for at least one-half day or according to the instructions listed on the carburetor cleaner) and reassemble it, which will usually take much longer than disassembly. When disassembling the carburetor, match each part with the illustration in the carburetor kit and lay the parts out in order on a clean work surface. Overhauls by inexperienced mechanics can result in an engine which runs poorly or not at all. To avoid this, use care and patience when disassembling the carburetor so you can reassemble it correctly.

4-10　Chapter 4　Fuel and exhaust systems

12　Because carburetor designs are constantly modified by the manufacturer in order to meet increasingly more stringent emissions regulations, the overhaul procedures in this Chapter may not apply exactly to your vehicle. You'll receive a detailed, well illustrated set of instructions with any carburetor overhaul kit; they'll apply in a more specific manner to the carburetor on your vehicle.

Adjustments

Note: *These carburetor adjustments are strictly in-vehicle adjustments. During overhaul, refer to the instructions included in the overhaul kit for complete procedures and any additional adjustments that are required.*

Fast idle speed

Refer to illustration 11.16

13　The ignition timing should be checked and adjusted before setting the fast idle speed. Adjustment should not be performed while the cooling fan is operating or during A/C operation. Make sure the fan is not operating and the air conditioning system is turned off before adjusting the fast idle speed.
14　Shift the transmission to Neutral and set the parking brake. Connect a tachometer according to the tool manufacturer's instructions.
15　With the engine at normal operating temperature and all the lights and accessories turned OFF, set the fast idle cam on the second step and observe the fast idle speed on the tachometer. The fast idle speed should be 1,650 to 2,150 rpm.
16　To adjust the fast idle speed, turn the fast idle speed screw **(see illustration)** until the fast idle speed listed in this Chapter's Specifications is obtained.

11.16　Location of the fast idle speed screw

Choke unloader

Refer to illustration 11.17

17　Open the throttle fully while holding the choke valve closed. Measure the gap between the upper edge of the choke plate **(see illustration)**. A drill bit or Allen wrench can be used.
18　Compare the measurement to the Specifications listed in this Chapter.
19　To adjust the choke unloader, bend the choke unloader lever until the correct clearance is attained.

Choke breaker

Refer to illustration 11.22

20　Set the choke to the fully closed position and, using a hand-held vacuum pump, apply 16 in-Hg to the choke breaker diaphragm. Check the choke valve-to-air horn clearance.
21　Make sure the automatic choke housing cover index mark is opposite the center mark on the carburetor body.
22　Adjust the clearance by bending the tab on the lever **(see illustration)**.
23　Adjust the engine idle speed (see Chapter 1).

11.17　Bend the choke unloader lever (arrow) until the gap between the choke plate and air horn is as specified

12　Carburetor - removal and installation

Warning: *Gasoline is extremely flammable, so take extra precautions when you work on any part of the fuel system. See the **Warning** at the beginning of Section 2.*

Removal

1　Remove the fuel tank cap to relieve the tank pressure. Disconnect the cable from the negative terminal of the battery.
2　Remove the air cleaner from the carburetor. Be sure to label all vacuum hoses attached to the air cleaner housing.
3　Disconnect the accelerator cable from the throttle lever (see Section 10).
4　If the vehicle is equipped with an automatic transmission, disconnect the Throttle Valve (TV) cable from the throttle lever.
5　Clearly label all vacuum hoses and fittings, then disconnect the hoses.
6　Disconnect the fuel line from the carburetor. Plug the line to pre-

11.22　Choke breaker adjustment

vent fuel spillage.
7　Label the wires and terminals, then unplug all electrical connectors.
8　Remove the mounting fasteners and lift the carburetor from the intake manifold. Remove the gasket. Do not remove the EFE heater (see Chapter 6). Stuff a shop rag into the intake manifold openings to prevent the entry of debris.

Installation

9　Use a gasket scraper to remove all traces of gasket material and sealant from the EFE heater, being careful not to damage the heater, (and the carburetor, if it's being reinstalled), then remove the shop rag from the manifold openings. Clean the mating surfaces with lacquer thinner or acetone.
10　Place a new gasket on the EFE heater.
11　Position the carburetor on the gasket then install the mounting fasteners.

Chapter 4 Fuel and exhaust systems

14.6 The area inside the throttle body near the throttle plate usually gets a lot of sludge build-up due to the crankcase vapor venting into the PCV system. With the engine off, use carburetor cleaner (make sure it is safe for use with catalytic converters and oxygen sensors) and a toothbrush to clean the throttle body - open the throttle plate so you can clean behind it

12 To prevent carburetor distortion or damage, tighten the fasteners, in a criss-cross pattern, a little at a time until they are secure.
13 The remaining installation steps are the reverse of removal.
14 Check and, if necessary, adjust the idle speed (see Chapter 1).
15 If the vehicle is equipped with an automatic transmission, refer to Chapter 7, Part B, for the TV cable adjustment procedure.
16 Start the engine and check carefully for fuel leaks.

13 Electronic Fuel Injection (EFI) system - general information

1 1990 and later models are equipped with an Electronic Fuel Injection (EFI) system. The EFI system is composed of three basic sub systems: the fuel system, the air induction system and the electronic control system.

Fuel system

2 An electric fuel pump, located inside the fuel tank, supplies fuel under constant pressure to the fuel rail, which distributes fuel evenly to all injectors. From the fuel rail, fuel is injected into the intake ports, just above the intake valves, by four fuel injectors. The amount of fuel supplied by the injectors is precisely controlled by an Electronic Control Module (ECM). A pressure regulator controls system pressure in relation to intake manifold vacuum. A fuel filter between the fuel pump and the fuel rail filters fuel to protect the components of the system.

Air induction system

3 The air induction system consists of an air filter housing, an airflow meter (1990 through 1993 models) or a mass airflow sensor (1994 and later models) and a throttle body. The airflow meter (VAF) is an information gathering device for the ECM and it is located in the air intake duct near the air cleaner assembly. The VAF assembly contains a movable vane that is connected to a potentiometer. As the air flows through the airflow meter the vane changes position, thereby providing an input signal to the computer in respect to air volume changes. The mass airflow sensor (MAF) is an information gathering device for the ECM. The Mass Airflow (MAF) Sensor is located on the air intake duct. This sensor uses a hot wire sensing element to measure the amount of air entering the engine. The air passing over the hot wire causes it to cool. Consequently, this change in temperature can be converted into an analog voltage signal to the ECM which in turn calculates the required fuel injector pulse width. The On-Board Diagnostic (OBD) system can detect a variety of different VAF and MAF sensor problems and set codes to indicate the specific trouble area. Refer to Chapter 6 for additional information.

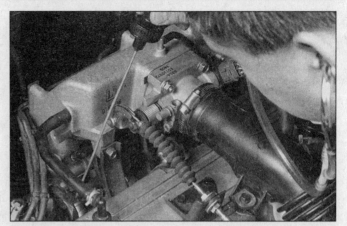

14.7 Use a stethoscope to determine if the injectors are working properly - they should make a steady clicking sound that rises and falls with engine speed changes

Electronic control system

4 The Computer Control System controls the EFI and other systems by means of an Electronic Control Module (ECM), which employs a microcomputer. The ECM receives signals from a number of information sensors which monitor such variables as intake air volume, intake air temperature, coolant temperature, engine rpm, acceleration/deceleration and exhaust oxygen content. These signals help the ECM determine the injection duration necessary for the optimum air/fuel ratio. Some of these sensors and their corresponding ECM-controlled relays are not contained within EFI components, but are located throughout the engine compartment. For further information regarding the ECM and its relationship to the engine electrical and ignition system, see Chapter 6.

14 Electronic Fuel Injection (EFI) system - check

Refer to illustrations 14.6, 14.7, 14.8 and 14.9
Warning: *Gasoline is extremely flammable, so take extra precautions when you work on any part of the fuel system. See the* **Warning** *at the beginning of Section 2.*

1 Check the ground wire connections for tightness. Check all wiring and electrical connectors that are related to the system. Loose electrical connectors and poor grounds can cause many problems that resemble more serious malfunctions.
2 Check to see that the battery is fully charged, as the control unit and sensors depend on an accurate supply voltage in order to properly meter the fuel.
3 Check the air filter element - a dirty or partially blocked filter will severely impede performance and economy (see Chapter 1).
4 If a blown fuse is found, replace it and see if it blows again. If it does, search for a grounded wire in the harness related to the system.
5 Check the air intake duct from the airflow meter to the intake manifold for leaks, which will result in an excessively lean mixture. Also check the condition of the vacuum hoses connected to the intake manifold.
6 Remove the air intake duct from the throttle body and check for dirt, carbon or other residue build-up. If it's dirty, clean it with carburetor cleaner (make sure the can says it's safe for use with oxygen sensors and catalytic converters) and a toothbrush **(see illustration)**.
Caution: *The throttle body on 1994 and later models is coated with a sludge-resistant material designed to protect the bore and throttle plate. Do not attempt to clean the interior of the throttle body with carburetor or other spray cleaners. This throttle body is designed to resist sludge accumulation and cleaning will impair the performance of the engine.*
7 With the engine running, place a stethoscope against each injector, one at a time, and listen for a clicking sound, indicating operation

14.8 Install the "noid" light into the fuel injector electrical connector and check to see that it blinks when the engine is running

14.9 Using an ohmmeter, measure the resistance across the terminals of the injector (arrows)

15.11 Remove the throttle body bolts (arrows)

(see illustration). If you don't have a stethoscope, place the tip of a long screwdriver against the body of the injector and listen through the handle.

8   If there is a problem with an injector, purchase a special injector test light (sometimes called a "noid" light) and install it into the injector electrical connector **(see illustration)**. Start the engine and make sure that each injector connector flashes the noid light. This will test for the proper voltage signal to the injectors.

9   With the engine OFF and the fuel injector electrical connectors disconnected, measure the resistance of each injector **(see illustration)**. Each injector should measure about 11 to 18 ohms. If not, the injector is probably faulty.

10 The remainder of the system checks should be left to a dealer service department or other qualified repair shop, as there is a chance that the control unit may be damaged if not performed properly.

15 Electronic Fuel Injection (EFI) system - component check and replacement

Warning: *Gasoline is extremely flammable, so take extra precautions when you work on any part of the fuel system. See the Warning at the beginning of Section 2.*

Throttle body

Check

1   Verify that the throttle linkage operates smoothly.
2   Start the engine, detach each vacuum hose and, using your finger, check the vacuum at each port on the throttle body with the engine at idle and above idle. Vacuum signals will vary depending upon location of the vacuum port (port or manifold vacuum).

Replacement

Refer to illustration 15.11
Warning: *The engine must be completely cool before beginning this procedure.*
3   Detach the cable from the negative terminal of the battery.
4   Drain the radiator (see Chapter 1).
5   Loosen the hose clamps and remove the air intake duct.
6   Detach the accelerator cable from the throttle lever arm (see Section 10), then detach the cable bracket and set it aside (it's not necessary to detach the cable from the bracket).
7   If your vehicle is equipped with an automatic transmission, detach the throttle valve (TV) cable from the throttle linkage (see Chapter 7B), detach the TV cable brackets from the engine and set the cable and brackets aside.

8   Clearly label, then detach, all vacuum hoses from the throttle body.
9   Mark and detach the coolant hoses from the throttle body.
10 Unplug the electrical connectors from the idle switch and the Throttle Position Sensor (TPS).
11 Remove the throttle body mounting bolts and detach the throttle body and gasket **(see illustration)** from the intake manifold.
12 Using a soft brush and carburetor cleaner, thoroughly clean the throttle body casting, then blow out all passages with compressed air. **Caution:** *Do not clean the throttle position sensor with anything. Just wipe it off carefully with a clean soft cloth.*
13 Installation of the throttle body is the reverse of removal. Be sure to tighten the throttle body mounting bolts to the torque listed in the Specifications Section at the beginning of this Chapter.

Throttle Position Sensor (TPS)

General description

14 The Throttle Position Sensor (TPS) is located on the end of the throttle shaft on the throttle body. By monitoring the output voltage from the TPS, the ECM can determine fuel delivery based on throttle valve angle (driver demand). A broken or loose TPS can cause intermittent bursts of fuel from the injector and an unstable idle because the ECM thinks the throttle is moving. The OBD system can detect a variety of different TPS problems and set codes to indicate the specific trouble area. Trouble code 12 for the OBD-I system designated for the TPS and its circuit control. PO122, PO123 and PO222, PO223 are designated TPS and circuit control codes for OBD-II systems. If an OBD-II SCAN tool is not available, have the codes extracted from the ECM by a dealer service department or other qualified automotive repair facility. Refer to Chapter 6 for additional information.

Check

15 To check the TPS, connect the probes of the voltmeter to the ground wire (-) (GND) and signal wire (+) (SIG) terminals on the backside of the electrical connector **(see illustration 15.18** for the terminal designations probe the wires that correspond to these terminals**)** and turn the ignition switch to ON (engine not running). This test checks for the proper signal voltage from the TPS. **Note:** *It will be necessary to backprobe the harness connector with pins or paper clips. Be careful when backprobing the electrical connector. Do not damage the wiring harness or pull on any connectors to make clean contact.*
16 The sensor should read 0.50 to 1.0-volt at idle. Have an assistant depress the accelerator pedal to simulate full throttle and the sensor should increase voltage to 4.0 to 5.0-volts. If the TPS voltage readings are incorrect, replace it with a new unit.
17 Also, check the TPS reference voltage. Install the positive probe of the voltmeter onto the voltage reference wire (+) (REF) and the nega-

Chapter 4 Fuel and exhaust systems

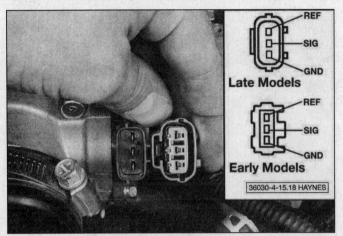

15.18 To check the throttle position sensor (TPS), insert a feeler gauge of the specified thickness between the throttle stop screw and the stop lever, then measure the resistance between the proper terminals

15.25 Fuel pressure regulator mounting bolts

tive probe of the voltmeter onto the ground wire (-) (GND) . There should be approximately 5.0 volts sent from the ECM to the TPS.

Adjustment

Refer to illustration 15.18

18 Before adjusting the TPS, make sure the idle speed (see Chapter 1) and the ignition timing (see Chapter 5) are correct. Disconnect the TPS harness connector and install an ohmmeter onto the following terminals **(see illustration)**.

	GND - SIG	REF - SIG
0.02 inch (0.5 mm)	Continuity	No continuity
0.027 inch (0.7 mm)	No continuity	No continuity
Wide open throttle	No continuity	Continuity

19 Install a feeler gauge between the throttle lever and the adjustment screw near the dashpot and make sure the TPS switch is adjusted properly. Loosen the TPS adjustment screws and move the TPS into the correct range (feeler gauge) and tighten the screws if necessary.

Fuel pressure regulator

Check

20 Refer to the fuel pump/fuel pressure check procedure (see Section 3).

Removal

Refer to illustration 15.25

21 Relieve the fuel system pressure (see Section 2), then detach the cable from the negative terminal of the battery.
22 Detach the vacuum sensing hose from the regulator.
23 Place a metal container or shop towel under the fuel return hose.
24 Loosen the clamp and detach the fuel return hose from the regulator.
25 Remove the mounting bolts **(see illustration)** and detach the pressure regulator from the fuel rail.

Installation

26 Installation is the reverse of removal. Be sure to use a new O-ring and make sure that the pressure regulator is installed properly on the fuel rail without the seal pinching or angled to prevent fuel leaks.
27 The remainder of installation is the reverse of removal.

Air intake plenum

Removal

Refer to illustration 15.32

28 Relieve the fuel system pressure (see Section 2), then detach the

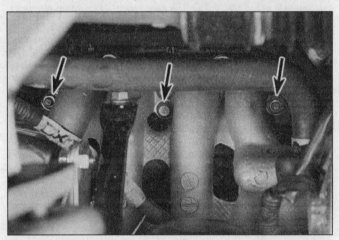

15.32 Location of the air intake plenum mounting bolts (arrows)

cable from the negative terminal of the battery.
29 Detach the vacuum lines from the air intake plenum. Be sure to mark them to insure proper installation.
30 Disconnect the accelerator cable (see Section 10) and the bracket assembly and separate them from the plenum.
31 Loosen the clamp and detach the air intake duct from the throttle body.
32 Raise the vehicle and support it securely on jackstands. Remove the air intake plenum mounting bolts working from underneath **(see illustration)**.

Installation

33 Installation is the reverse of removal. Be sure to use a new gasket between the air intake plenum and the intake manifold. Tighten the bolts to the Specifications listed in this Chapter.
34 The remainder of installation is the reverse of removal.

Idle Speed Control-Bypass Air (ISC-BPA)

Note: *The minimum idle speed is pre-set at the factory and should not require adjustment under normal operating conditions; however if the throttle body has been replaced or you suspect the minimum idle speed has been tampered with (for example, if the idle speed screw was removed from the throttle body), refer to the idle speed procedure in Chapter 1.*

Check

Refer to illustration 15.36

35 The ISC-BPA has two functions. During cold operating conditions, the ISC-BPA increases idle speed to allow quicker warm-up and

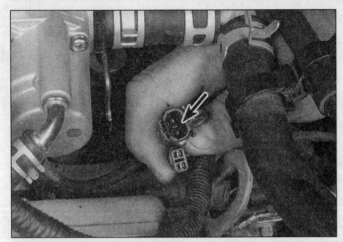

15.36 Disconnect the harness connector (arrow) and check the resistance of the ISC-BPA valve (late model EFI system shown)

15.51 Remove the fuel rail mounting bolts (arrows)

it also functions as a controller for idle speed during normal operating conditions. This device controls the amount of air into the intake system by funneling air from ahead of the throttle body into the air intake plenum. The ISC-BPA is a cylindrical shaped solenoid located next to the intake manifold below the air intake plenum.

36 Check the resistance of the ISC-BPA. Disconnect the electrical connector and check the resistance between the two terminals. It should be 7.7 to 9.3 ohms **(see illustration)**.

37 Also, check the vacuum hoses from the throttle body and the intake plenum for leaks, cracks, tight fittings and damage that could cause an air leak. Replace any hoses if necessary.

38 If the engine continues to have erratic idling problems, check for vacuum leaks, check the emission control systems (see Chapter 6) and the On Board Diagnostic (OBD) system. In the event of a hard-to-find idling system problem, have the system checked by a dealer service department or other qualified repair shop.

Removal

39 Disconnect the ISC-BPA electrical connector.
40 Disconnect the vacuum hoses from the ISC-BPA.
41 Remove the mounting screws and detach the ISC-BPA from its bracket.
42 Installation is the reverse of removal.

Fuel rail and fuel injectors

Check

43 Refer to Section 14 for the fuel injector checking procedure.

Replacement

Refer to illustrations 15.51, 15.52, 15.53, 15.54 and 15.56

44 Relieve the fuel system pressure (see Section 2).
45 Detach the cable from the negative terminal of the battery.
46 Remove the PCV hose from the cylinder head and intake manifold.
47 Unplug the electrical connectors from the fuel injectors and set the injector wiring harness aside.
48 Remove the air intake plenum (see Steps 28 through 32).
49 Detach the vacuum hose from the fuel pressure regulator.
50 Disconnect the fuel lines from the fuel pressure regulator and the fuel rail.
51 Remove the fuel rail mounting bolts **(see illustration)**.
52 Remove the fuel rail and retrieve the two plastic spacers from the manifold **(see illustration)**.
53 Remove the fuel injector(s) from the fuel rail **(see illustration)**.
54 Remove the rubber grommets from the injector bores in the intake manifold **(see illustration)**.
55 If the injectors remain in the intake manifold, remove them using a twisting motion.
56 If you are replacing the injector(s), discard the old injector, the grommet and the O-ring. If you are simply replacing leaking injector O-rings and intend to re-use the same injectors, remove the old grommet and O-ring **(see illustration)** and discard them.
57 Further testing of the injector(s) is beyond the scope of the home mechanic. If you are in doubt as to the status of any injector(s), it can be bench tested for volume and leakage at a dealer service department or other repair shop.

15.52 Retrieve the fuel rail mounting spacers

15.53 Lift the fuel injectors from the fuel rail, being careful not to damage the tip(s)

Chapter 4 Fuel and exhaust systems

4-15

15.54 Remove the rubber grommets from the cylinder head

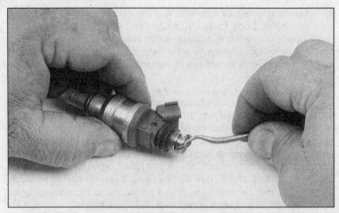

15.56 If you plan to reinstall the same injectors, be sure to remove and discard the old O-rings and replace them with new ones

58 Installation of the fuel injectors is the reverse of removal. Be sure to use new grommets and O-rings on the injector(s) and new sealing washers on the fuel delivery pipe. Lubricate the O-rings with gasoline. Tighten the fuel rail mounting bolts to the torque listed in this Chapter's Specifications.

Dashpot

Adjustment

Refer to illustrations 15.63a and 15.63b

59 Warm the engine to normal operating temperature and then allow it to idle.
60 Place the shift lever in PARK for an automatic or NEUTRAL for a manual transaxle vehicle.
61 Install a tachometer according to the manufacturer's specifications.
62 Increase the engine speed to approximately 3,500 rpm. Release the throttle and confirm that the dashpot rod touches the throttle lever at 2,600 to 2,900 rpm on models with a manual transaxle or 2,500 to 3,100 for models with an automatic transaxle.
63 If the rod touches the throttle lever out of the specified range, adjust the contacting point of the dashpot by turning the adjusting bolt **(see illustrations)**.
64 Recheck the dashpot adjustment.

16 Exhaust system servicing - general information

Refer to illustrations 16.1a and 16.1b
Warning: *Inspection and repair of exhaust system components should*

15.63a Remove the air reservoir chamber to gain access to the dashpot

be done only after enough time has elapsed after driving the vehicle to allow the system components to cool completely. Also, when working under the vehicle, make sure it is securely supported on jackstands.

1 The exhaust system consists of the exhaust manifold, catalytic converter, the muffler, the tailpipe and all connecting pipes, brackets, hangers and clamps **(see illustrations)**. The exhaust system is attached to the body with mounting brackets and rubber hangers. If any of these parts are damaged or deteriorated, excessive noise and vibration will be transmitted to the body.

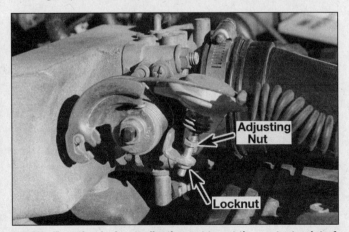

15.63b Turn the dashpot adjusting nut to set the contact point of the throttle lever

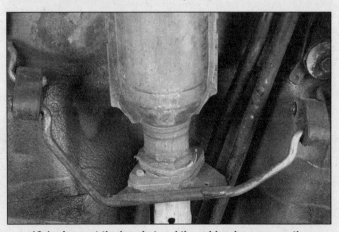

16.1a Inspect the bracket and the rubber hangers on the catalytic converter

2 Conducting regular inspections of the exhaust system will keep it safe and quiet. Look for any damaged or bent parts, open seams, holes, loose connections, excessive corrosion or other defects which could allow exhaust fumes to enter the vehicle. Deteriorated exhaust system components should not be repaired - they should be replaced with new parts.

3 If the exhaust system components are extremely corroded or rusted together, they will probably have to be cut from the exhaust system. The convenient way to accomplish this is to have a muffler repair shop remove the corroded sections with a cutting torch. If, however, you want to save money by doing it yourself and you don't have an oxy-acetylene welding outfit with a cutting torch), simply cut off the old components with a hack-saw. If you have compressed air, special pneumatic cutting chisels can also be used. If you do decide to tackle the job at home, be sure to wear eye protection to protect your eyes from metal chips and work gloves to protect your hands.

4 Here are some simple guidelines to apply when repairing the exhaust system:

   a) *Work from the back to the front when removing exhaust system components.*
   b) *Apply penetrating oil to the exhaust system component fasteners to make them easier to remove.*
   c) *Use new gaskets, hangers and clamps when installing exhaust system components.*
   d) *Apply anti-seize compound to the threads of all exhaust system fasteners during reassembly.*
   e) *Be sure to allow sufficient clearance between newly installed parts and all points on the underbody to avoid overheating the floor pan*

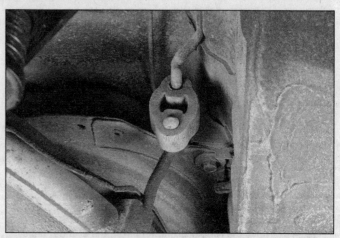

16.1b Also inspect the rubber hangers that support the tailpipe near the rear of the vehicle

and possibly damaging the interior carpet and insulation. Pay particularly close attention to the catalytic converter and its heat shield. **Warning:** *The catalytic converter operates at very high temperatures and takes a long time to cool. Wait until it's completely cool before attempting to remove the converter. Failure to do so could result in serious burns.*

Chapter 5
Engine electrical systems

Contents

	Section		Section
Alternator - removal and installation	13	Ignition coil - check and replacement	7
Battery cables - check and replacement	4	Ignition system - check	6
Battery check, maintenance and charging	See Chapter 1	Ignition system - general information and precautions	5
Battery - emergency jump starting	2	Ignition timing - check and adjustment	10
Battery - removal and installation	3	Spark plug replacement	See Chapter 1
Charging system - check	12	Spark plug wire, distributor cap and rotor check	
Charging system - general information and precautions	11	and replacement	See Chapter 1
CHECK ENGINE light	See Chapter 6	Starter motor - removal and installation	16
Distributor - removal and installation	9	Starter motor - testing in vehicle	15
Drivebelt check, adjustment and replacement	See Chapter 1	Starting system - general information and precautions	14
General information	1		
Ignition control module (1990 through 1993 models) - replacement	8		

Specifications

Ignition timing
1988 and 1989 models	TDC
1990 through 1995 models	10-degrees BTDC (white mark on pulley)
1996 and 1997 models	10-degrees BTDC (yellow mark on pulley)

Ignition coil resistance (cold)
Primary resistance
1988 through 1993 models	0.8 to 1.6 ohms
1994 and later	Not available

Secondary resistance
1988 through 1993 models	6 to 30 K-ohms
1994 and later	Not available

Charging system
Charging voltage	14.1 to 14.7 volts

Torque specifications
	Ft-lbs
Alternator mounting bolts	
Adjusting bolt	12 to 16
Pivot bolt	24 to 33
Distributor mounting bolt	14 to 18
Starter mounting bolts	24 to 33

Chapter 5 Engine electrical systems

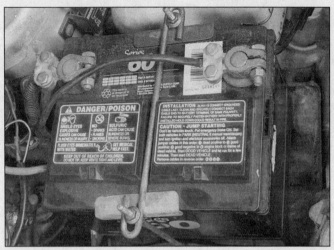

3.1 To remove the battery, detach the negative battery cable first, then the positive cable; remove the hold-down strap nuts and bolts and remove the hold-down strap

3.4 If necessary, remove the bolts (arrows) that retain the battery tray to the engine compartment chassis

1 General information

The engine electrical systems include all ignition, charging and starting components. Because of their engine related functions, these components are discussed separately from chassis electrical devices such as the lights, the instruments, etc. (which are included in Chapter 12).

Always observe the following precautions when working on the electrical systems:

a) *Be extremely careful when servicing engine electrical components. They are easily damaged if checked, connected or handled improperly.*
b) *Never leave the ignition switch on for long periods of time (10 minutes maximum) with the engine off.*
c) *Don't disconnect the battery cables while the engine is running.*
d) *Maintain correct polarity when connecting a battery cable from another vehicle during jump starting.*

Always disconnect the negative cable first and hook it up last or the battery may be shorted by the tool being used to loosen the cable clamps.

It's also a good idea to review the safety-related information regarding the engine electrical systems located in the *Safety First!* Section near the front of this manual before beginning any operation included in this Chapter.

2 Battery - emergency jump starting

Refer to the *Booster battery (jump) starting* procedure at the front of this manual.

3 Battery - removal and installation

Refer to illustration 3.1 and 3.4

1   Disconnect the negative battery cable, then the positive battery cable from the battery terminals **(see illustration)**.
2   Remove the battery hold-down clamp.
3   Lift out the battery. Be careful, it's heavy.
4   While the battery is out, inspect the carrier (tray) for corrosion **(see illustration)**.
5   If you are replacing the battery, make sure that you get one that is identical, with the same dimensions, amperage rating, cold cranking amperage rating, etc., as the original.
6   Installation is the reverse of removal.

4 Battery cables - check and replacement

1   Periodically inspect the entire length of each battery cable for damage, cracked or burned insulation and corrosion. Poor battery cable connections can cause starting problems and decreased engine performance.
2   Check the cable-to-terminal connections at the ends of the cables for cracks, loose wire strands and corrosion. The presence of white, fluffy deposits under the insulation at the cable terminal connection is a sign that the cable is corroded and should be replaced. Check the terminals for distortion, missing mounting bolts and corrosion.
3   When removing the cables, always disconnect the negative cable first and hook it up last or the battery may be shorted by the tool used to loosen the cable clamps. Even if only the positive cable is being replaced, be sure to disconnect the negative cable from the battery first (see Chapter 1 for further information regarding battery cable removal).
4   Disconnect the old cables from the battery, then trace each of them to their opposite ends and detach them from the starter solenoid and ground terminals. Note the routing of each cable to ensure correct installation.
5   If you are replacing either or both of the old cables, take them with you when buying new cables. It is vitally important that you replace the cables with identical parts. Cables have characteristics that make them easy to identify: positive cables are usually red, larger in cross-section and have a larger diameter battery post clamp; ground cables are usually black, smaller in cross-section and have a slightly smaller diameter clamp for the negative post.
6   Clean the threads of the solenoid or ground connection with a wire brush to remove rust and corrosion. Apply a light coat of battery terminal corrosion inhibitor, or petroleum jelly, to the threads to prevent future corrosion.
7   Attach the cable to the solenoid or ground connection and tighten the mounting nut/bolt securely.
8   Before connecting a new cable to the battery, make sure that it reaches the battery post without having to be stretched.
9   Connect the positive cable first, followed by the negative cable.

5 Ignition system - general information and precautions

1   The electronic ignition system includes the ignition switch, the battery, the pick-up coil, the ignition module, the ignition coil, the primary (low voltage) and secondary (high voltage) circuits, the distributor and the spark plugs. The ignition system is controlled by the Electronic Control Module (ECM). Using data provided by information sensors

Chapter 5 Engine electrical systems

6.1 To use a calibrated ignition tester (available at most auto parts stores), remove an ignition wire from a cylinder, connect the wire to the tester and clip the tester to a good ground - if there is enough voltage to fire the plug, sparks will be clearly visible between the electrode tip and the tester body as the engine is cranked

which monitor various engine functions (such as rpm, intake air volume, engine temperature, etc.), the ECM ensures a perfectly timed spark under all conditions. There are three different ignition systems on these models depending upon the location of the ignition module and the ignition coil. **Note:** *On 1988 and 1989 models, the ignition module and the pick-up coil are mounted in the distributor. This assembly is called the igniter. On 1990 through 1993 models, the ignition module is mounted externally next to the ignition coil with the pick-up coil and crankshaft sensor mounted inside the distributor. 1994 and later models, the ignition coil, the ignition module and the crankshaft position sensors are all mounted in the distributor body, now called the ignition control assembly.*

2 On 1988 and 1989 models, the ignition system is equipped with an external coil, an ignition module and pick-up coil assembly that is mounted in the distributor called the igniter assembly, secondary ignition wires, a cap and rotor. The distributor is mounted in the flywheel end of the cylinder head and it is direct driven by the camshaft at one-half engine speed. Crankshaft position is sensed by a pulse generator consisting of a reluctor and pick-up coil assembly. The pick-up coil, which is part of the ignition module, controls the current through the ignition coil primary winding. Switching is achieved through the magnetic field collapsing between the reluctor and the two pole magnetic stator or pick-up coil. The ignition module and pick-up coil should be replaced as a complete unit in the event of failure of one component (or both components). This ignition system is also equipped with a High Altitude Spark Advance correction system. This electronic system switches into high altitude mode when the barometric pressure drops below 26.4 in-Hg. The barometric pressure switch, which is mounted on the firewall on the right side of the engine compartment, senses the changes in altitude pressures. The delay circuit in the ignition control module and the EMC is bypassed to allow extra spark advance to compensate for the thinner (less oxygen) air and the richer fuel mixture.

3 On 1990 through 1993 models, the ignition system uses a transistorized ignition module that relays spark control signals from the ECM to the ignition coil. The ignition module receives a Profile Ignition Pulse (PIP) from the pick-up coil located in the distributor. The module also receives a Spark Output (SPOUT) signal from the ECM which allows the computer to correct deviations in ignition timing. The ignition module is located next to the ignition coil mounted on the fenderwell near the strut tower. The pick-up coil is mounted inside the distributor. The crankshaft position sensor is mounted in the distributor along with the pick-up coil. This type of crankshaft sensor uses a vane with four equally spaced slots which rotates with the distributor shaft. The vane slots pass through a photo optic sensor which senses the slots and converts this interval into an electronic signal that relays crankshaft position to the ECM.

4 On 1994 and later models, the ignition system is equipped with a distributor that houses the crankshaft position sensors (CID and CKP), ignition coil, the ignition module, the ignition primary circuit and the secondary ignition wires. The computer (ECM) controls the ignition timing. Initial timing adjustments are not necessary unless the distributor has been moved or removed from the engine. This system incorporates a Hall Effect type switch assembly along with a vane cup to trigger the signal ON and OFF. High altitude spark correction is entirely electronic and the device is contained within the ECM. The ECM will alter the SPOUT signal to correct ignition timing for more advance during high altitude driving. **Note:** *The crankshaft position sensors (CKP and CID) read crankshaft position and engine rpm and relay the signals electronically to the computer. These two crankshaft sensors are contained within the distributor on 1994 and 1995 models. The 1996 and 1997 models locate the crankshaft sensor near the front oil pump. It reads crankshaft position off the timing pulley. Cylinder detection is detected by the camshaft sensor which is located in the distributor. These sensors are difficult to check, so have the ignition system tested by a dealer service department or other qualified automotive repair facility.*

5 When working on the ignition system, take the following precautions:

a) *Do not keep the ignition switch on for more than 10 seconds if the engine will not start.*
b) *Always connect a tachometer in accordance with the manufacturer's instructions. Some tachometers may be incompatible with this ignition system. Consult a dealer service department before buying a tachometer for use with this vehicle.*
c) *Never allow the ignition coil terminals to touch ground. Grounding the coil could result in damage to the igniter and/or the ignition coil.*
d) *Do not disconnect the battery when the engine is running.*
e) *On 1994 and earlier models, make sure the igniter is properly grounded.*

6 Ignition system - check

Refer to illustrations 6.1, 6.7 and 6.9
Warning: *Because of the high voltage generated by the ignition system, extreme care should be taken whenever an operation is performed involving ignition components. This not only includes the igniter, coil, distributor and spark plug wires, but related components such as plug connectors, tachometer and other test equipment also.*

1 If the engine turns over but will not start, disconnect the spark plug wire from any spark plug and attach it to a calibrated tester available at most auto parts stores **(see illustration)**. Connect the clip on the tester to a bolt or metal bracket on the engine. If you're unable to obtain a calibrated ignition tester, remove the wire from one of the spark plugs and using an insulated tool, pull back the boot and hold the end of the wire about 1/4-inch from a good ground.

2 Crank the engine and watch the end of the tester or spark plug wire to see if bright blue, well-defined sparks occur.

3 If sparks occur, sufficient voltage is reaching the plug to fire it (repeat the check at the remaining plug wires to verify that the distributor cap and rotor are OK). However, the plugs themselves may be fouled, so remove and check them as described in Chapter 1.

4 If no sparks or intermittent sparks occur, remove the distributor cap and check the cap and rotor as described in Chapter 1. If moisture is present, dry out the cap and rotor, then reinstall the cap and repeat the spark test.

5 If there is still no spark, detach the coil secondary wire from the distributor cap and hook it up to the tester (re-attach the plug wire to the spark plug), then repeat the spark check. Again, if you don't have a tester, hold the end of the wire about 1/4-inch from a good ground.

6 If sparks now occur, the distributor cap, rotor or plug wire(s) may be defective.

Chapter 5 Engine electrical systems

6.7 Check for battery voltage to the positive (+) terminal on the ignition coil with the key ON (engine not running) (carbureted model shown)

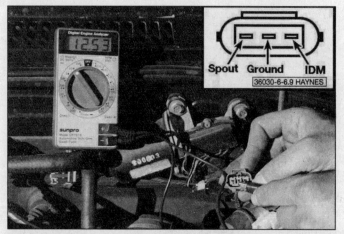

6.9 Disconnect the ignition module harness connector and check for battery voltage to the module on the IDM terminal with the ignition key ON (engine not running)

7   If no sparks occur, check the primary wire connections at the coil to make sure they're clean and tight. Check for voltage to the coil on the primary circuit from the ignition switch **(see illustration)**. Check the ignition coil (see Section 7). Make any necessary repairs, then repeat the check again.

8   If there's still no spark, the coil-to-cap wire may be bad (check the resistance with an ohmmeter and compare it to the spark plug wire resistance Specifications found in Chapter 1). If a known good wire doesn't make any difference in the test results, the igniter may be defective.

9   Check for battery voltage to the module **(see illustration)** with the ignition key ON, engine not running. If voltage is available and there is still no spark, check for a switching signal (next step). **Note:** *Because the ignition module is housed within the distributor on 1988 and 1989 models and also on 1994 and later models, it will not be possible to separate the testing procedures on these particular vehicles. This test procedure covers only the 1990 through 1993 models that are equipped with an external ignition module. In the event the other types of distributor modules are defective or if the electronic distributor is not functioning properly, consult a dealer parts department or auto parts store. These distributors will have to be replaced as a complete assembly in order to insure against further ignition system malfunctions.*

10 Disconnect the ignition coil harness connector. Connect an LED test light between the IDM wire and the battery positive terminal (+) **(see illustration 5.3)**. Confirm that the test light flashes when an assistant cranks the engine over. This indicates a switching signal from the ignition module.

11 If no flashing LED test light is observed, replace the ignition module.

12 Also on external ignition module systems, check the pick-up coil resistance. Disconnect the distributor harness connector and check the resistance between the two terminals. It should be between 900 to 1,200 ohms.

13 If the resistance is out of range, replace the distributor with a complete assembly.

7 Ignition coil - check and replacement

Note: The coil is housed within the distributor on 1994 and later models. In the event of ignition system problems, replace the distributor as a complete assembly.

Check

Refer to illustrations 7.5 and 7.6

1   Perform the ignition system checks as described in Section 6.

2   Crank the engine and verify that a strong blue spark is visible at the coil wire or spark plug wire.

3   If there is no spark, check for voltage at the positive (+) terminal of the coil with the ignition switch in the ON position **(see illustration 6.7)**.

4   If there is no battery voltage, check the main fuse, ignition switch, and wiring harness.

7.5 Connect an ohmmeter between the two terminals of the ignition coil to measure the coil primary winding resistance

7.6 Connect an ohmmeter between one terminal of the connector and the high-tension tower to measure the resistance of the coil secondary windings

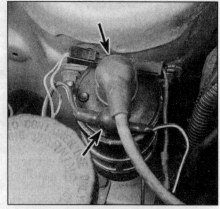

7.9 Remove the rubber protectors and disconnect the coil terminal nuts to remove the connectors

Chapter 5 Engine electrical systems

9.5a Make a mark on the edge of the distributor (arrow) directly below the rotor tip and in line with it

5 Use an ohmmeter to measure the resistance of the coil primary winding **(see illustration)**. Refer to the Specifications listed in this Chapter. If the readings are incorrect, replace the coil.
6 Use an ohmmeter to measure the resistance of the coil secondary winding **(see illustration)**. Refer to the Specifications listed in this Chapter. If the readings are incorrect, replace the coil.

Replacement

Refer to illustration 7.9

7 Detach the cable from the negative terminal of the battery.
8 Remove the heat shield from the coil, if equipped.
9 Label and disconnect the wires from the coil terminals **(see illustration)**.
10 Remove the coil mounting screws.
11 Installation is the reverse of removal.

8 Ignition control module (1990 through 1993 models) - replacement

Note: *Because the ignition module is housed within the distributor on 1988 and 1989 models and also on 1994 and later models, it will not be possible to test these particular models. This test procedure covers only the 1990 through 1993 models equipped with an external ignition module. In the event the other types of ignition modules are defective or if the electronic distributor is not functioning properly, consult with a dealer service department or other qualified repair facility.*

1 Detach the cable from the negative terminal of the battery.
2 Disconnect the electrical connector from the module. The ignition module is located next to the ignition coil near the front strut tower in the corner of the engine compartment.
3 Remove the screws from the engine compartment sheet metal and remove the module from the engine compartment.
4 Installation is the reverse of removal.

9 Distributor - removal and installation

Removal

Refer to illustrations 9.5a and 9.5b

1 Detach the cable from the negative battery terminal.
2 Look for a raised "1" on the distributor cap. This marks the location for the number one cylinder spark plug wire terminal. If the cap does not have a mark for the number one terminal, locate the number one spark plug and trace the wire back to the terminal on the cap.
3 Remove the distributor cap (see Chapter 1) and turn the engine over until the rotor is pointing toward the number one spark plug terminal (see locating TDC procedure in Chapter 2A).
4 Disconnect and label the electrical connectors from the distributor.
5 Make a mark on the edge of the distributor base directly below the rotor tip **(see illustration)** and in line with it. Also, mark the distributor base and the engine block to ensure that the distributor is installed correctly **(see illustration)**.
6 Remove the distributor hold-down bolt, then pull the distributor straight out to remove it. **Caution:** *DO NOT turn the crankshaft while the distributor is out of the engine, or the alignment marks will be useless.*

Installation

Refer to illustrations 9.7 and 9.8

Note: *If the crankshaft has been moved while the distributor is out, locate Top Dead Center (TDC) for the number one piston (see Chapter 2A) and position the distributor and the rotor accordingly.*

7 Install a new O-ring onto the distributor housing **(see illustration)**.
8 Align the off-set portion of the coupling with the groove in the housing **(see illustration)**.
9 Insert the distributor into the engine in exactly the same relationship to the block that it was in when removed.
10 If the distributor does not seat completely, recheck the alignment marks between the distributor base and the block to verify that the distributor is in the same position it was in before removal. Also check the rotor to see if it's aligned with the mark you made on the edge of the distributor base.
11 Loosely install the distributor hold-down bolt(s).
12 Install the distributor cap and connect the electrical connectors.
13 Check the ignition timing (see Section 10) and tighten the distributor hold-down bolt securely.

9.5b Also, mark the distributor base and the engine block (arrow) to ensure that the distributor can be reinstalled correctly

9.7 Install a new O-ring onto the distributor housing

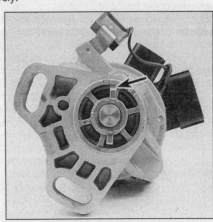

9.8 Align the off-set portion (arrow) of the coupling with the groove in the cylinder head housing

10 Ignition timing - check and adjustment

Refer to illustrations 10.2 and 10.3

Note: *The following ignition timing procedure applies to most models covered by this manual. However, if the procedure specified on the VECI label of your vehicle differs from this one, use the procedure found on the VECI label.*

1 Connect a tachometer according to the manufacturer's specifications.
2 If you're working on a 1988 or 1989 model, disconnect and plug the hose(s) from the vacuum advance mechanism. If you're working on a 1990 or later model, locate the diagnostic connector and ground the STI terminal with a jumper wire. **Note:** *On 1990 through 1993 models, ground the STI terminal (the one-terminal connector near the power brake booster) to the engine with a jumper wire. On 1994 and later models, connect the STI and GND terminals in the Test Connector with a jumper wire or paper clip* **(see illustration)**.
3 With the ignition switch off, connect a timing light according to the tool manufacturer's instructions **(see illustration)**. Most timing lights are powered by the battery. Also, an inductive style pick-up is connected to the number one cylinder spark plug wire.
4 On carbureted models, disconnect the barometric switch from the distributor.
5 Turn off all the electrical accessories (A/C, radio, headlights, etc.)
6 Locate the timing marks on the pointer index and the crankshaft pulley.
7 Start the engine and allow it to warm up to normal operating temperature (upper radiator hose hot). Verify that the engine idle is correct (see Chapter 1). Aim the timing light at the timing scale (see Chapter 2A, **illustration 3.8**). The mark on the crankshaft pulley should line up with the timing indicator. Refer to the Specifications listed at the beginning of this Chapter. If necessary, loosen the distributor hold-down bolt and slowly rotate the distributor until the timing marks align. Tighten the hold-down bolt and recheck the timing.
8 On 1989 and earlier models, reconnect the vacuum hose(s) to the vacuum advance unit. Remove the jumper wire from the diagnostic connector on later systems.
9 Turn the engine off and remove the tachometer and the timing light.

11 Charging system - general information and precautions

The charging system includes the alternator, an internal voltage regulator, a charge indicator, the battery and the wiring between all the components. The charging system supplies electrical power for the ignition system, the lights, the radio, etc. The alternator is driven by a drivebelt at the front of the engine.

The purpose of the voltage regulator is to limit the alternator's voltage to a preset value. This prevents power surges, circuit overloads, etc., during peak voltage output.

The charging system doesn't ordinarily require periodic maintenance. However, the drivebelt, battery and wires and connections should be inspected at the intervals outlined in Chapter 1.

The dashboard warning light should come on when the ignition key is turned to Start, then should go off immediately. If it remains on, there is a malfunction in the charging system. Some vehicles are also equipped with a voltage gauge. If the voltage gauge indicates abnormally high or low voltage, check the charging system (see Section 12).

Be very careful when making electrical circuit connections to a vehicle equipped with an alternator and note the following:

a) When reconnecting wires to the alternator from the battery, be sure to note the polarity.
b) Before using arc welding equipment to repair any part of the vehicle, disconnect the wires from the alternator and the battery terminals.
c) Never start the engine with a battery charger connected.

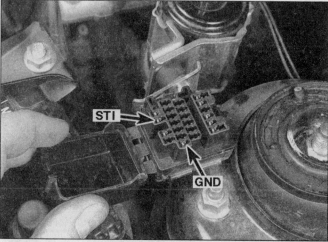

10.2 Attach a jumper wire between terminals STI and GND of the test connector before checking the ignition timing

d) Always disconnect both battery leads before using a battery charger.
e) The alternator is driven by an engine drivebelt which could cause serious injury if your hand, hair or clothes become entangled in it with the engine running.
f) Because the alternator is connected directly to the battery, it could arc or cause a fire if overloaded or shorted out.
g) Wrap a plastic bag over the alternator and secure it with rubber bands before steam cleaning the engine.

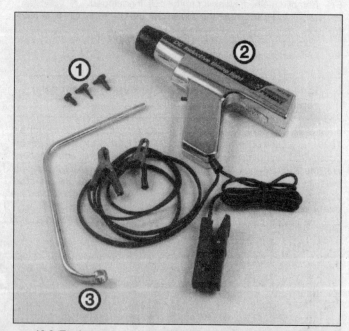

10.3 Tools needed to check and adjust the ignition timing

1 **Vacuum plugs** - Vacuum hoses will, in most cases, have to be disconnected and plugged. Molded plugs in various shapes and sizes are available for this
2 **Inductive pick-up timing light** - Flashes a bright, concentrated beam of light when the number one spark plug fires. Connect the leads according to the instructions supplied with the light
3 **Distributor wrench** - On some models, the hold-down bolt for the distributor is difficult to reach and turn with conventional wrenches or sockets. A special wrench like this must be used

Chapter 5 Engine electrical systems

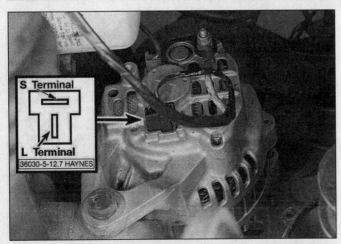

12.7 Backprobe the S and L terminals and check the voltage with the ignition key On and then again with the engine running

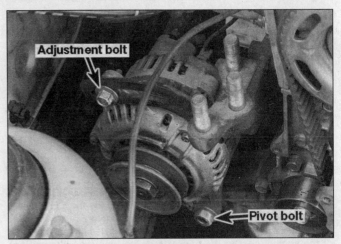

13.3 First, loosen the pivot bolt and then remove the adjustment bolt and pivot bolts from the alternator brackets

12 Charging system - check

Refer to illustrations 12.7

1 If a malfunction occurs in the charging circuit, don't automatically assume that the alternator is causing the problem. First check the following items:

  a) *Check the drivebelt tension and its condition. Replace it if worn or deteriorated.*
  b) *Make sure the alternator mounting and adjustment bolts are tight.*
  c) *Inspect the alternator wiring harness and the electrical connectors at the alternator and voltage regulator. They must be in good condition and tight.*
  d) *Check the large main fusible link in the engine compartment next to the shock tower (see Chapter 12). If it's burned, determine the cause, repair the circuit and replace the fuse (the vehicle won't start and/or the accessories won't work if the fuse blows).*
  e) *Check all the fuses that are in series with the charging system circuit. The location of these fuses may vary from year and model but the designations are the same.*
  f) *Start the engine and check the alternator for abnormal noises (a shrieking or squealing sound indicates a bad bushing).*
  g) *Check the specific gravity of the battery electrolyte. If it's low, charge the battery (doesn't apply to maintenance free batteries).*
  h) *Make sure that the battery is fully charged (one bad cell in a battery can cause overcharging by the alternator).*
  i) *Disconnect the battery cables (negative first, then positive). Inspect the battery posts and the cable clamps for corrosion. Clean them thoroughly if necessary (see Chapter 1). Reconnect the positive cable, then the negative cable.*

2 Using a voltmeter, check the battery voltage with the engine off. It should be approximately 12 volts.

3 Start the engine and check the battery voltage again. It should now be approximately 13.5 to 14.7 volts.

4 Turn on the headlights. The voltage should drop and then come back up, if the charging system is working properly.

5 If the voltage reading is greater than the specified charging voltage, the regulator is faulty. Replace the alternator as a complete assembly.

6 If the voltmeter reading is less than standard voltage, check the alternator as follows:

7 Using a voltmeter and working at the backside of the alternator, backprobe the S and L terminals of the alternator connector **(see illustration)**. Also check the voltage at the large alternator output (B+) terminal. With the key On (engine not running), there should be 12 volts at the S terminal, one volt at the L terminal and 12 volts at the B+ terminal. Now start the engine and check again - there should be 14.0 to 14.7 volts at the B+ terminal, 14.0 to 14.7 volts at the S terminal and 13.0 to 14.0 volts at the L terminal with the engine idling. If the voltages are not as specified, check the wiring harness. If the wiring harness is not defective, replace the alternator.

8 If you suspect that there is a voltage drain on the battery while the vehicle is sitting in the driveway, remove the battery positive terminal, install a test light between the battery terminal and the cable (series connection) and observe the test light. If the light is brightly illuminated, there is a circuit that continues to be activated. **Note:** *The test light will glow dimly because of the parasitic drain of the ECM, radio, clock, etc.*

9 Carefully remove the fuses one-by-one that govern accessories such as radio, blower motor, trunk lights, etc. until the test light goes out. Track down the short circuit in the particular fused circuit and repair the problem. Recheck the electrical system as described.

10 If all the fuses are pulled out and the test light remains lit, remove the output cable at the rear of the alternator then unplug all the connectors from the backside of the alternator. If the test light goes out, then there is an internal drain in the alternator or voltage regulator. Replace the alternator.

13 Alternator - removal and installation

Removal

Refer to illustrations 13.3

1 Detach the cable from the negative terminal of the battery.
2 Detach the electrical connectors from the alternator.
3 Loosen the alternator adjustment, pivot and lock bolts **(see illustration)** and detach the drivebelt.
4 Remove the adjustment and lock bolts from the alternator adjustment bracket.
5 Separate the alternator and bracket from the engine. **Note:** *It may be necessary to bend the catalytic converter shield slightly to be able to drop the alternator past the catalytic converter and out the bottom of the engine compartment.*
6 If you are replacing the alternator, take the old alternator with you when purchasing a replacement unit. Make sure that the new/rebuilt unit is identical to the old alternator. Look at the terminals - they should be the same in number, size and locations as the terminals on the old alternator. Finally, look at the identification markings - they will be stamped in the housing or printed on a tag or plaque affixed to the housing. Make sure that these numbers are the same on both alternators.
7 Many new/rebuilt alternators do not have a pulley installed, so you may have to switch the pulley from the old unit to the new/rebuilt one. When buying an alternator, find out the store policy regarding installation of pulleys - some stores will perform this service free of charge.

Installation

8   Install in the reverse order of removal.
9   After the alternator is installed, adjust the drivebelt tension (see Chapter 1).
10 Check the charging voltage to verify proper operation of the alternator (see Section 12).

14 Starting system - general information and precautions

Refer to illustration 14.3

1   The sole function of the starting system is to turn over the engine quickly enough to allow it to start.
2   The starting system consists of the battery, the starter motor, the starter solenoid and the electrical circuit connecting the components. The solenoid is mounted directly on the starter motor.
3   The starter/solenoid motor assembly is installed on the upper part of the engine, next to the transaxle bellhousing **(see illustration)**. **Note:** *The starter/solenoid assembly brackets are mounted in different locations according to automatic and manual transmission options.*
4   When the ignition key is turned to the Start position, the starter solenoid is actuated through the starter control circuit. The starter solenoid then connects the battery to the starter. The battery supplies the electrical energy to the starter motor, which does the actual work of cranking the engine.
5   The starter motor on a vehicle equipped with a manual transaxle can be operated only when the clutch pedal is depressed; the starter on a vehicle equipped with an automatic transaxle can be operated only when the transaxle selector lever is in Park or Neutral.
6   Always observe the following precautions when working on the starting system:

a) *Excessive cranking of the starter motor can overheat it and cause serious damage. Never operate the starter motor for more than 15 seconds at a time without pausing to allow it to cool for at least two minutes.*
b) *The starter is connected directly to the battery and could arc or cause a fire if mishandled, overloaded or short circuited.*
c) *Always detach the cable from the negative terminal of the battery before working on the starting system.*

15 Starter motor - testing in vehicle

Note: *Before diagnosing starter problems, make sure the battery is fully charged.*

1   If the starter motor does not turn at all when the switch is operated, make sure that the shift lever is in Neutral or Park (automatic transaxle) or that the clutch pedal is depressed (manual transaxle).
2   Make sure that the battery is charged and that all cables, both at the battery and starter solenoid terminals, are clean and secure.
3   If the starter motor spins but the engine is not cranking, the over-running clutch in the starter motor is slipping and the starter motor must be replaced.
4   If, when the switch is actuated, the starter motor does not operate at all but the solenoid clicks, then the problem lies with either the battery, the main solenoid contacts or the starter motor itself (or the engine is seized).
5   If the solenoid plunger cannot be heard when the switch is actuated, the battery is bad, the circuit is open, or the starter solenoid itself is defective.

14.3 The starter/solenoid assembly is mounted on the transaxle bellhousing, below the distributor

6   To check the solenoid, connect a jumper lead between the battery (+) and the ignition switch terminal (the small terminal) on the solenoid. If the starter motor now operates, the solenoid is OK and the problem is in the ignition switch, Neutral start switch or in the wiring.
7   If the starter motor still does not operate, remove the starter/solenoid assembly for disassembly, testing and repair.
8   If the starter motor cranks the engine at an abnormally slow speed, first make sure that the battery is charged and that all terminal connections are tight. If the engine is partially seized, or has the wrong viscosity oil in it, it will crank slowly.
9   Run the engine until normal operating temperature is reached, then disconnect the coil wire from the distributor cap and ground it on the engine.
10 Connect a voltmeter positive lead to the battery positive post and connect the negative lead to the negative post.
11 Crank the engine and take the voltmeter readings as soon as a steady figure is indicated. Do not allow the starter motor to turn for more than 15 seconds at a time. A reading of nine volts or more, with the starter motor turning at normal cranking speed, is normal. If the reading is nine volts or more but the cranking speed is slow, the motor is faulty. If the reading is less than nine volts and the cranking speed is slow, the solenoid contacts are probably burned, the starter motor is bad, the battery is discharged or there is a bad connection.

16 Starter motor - removal and installation

Note: *The starter/solenoid assembly cannot be repaired using separate components. In the event of failure, exchange the starter/solenoid assembly for a new or rebuilt unit.*

1   Detach the cable from the negative terminal of the battery.
2   Remove the battery and the battery tray (see Section 3).
3   Remove the starter motor assembly brace and bracket from the transaxle. **Note:** *The starter/solenoid assembly brackets are mounted in different locations according to automatic and manual transmission options.*
4   Disconnect the electrical connector from the solenoid, remove the nut and disconnect the battery cable from the starter.
5   Remove starter mounting bolts and remove the starter.
6   Installation is the reverse of removal.

Chapter 6
Emissions and engine control systems

Contents

	Section		Section
Catalytic converter	14	General information	1
CHECK ENGINE light	See Section 7	Heated Air Inlet (HAI) system	5
Deceleration Control system	4	High Altitude Compensation (HAC) system	6
Early Fuel Evaporation (EFE) system	2	Information sensors	8
Electronic Control Module (ECM) - removal and installation	9	On-Board Diagnostic (OBD) System and trouble codes	7
Evaporative Emission Control (EVAP) system	10	Positive Crankcase Ventilation (PCV) system	12
Exhaust Gas Recirculation (EGR) system	11	Pulse Air system	3
Feedback carburetor system	13		

1 General Information

Refer to illustrations 1.1 and 1.6

1 To minimize pollution of the atmosphere from incompletely burned and evaporating gases and to maintain good driveability and fuel economy, a number of emission control systems are used on these vehicles **(see illustration)**. They include the:

    Early Fuel Evaporation system (carbureted engines)
    Pulse Air system (carbureted engines)
    Deceleration Control system (carbureted engines)
    Heated Air Inlet (HAI) (carbureted engines)
    High Altitude Compensation (HAC) system
    Positive Crankcase Ventilation (PCV) system
    Evaporative Emission Control (EVAP) system
    Exhaust Gas Recirculation (EGR) system
    Three-way catalytic converter (TWC) system
    Electronic Control Module (ECM)
    Idle Speed Control (ISC) system (fuel-injected engines)
        (Chapter 4)

1.1 Typical emission and engine control components (fuel-injected model) - items may vary with the model year of the vehicle

1	MAF sensor	4	ISC-BPA valve	6	Coolant temperature sensor (threaded into the intake manifold)	7	TPS
2	PCV valve	5	Test connector			8	Intake Air Temperature sensor
3	Charcoal canister						

Chapter 6 Emissions and engine control systems

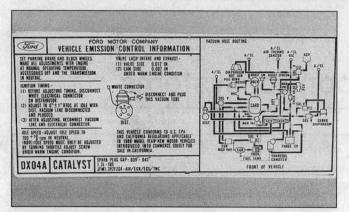

1.6 The Vehicle Emissions Control Information (VECI) label contains such essential information as the types of emission control systems installed on the engine, the idle speed and ignition timing specifications (early carbureted model shown)

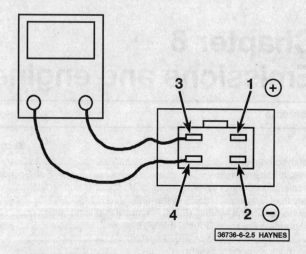

2.5 Continuity should not exist between terminals 3 and 4 unless battery voltage is applied to terminals 1 (+) and 2 (-)

2 The Sections in this Chapter include general descriptions, checking procedures within the scope of the home mechanic and component replacement procedures (when possible) for each of the systems listed above.

3 Before assuming an emissions control system is malfunctioning, check the fuel and ignition systems carefully (see Chapters 4 and 5). The diagnosis of some emission control devices requires specialized tools, equipment and training. If checking and servicing become too difficult or if a procedure is beyond the scope of your skills, consult your dealer service department or other repair shop.

4 This doesn't mean, however, that emission control systems are particularly difficult to maintain and repair. You can quickly and easily perform many checks and do most of the regular maintenance at home with common tune-up and hand tools. **Note:** *The most frequent cause of emissions problems is simply a loose or broken electrical connector or vacuum hose, so always check the vacuum hoses and electrical connectors.*

5 Pay close attention to any special precautions outlined in this Chapter. It should be noted that the illustrations of the various systems may not exactly match the system installed on your vehicle because of changes made by the manufacturer during production or from year-to-year.

6 The Vehicle Emissions Control Information (VECI) label and a vacuum hose diagram are located on the hood **(see illustration)**. These contain important emissions specifications and setting procedures, and a vacuum hose schematic with emissions components identified. When servicing the engine or emissions systems, the VECI label in your particular vehicle should always be checked for up-to-date information.

2 Early Fuel Evaporation (EFE) system

Refer to illustration 2.5
Note: *This system applies to carbureted models only. Not all models are equipped with an EFE system.*

General information

1 The Early Fuel Evaporation (EFE) system is designed to preheat the fuel mixture to improve driveability and fuel economy while the vehicle is at less than normal operating temperature. The components of this system include the heater (located at the base of the carburetor), the coolant temperature sensor, the heater relay and the ECM. When the engine is cold, the ECM will activate the EFE heater by energizing the EFE heater relay. Any malfunction in the heater relay or heater will set a code 38. **Note:** *The coolant temperature sensor check and replacement procedure is located in Section 8.*

Check

Heater element

2 Disconnect the EFE heater electrical connector and check for continuity between terminals 1 and 2, then between 1 and 3. Continuity should exist. If continuity does not exist, replace the heater. **Note:** *The heater assembly is located at the base of the carburetor. It will be necessary to remove the carburetor (see Chapter 4) to replace it.*

Heater relay

3 Locate the heater relay under the dash on the driver's side above the steering column. Disconnect the heater relay harness connector. With the ignition key ON (engine not running), check for battery voltage to the relay using a voltmeter.
4 If voltage is not available, check the circuit from the battery to the relay. Refer to the wiring diagrams at the end of Chapter 12.
5 Test the relay by checking for continuity across terminals 3 and 4 **(see illustration)**. Continuity should not exist.
6 Apply battery voltage to terminals 1 (+) and 2 (-) and check for continuity across terminals 3 and 4. Continuity should exist.
7 If the test results are incorrect, replace the heater relay.

3 Pulse Air system

Note: *This system applies to carbureted models only.*

General description

1 The pulse air system reduces carbon monoxide and hydrocarbon content in the exhaust gases by directing fresh air (intake air) into the hot exhaust gases leaving the exhaust ports during cold engine operation. This system uses negative pressure pulses from the exhaust system to draw air into the catalytic converter and the exhaust manifold. When fresh air is mixed with hot exhaust gases, oxidation is increased, reducing the concentration of hydrocarbons and carbon monoxide and converting them into harmless carbon dioxide and water.
2 The pulse air system consists of an air control valve, an air control valve solenoid, two reed valves located in the air cleaner assembly, vacuum signal lines and pipes.

Check

Refer to illustration 3.6
3 Inspect all the vacuum lines, the hoses, the electrical connectors and the wiring for damage. Check for kinks, opens or broken wires. Replace any damaged components with new ones.

Chapter 6 Emissions and engine control systems

3.6 Use a hand-held vacuum pump and apply a vacuum of 5 in-Hg to the air control valve and see if the valve opens (it should)

4   Warm the engine to operating temperature. Disconnect the neutral switch.
5   Remove the vacuum hose from the air control valve vacuum diaphragm. Increase the engine speed to 1,200 rpm and verify that the vacuum signal is present below 1,200 rpm and bleeds off above 1,200 rpm.
6   Using a hand-held vacuum pump, apply vacuum to the air control valve diaphragm **(see illustration)**. The valve should open at 5.5 in-Hg of vacuum. If the valve does not open, replace it with a new one.
7   Check the voltage on the air control solenoid valve. Below 1,200 rpm, voltage should read 12 to 14 volts. Increase the engine rpm to 2,500 and if voltage is still not present, check the black/white wire to the solenoid for battery voltage. Check the entire circuit back to the ignition switch if voltage is not present.
8   Disconnect the vacuum hose from port A and apply air pressure (slight amount) through the solenoid valve. **Note:** *This can be done my mouth if you connect a clean vacuum hose to the valve.* Air should flow through the valve and out port C. If not, replace the solenoid valve.
9   Briefly apply battery voltage and ground to the solenoid terminals and see if the valve clicks. Air should bleed from port A through port B when battery voltage is applied.

Component replacement

10 To replace the air control valve, solenoid valve, the reed valves or air suction pipe, clearly label, then disconnect, the hoses leading to them, replace the faulty component and reattach the hoses to the proper ports. Make sure the hoses are in good condition. If not, replace them with new ones.

4 Deceleration Control system

Note: *This system applies to carbureted models only.*

General description

1   During rapid deceleration, the fuel/air mixture becomes heavily concentrated for a short period of time. As the vacuum in the intake manifold increases, it causes the fuel residue on the inside wall of the manifold to vaporize and to enter the combustion chamber.
2   The anti-backfire system prevents this heavy concentration from occurring by blocking the fuel flow in the primary slow fuel delivery circuit in the carburetor when the vehicle is decelerating or the ignition switch is OFF. This is accomplished by the deceleration fuel shutoff valve.
3   In conjunction with fuel shut-off valve, the anti-backfire valve or "gulp" valve allows additional fresh air to be drawn into the combustion chamber. This provides extra air to mix with any unburned fuel that may be drawn through the metering circuits due to the high manifold vacuum (closed throttle) during deceleration. The fuel shut-off valve is located next to the intake manifold.
4   The main components of this system are the anti-backfire valve (gulp valve), the deceleration fuel shut-off valve, the neutral switch, the clutch switch, the brake switch, the idle switch, the A/C switch, the MAP sensor and the BP switch.

Check

5   First check the fuel shut-off valve. **Note:** *The fuel shut-off valve is a solenoid type device mounted on the side of the carburetor.* Have an assistant turn the ignition key ON and OFF while checking for a distinct click from the fuel shut-off valve. Start the engine and disconnect the valve harness connector. The engine should stop. If the engine continues to run, replace the shut-off valve.
6   Check the anti-backfire valve **(see illustration 13.1 for the location of the anti-backfire valve)**. With the engine idling at normal operating temperature, check the vacuum line at the valve for the presence of manifold vacuum. Check the line for damage if there is no vacuum signal.
7   Remove the vacuum line from the anti-backfire valve for five seconds and then reconnect the vacuum line. There should be a change in rpm.
8   Place your hand at the bottom of the anti-backfire valve and increase the engine speed to 2500 rpm. Quickly release the throttle and note that the valve draws in air for approximately 2 to 3 seconds. If the valve does not respond properly, replace it with a new part.

Component replacement

9   Remove the anti-backfire valve by disconnecting the hoses, then unclip the valve from its bracket. Installation is the reverse of removal.

5 Heated Air Inlet (HAI) system

Note: *This system applies to carbureted models only.*

General description

1   The inlet air temperature control system provides heated intake air during warm-up, then maintains the inlet air temperature within a 100-degrees F to 130-degrees F (30 to 40-degrees C) operating range by mixing warm and cool air. This allows leaner fuel/air mixture settings for the carburetor system, which reduces emissions and improves driveability.
2   Two fresh air inlets - one warm and one cold - are used. The balance between the two is controlled by intake manifold vacuum, a temperature sensor and a vacuum motor. A vacuum motor, which operates a heat duct valve in the air cleaner, is controlled by the temperature sensor.
3   When the underhood temperature is cold, warm air radiating off the exhaust manifold is routed by a shroud, which fits over the manifold, up through a hot air inlet tube and into the air cleaner. This provides warm air for the carburetor, resulting in better driveability and faster warm-up. As the underhood temperature rises, a heat duct valve is gradually closed by a vacuum motor and the air cleaner draws air through a cold air duct instead. The result is a consistent intake air temperature.
4   A temperature vacuum sensor mounted inside the air intake snorkel, monitors the temperature of the inlet air heated by the exhaust manifold.
5   The vacuum motor regulates the duct valve allowing warm air to pass through when the engine is cold and conversely shuts the flow of warm air and opens the cold air duct after the engine has reached operating temperature.

Check

Note: *Make sure that the engine is cold before beginning this test.*
6   Check the vacuum source and the integrity of all vacuum hoses between the source and the vacuum motor before beginning the test. Do not proceed until they're okay.

7 Apply the parking brake and block the wheels.
8 Remove the air filter (see Chapter 1).
9 Use a hand-held vacuum pump and apply vacuum to the vacuum motor. The duct valve should move. Watch carefully - it may be binding or sticking. Make sure that it's not rusted in an open or closed position by attempting to move it by hand. If it's rusted, it can usually be freed by cleaning and oiling the hinge. If it fails to work properly after servicing, replace it.
10 If the vacuum motor door is okay but the motor still fails to operate correctly, check carefully for a leak in the hose leading to it. Check the vacuum source to and from the sensor and the vacuum motor as well. If no leak is found, replace the vacuum motor.
11 Check the temperature sensor. Monitor the vacuum using a vacuum gauge first, when the engine is cold (start-up) and then finally when the engine has reached operating temperature. There should be vacuum when the engine is cold and then gradually shut down when the engine warms up.

Component replacement

Temperature sensor
12 Clearly label, then detach both vacuum hoses from the temperature sensor (one is coming from the vacuum source at the manifold and the other is going to the vacuum motor underneath the air cleaner housing).
13 Remove the cover from the air cleaner housing (see Chapter 4).
14 Remove the bolts that retain the sensor to the bottom of the air cleaner housing.
15 Remove the temperature sensor.
16 Installation is the reverse of removal.

Vacuum motor
17 Remove the air cleaner housing (see Chapter 4) and place it on a workbench.
18 Detach the vacuum hose from the vacuum motor.
19 Drill out the vacuum motor retaining strap rivet.
20 Remove the vacuum motor.
21 Installation is the reverse of removal. Use a sheet metal screw of the appropriate size to replace the rivet.

6 **High-Altitude Compensation (HAC) system**

Note: *These testing procedures apply to carbureted engines equipped with the High Altitude Compensation system. Later EFI engines are equipped with a barometric pressure sensor that is incorporated into the ECM. This sensor relays pressure and altitude changes to the circuits within the ECM that directly control the air/fuel mixture. Any diagnosis or repairs to this later system will require special testing equipment. Have the vehicle diagnosed by a dealer service department or other qualified repair shop.*

General Information
1 Vehicles that are operated at high altitude, where the air is thinner (and the air/fuel mixture ratio and manifold vacuum boost vary and the amount of emissions increase) require a special system to compensate for the atmospheric changes.
2 The altitude compensator is located in the corner of the engine compartment. Three rubber hoses attach to the unit.

Check
3 Check all the vacuum hoses and fittings for leaks, cuts or damage that would hinder the proper amount of engine vacuum to reach the components.
4 The primary symptom of a defective HAC valve at low altitudes (below 3300 feet) is a rough idle condition caused by an excessively lean fuel mixture (the result of the HAC valve allowing air to pass through when it shouldn't; it's actually a vacuum leak). To check the valve at low altitudes, detach the three hoses from the HAC valve (see

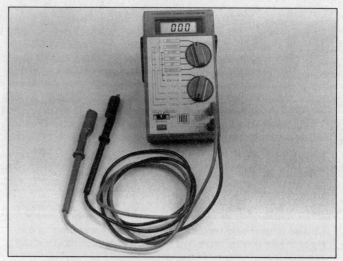

7.1 Digital multimeters can be used for testing all types of circuits; because of their high impedance, they are much more accurate than analog meters for measuring voltage in low-voltage circuits

illustration 13.1) and note the idle quality (don't plug the hoses). If the HAC valve is defective, the idle quality won't change (it'll still be rough, as there is no difference between vacuum leaking from the defective valve or from the disconnected hoses). If the idle quality doesn't change, the valve is probably defective. Now plug the hoses - the idle should smooth out. If it does, this indicates that the HAC valve was allowing extra air to pass through when it shouldn't. Replace the valve.
5 If the vehicle runs rough and emits black exhaust (caused by an overly rich air/fuel ratio) at high altitudes (above 3,300 feet), disconnect all the hoses from the HAC valve and *plug* the hoses (this simulates a HAC valve that *won't* allow the extra air to pass through). If the idle doesn't change, the HAC valve is probably defective. Now unplug the hoses - if the idle smoothes out, this indicates that the HAC valve wasn't allowing the additional air to pass through to lean-out the fuel mixture. Replace the valve.
6 If the test results are incorrect, replace the high altitude compensator.

7 **On-Board Diagnosis (OBD) System and trouble codes**

Note: *This procedure does not include the complete list of diagnostic codes or the code extracting procedure for 1996 or later models equipped with the OBD II system. The diagnostic system and trouble codes are only accessible using expensive specialized equipment. The OBD II codes indicated in the text are designed and mandated by the EPA for all 1996 and later OBD II vehicles produced by automobile manufacturers. These generic trouble codes do not include the manufacturer's specific trouble codes. Consult a dealer service department or other qualified repair shop for additional information on manufacturer-specific trouble codes. General information on the system sensors and actuators for all models is described in the following text. See the Troubleshooting section at the beginning of this manual for some basic diagnostic aids. Because the OBD II system requires a special SCAN tool to access the trouble codes, have the vehicle diagnosed by a dealer service department or other qualified automotive repair facility if the proper SCAN tool is not available.*

Diagnostic tool information
Refer to illustrations 7.1, 7.2 and 7.4
1 A digital multimeter is necessary for checking fuel injection and emission related components **(see illustration)**. A digital volt-ohmmeter is preferred over the older style analog multimeter for several reasons. An analog multimeter cannot display the measured values pre-

Chapter 6 Emissions and engine control systems

7.2 Scanners like the Actron Scantool and the AutoXray XP240 are powerful diagnostic aids - programmed with comprehensive diagnostic information, they can tell you just about anything you want to know about your engine management system

7.4 Trouble code tools simplify the task of extracting the trouble codes

cisely enough. When working with electronic circuits which are often very low voltage, this accurate reading is most important. Another reason for the digital multimeter is the high-impedance circuitry. Most digital multimeters are equipped with a high resistance internal circuitry (10 million ohms). Because a voltmeter is hooked up in parallel with the circuit when testing, it is vital that none of the voltage being measured should be allowed to travel the parallel path set up by the meter itself. This dilemma does not show itself when measuring greater voltages (9 to 12 volt circuits) but if you are measuring a low voltage circuit such as the oxygen sensor signal voltage, a fraction of a volt may be a significant amount when diagnosing a problem. Obtaining the diagnostic trouble codes is one exception where using an analog voltmeter is necessary.

2 Hand-held scanners are the most powerful and versatile tools for analyzing engine management systems used on later model vehicles **(see illustration)**. Early model scanners handle codes and some diagnostics for many OBD I systems. Each brand scan tool must be examined carefully to match the year, make and model of the vehicle you are working on. Interchangeable cartridges are often available to access the particular manufacturer (Ford, GM, Chrysler, etc.). Some manufacturers are specified by continent (Asia, Europe, USA, etc.).

3 With the arrival of the Federally mandated emission control system (OBD II), a specially designed scanner must be used. At this time, several manufacturers plan to release OBD II scan tools for the home mechanic. Ask the parts salesman at a local auto parts store for additional information concerning these tools.

4 Another type of code reader, which is far less expensive, is available at most auto parts stores **(see illustration)**. These tools simplify the procedure for extracting codes from the engine management computer by simply plugging into the diagnostic connector on the vehicle wiring harness. **Note:** *Some diagnostic connectors are located under the dash, kick panel or glovebox, while others are located in the engine compartment.*

OBD general information

Refer to illustrations 7.7a, 7.7b and 7.7c

5 The ECM (computer) has a built-in self-diagnosis system, or On Board Diagnosis (OBD) system, which detects malfunctions in the system sensors and alerts the driver by illuminating a CHECK ENGINE warning light on the instrument panel. The On Board Diagnosis (OBD) system is equipped with a computer which stores the failure code until the diagnostic system is cleared by disconnecting the negative battery cable then depressing the brake pedal for a period of five seconds or longer. The warning light goes out automatically when the malfunction is repaired.

6 The CHECK ENGINE warning light should come on when the ignition switch is placed in the ON position, this checks the bulb for proper operation. When the engine is started the warning light should go out. If the light remains on, the diagnostic system has detected a malfunction or abnormality in the system.

7 To access the self diagnosis system (OBD), connect a jumper wire to the correct terminals of the diagnostic connector:

A) *On feedback carburetor systems, ground the one-pin terminal connector* **(see illustration)** *with a jumper wire to chassis ground. Also, connect the positive lead of an analog voltmeter to terminal 6 (green/red wire) of the 6-pin test connector under the dash. Connect the negative lead of the voltmeter to ground. Watch the sweeps of the voltmeter needle. The voltmeter must be analog to see the movement of the needle as it sweeps across the face of the meter.*

B) *On EFI systems through 1995, connect a jumper wire from terminal STI to GND* **(see illustrations)**. *Also, connect a voltmeter from*

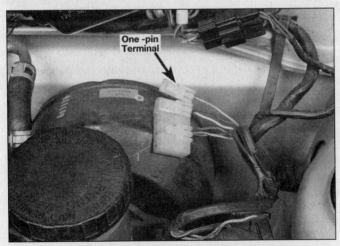

7.7a The one-pin check terminal on feedback carburetor models is near the power brake booster

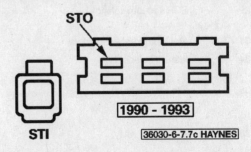

7.7b Diagnostic terminal designations (1990 to 1993 models)

Chapter 6 Emissions and engine control systems

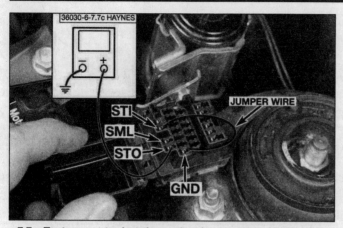

7.7c Test connector location and code extracting details (1994 and 1995 models)

the STO terminal and engine ground. With the ignition key ON (engine not running), watch the sweeps of the voltmeter needle. The voltmeter must be analog to see the movement of the needle as it sweeps across the face of the meter. **Note:** *It is also possible to read the trouble codes from the CHECK ENGINE light on the dash. Simply ground the STI and GND terminals with a jumper wire and watch the light on the dash as it flashes the stored trouble code(s).*

C) On EFI systems equipped with the OBD II system (1996 and later models), it will be necessary to use a specialized SCAN tool to extract any stored trouble codes.

Make sure the battery voltage is greater than 11 volts, the transaxle is in Neutral, the accessories are off, the throttle valve is closed and the engine is at normal operating temperature, then turn the ignition switch to the ON position but do not start the engine.

8 The diagnostic code is the number of flashes indicated on the CHECK ENGINE light or sweeps on the voltmeter needle. If no codes are stored, the CHECK ENGINE light will come on for a few moments, then go out. If any malfunction has been detected, the light will blink the first digit(s) of the code at a long interval(s) and then blink the second digit of the code at short interval(s). For example, a code 34 (IAC valve) will first blink three long flashes and then pause and blink four quick flashes. **Note:** *If the code is simply a single digit number, the CHECK ENGINE light will flash in the quick mode.*

9 The accompanying tables indicate the diagnostic code, the circuit or system, and the corrective action to take to rectify the problem.

Feedback carbureted engines

Trouble code	Circuit or system	Corrective action
01	Ignition coil signal	Check the ignition coil for loose or damaged wire harness (see Chapter 5)
09	Coolant temperature sensor	Check the coolant temperature sensor circuit for an open or short circuit or a sensor malfunction (see Section 8)
13	Manifold Absolute Pressure (MAP) sensor or circuit	Check the MAP sensor and circuit (see Section 8)
15	Oxygen sensor malfunction	Check for a defective oxygen sensor and the oxygen sensor circuit for an open or short circuit condition (see Section 8)
16	EGR valve position sensor or	Check the EGR valve position sensor and the circuit for an open or short circuit (see Section 8)
17	Feedback system	Check the feedback system for component malfunctions (see Section 13)
18	Air Bleed Control Valve Number 1	Check for loose wires, harness connectors that are damaged or defective ABCV (see Section 13)
19	Air Bleed Control Valve Number 2	Check for loose wires, harness connectors that are damaged or defective ABCV (see Section 13)
20	Air Bleed Control Valve Number 3	Check for loose wires, harness connectors that are damaged or defective ABCV (see Section 13)
21	Air Bleed Control Valve Number 4	Check for loose wires, harness connectors that are damaged or defective ABCV (see Section 13)
22	Fuel Shutoff solenoid or circuit	Possible defective harness connector or fuel shutoff solenoid. Have the vehicle diagnosed at a dealer service department or other qualified repair shop
28	EGR vacuum solenoid	Check for an open or short circuit in the EGR vacuum solenoid wire number 1 harness (see Section 11)
29	EGR vacuum solenoid	Check for an open or short circuit in the EGR vacuum solenoid wire number 2 harness (see Section 11)
31	Air Control Valve Solenoid	Check for an open or short circuit in the ACV solenoid (see Section 3)
34	Idle-up solenoids	Possible defective idle-up solenoid(s) that govern E/L solenoid and A/C solenoid. Have the vehicle tested by a dealer service department
35	Idle-up solenoids	Possible defective idle-up solenoid(s) that govern E/L solenoid and A/C solenoid. Have the vehicle tested by a dealer service department or other qualified repair shop
38	EFE heater relay	Check for an open or short circuit in the EFE heater relay and/or harness connector (see Section 2)
70	Vacuum switch	Check for an open or short circuit in the vacuum switch (see Section 13)

Chapter 6 Emissions and engine control systems

EFI engines (1990 through 1995)

Trouble code	Circuit or system	Corrective action
01	Ignition Diagnostic Monitor (1990 through 1993)	Check the distributor circuit to the IDM (see Chapter 5)
02	Crankshaft Position (CP) sensor (1990 through 1993)	Check the distributor CP sensor and circuit (see Section 8)
03	Cylinder Identification sensor (1990 through 1993)	Check the distributor CID sensor and circuit (see Section 8)
03	Camshaft Position Sensor (CMP) (1994 and later)	Check the distributor and ECM circuitry or components (see Chapter 5)
04	Crankshaft Position Sensor (CKP) (1994 and later)	Check the distributor and ECM circuitry or components (see Chapter 5)
06	Vehicle Speed Sensor	Check the VSS and circuit (see Section 8)
08	Airflow sensor (1990 through 1993)	Check the airflow sensor circuit from the sensor to the ECM for an open or short circuit or a sensor malfunction (see Section 8)
08	Mass Airflow (MAF) sensor (1994 and later)	Check the MAF sensor circuit from the sensor to the ECM for an open or short circuit or a sensor malfunction (see Section 8)
09	Coolant temperature sensor	Check the coolant temperature sensor circuit for an open or short circuit or a sensor malfunction (see Section 8)
10	Intake air temperature sensor	Check the intake air temperature sensor circuit for an open or short circuit or a sensor malfunction (see Section 8)
12	Throttle position sensor	Check the throttle position sensor circuit for an open or short circuit or a sensor malfunction (see Section 8)
14	Barometric pressure sensor	The barometric pressure sensor is integrated within the ECM. Check the ECM power and ground circuits. If no fault is found with the circuits, replace the ECM.
15	Oxygen sensor	Check the oxygen sensor circuit for an open or short circuit or a sensor malfunction (see Section 8)
16	EGR valve position sensor	Check the EGR valve position sensor and circuit (see Section 6)
17	Oxygen sensor	Check the fuel and ignition system performance (i.e. fuel pressure high or low, leaking fuel a rich or lean condition injectors, inoperative fuel injector, intake air leaks, ignition misfire, etc.) (see Chapter 4)
25	Fuel Pressure Regulator Control solenoid	Check for an open or short circuit from the solenoid valve to the ECM. Possible broken or (fuel pressure regulator) short circuit from the solenoid valve to the fuel injection main relay. Check for a defective solenoid valve (see Chapter 4)
26	Solenoid valve (purge control)	Check for an open or short circuit from the solenoid valve to the ECM. Check for an open or short circuit from the solenoid valve to the fuel injection main relay. Check for a defective solenoid valve.
28	EGR control solenoid (1994 and later)	Check for an open or short circuit from the solenoid valve to the ECM (see Section 11)
29	IAC solenoid (1994 and later)	Check for an open or short circuit from the IAC valve to the ECM. Check for an open or short circuit from the IAC valve to the fuel injection main relay. Check for a defective IAC valve.
34	IAC valve (1990 through 1993)	Check for an open or short circuit from the IAC valve to the ECM. Check for an open or short circuit from the IAC valve to the fuel injection main relay. Check for a defective IAC valve.

Clearing the codes

10 After the self-diagnosis check, remove the jumper wire and close the cover on the DIAGNOSTIC electrical connector. Check the indicated system or component or take the vehicle to a dealer service department or other qualified repair shop to have the malfunction repaired.

11 After repairs have been made, the diagnostic code must be canceled by detaching the cable from the negative terminal of the battery, then depressing the brake pedal for more than five seconds.

12 After cancellation, perform a road test and make sure the warning light does not come on. If the original trouble code is repeated, additional repairs are required.

Information sensors (OBD-I and II systems)

13 When battery voltage is applied to the air conditioning compressor solenoid, a signal is sent to the computer, which interprets the signal as an added load created by the compressor and increases engine idle speed accordingly to compensate.

14 The **Intake Air Temperature sensor (IAT)**, positioned in the air intake duct (see Section 8), provides the computer with fuel/air mixture temperature information. The computer uses this information to control fuel flow, ignition timing and EGR system operation.

15 The **Engine Coolant Temperature (ECT) sensor**, which is threaded into a coolant passage in the intake manifold, monitors engine coolant temperature. The ECT sends the computer a voltage

signal which influences control of the fuel mixture, ignition timing and EGR operation.

16 The **Heated Exhaust Gas Oxygen (HEGO) sensor**, which is threaded into the exhaust manifolds before (and, on OBD-II vehicles, after) the catalytic converter, constantly monitors the oxygen content of the exhaust gases. A voltage signal which varies in accordance with the difference between the oxygen content of the exhaust gases and the surrounding atmosphere is sent to the ECM/PCM. The computer converts this exhaust gas oxygen content signal to the fuel/air ratio, compares it to the ideal ratio for current engine operating conditions and alters the signal to the injectors accordingly.

17 The **Throttle Position Sensor (TPS)**, which is mounted on the side of the throttle body (see Section 8) and connected directly to the throttle shaft, senses throttle movement and position, then transmits an electrical signal to the computer. This signal enables the computer to determine when the throttle is closed, in its normal cruise condition or wide open.

18 The **Mass Air Flow (MAF) sensor**, which is mounted in the air cleaner intake passage, measures the mass of the air entering the engine (see Section 8). Because air mass varies with air temperature (cold air is denser than warm air), measuring air mass provides the computer with a very accurate way of determining the correct amount of fuel to obtain the ideal fuel/air mixture.

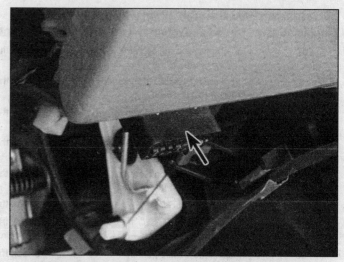

7.24 Location of the 16-pin Diagnostic Link Connector on OBD II vehicles

Output actuators (OBD-I and II systems)

19 The **power relay**, which is activated by the ignition switch, supplies battery voltage to the EEC system components when the switch is in the Start or Run position. Refer to Chapter 4 for additional information on the power relay and the fuel pump relay.

20 The **canister purge valve** switches manifold vacuum to operate the valve when a signal is received from the computer. The activates the valve, allowing fuel vapor to flow from the canister to the intake manifold to be burned in the combustion process.

21 The solenoid-operated **fuel injectors** are located above the intake ports (see Chapter 4). The computer controls the length of time the injectors are open. The "open" time of the injectors determines the amount of fuel delivered. For information regarding injector replacement, refer to Chapter 4.

22 The **fuel pump relay** is activated by the PCM with the ignition switch in the On position. When the ignition switch is turned to the On position, the relay is activated to supply initial line pressure to the system. For information regarding fuel pump check and replacement, refer to Chapter 4.

23 The **ignition module** (see Chapter 5) is incorporated into the distributor. The computer uses a signal from the camshaft sensor to determine piston position. Ignition timing is determined by the computer, which then signals the module to fire the coil. For further information regarding the ignition system, refer to the appropriate Section in Chapter 5.

OBD II general information

Refer to illustration 7.24

24 Beginning in 1994, the manufacturer began to manufacture a second generation self-diagnosis system specified by the CARB and EPA regulations called On Board Diagnosis (OBD) II. This system incorporates a series of diagnostic monitors that detect and identify emissions systems faults and store the information in the computer memory. This updated system also tests sensors and output actuators, diagnoses drive cycles, freezes data and clears codes. This powerful diagnostic computer must be accessed using the new OBD II SCAN tool and 16 pin Data Link Connector (DLC) located under the driver's dash area **(see illustration)**. This system consists of an onboard computer, known as the Powertrain Control Module (PCM), and information sensors, which monitors various functions of the engine and sends data to the PCM. Based on the data and the information programmed into the computer's memory, the PCM generates output signals to control various engine functions via control relays, solenoids and other output actuators. **Note:** *Although the manufacturer started producing OBD II in 1994, the first Aspire models equipped with this updated system came in 1996.*

25 The PCM, located under the instrument panel on the passenger's side, is the "brain" of the system. It receives data from a number of sensors and other electronic components (switches, relays, etc.). Based on the information it receives, the PCM generates output signals to control various relays, solenoids and other actuators. The PCM is specifically calibrated to optimize the emissions, fuel economy and driveability of the vehicle.

26 Because of a Federally mandated extended warranty which covers the OBD-II system components and because any owner-induced damage to the PCM, the sensors and/or the control devices may void the warranty, it isn't a good idea to attempt diagnosis or replacement of the PCM at home while the vehicle is under warranty. Take the vehicle to a dealer service department if the PCM or a system component malfunctions.

Obtaining OBD II system codes

27 On OBD II systems, the PCM will illuminate the Malfunction Indicator Light on the dash if it recognizes a component fault for two consecutive drive cycles. It will continue to set the light until the PCM does not detect any malfunction for three or more consecutive drive cycles. Because the OBD II system requires a SCAN tool to reset the light, if the tool is not available for diagnostics, have the system checked by a dealer service department or other qualified repair facility.

28 The diagnostic codes for the OBD-II systems can be extracted from the PCM using a special SCAN tool that is programmed to interface with this new system by plugging into the DLC. If the tool is not available, have the vehicle checked at a dealer service department or other qualified repair shop.

Clearing codes

29 To clear the codes from the PCM memory, install the OBD-II SCAN tool, scroll the menu for the function that describes "CLEARING CODES' and follow the prescribed method for that particular SCAN tool. If necessary, have the codes cleared by a dealer service department or other qualified repair facility. **Caution:** *Do not disconnect the battery from the vehicle to clear the codes. This will erase stored operating parameters from the KAM (Keep Alive Memory) and cause the engine to run rough for a period of time while the computer relearns the information.*

Chapter 6 Emissions and engine control systems

OBD-II Trouble Codes (not all codes apply to all models)

Code	Code Definition	Location
P0100, P0102, P0103	Mass Airflow (MAF) sensor circuit fault	See Section 8
P0110, P0112, P0113	Inlet Air Temperature (IAT) sensor circuit fault	See Section 8
P0115, P0117, P0118	Engine Coolant Temperature (ECT) sensor circuit fault	See Section 8
P0120, P0122, P0123	Throttle Position Sensor (TPS) circuit fault	See Section 8
P0125	Excessive time to enter closed loop	See Section 8
P0130, P0131, P0150	Heated O2 sensor circuit fault	See Section 8
P0133, P0134, P0140	Heated O2 sensor circuit slow response	See Section 8
P0154, P0160	Heated O2 sensor circuit slow response	See Section 8
P0135, P0141, P0155	Heated O2 sensor heater circuit fault	See Section 8
P0170, P0171	System fuel too lean	See Chapter 4
P0172, P0173	System fuel too rich	See Chapter 4
P0230, P0231, P0232	Fuel pump circuit fault	See Chapter 4
P0300	Random misfire detected	See Chapter 5
P0301	Cylinder number 1 misfire detected	See Chapter 5
P0302	Cylinder number 2 misfire detected	See Chapter 5
P0303	Cylinder number 3 misfire detected	See Chapter 5
P0304	Cylinder number 4 misfire detected	See Chapter 5
P0320	Ignition input circuit fault	See Chapter 5
P0335	Crankshaft position sensor circuit fault	See Section 8
P0340	Camshaft position sensor malfunction	See Section 8
P0400	EGR flow fault	See Section 11
P0420, P0430	Catalyst system efficiency below threshold	See Section 8
P0440	EVAP system fault	See Section 10
P0443	EVAP purge solenoid circuit fault	See Section 10
P0500, P0503	VSS fault	See Section 8
P0505	IAC valve system fault	See Chapter 4
P0510	Idle switch fault	See Chapter 4
P0552, P0553	Power steering pressure sensor circuit fault	See Section 8
P0603	ECM Keep Alive Memory test error	See Section 9
P0605	ECM Read Only Memory test error	See Section 9
P0703	Brake On/Off switch fault	See Chapter 9
P0704	Clutch pedal position switch fault	See Section 8
P0705	Transmission range sensor fault	See Chapter 7
P0710 thru 760	Electronic transmission control system fault	See Chapter 7

8 Information sensors

Note: *Most of the components described in this Section are protected by a Federally mandated extended warranty. See your dealer for the details regarding your vehicle. Refer to Chapters 4 and 5 for additional information on the location and the diagnostic procedures for the sensors that are not covered in this Section.*

Coolant temperature sensor

General description

1 The coolant temperature sensor is a thermistor (a resistor which varies the value of its voltage output in accordance with temperature changes). As the sensor temperature DECREASES, the resistance values will INCREASE. As the sensor temperature INCREASES, the resistance values will DECREASE. A failure in this sensor circuit should set a Code 09. This code indicates a failure in the coolant temperature sensor circuit, so in most cases the appropriate solution to the problem will be either repair of a connector or wire, or replacement of the sensor.

Check

Refer to illustrations 8.2a and 8.2b

Note: *The coolant temperature sensor on EFI models is located in the top center section of the intake manifold under the throttle controls. On carbureted models, it's threaded into the underside of the intake manifold.*

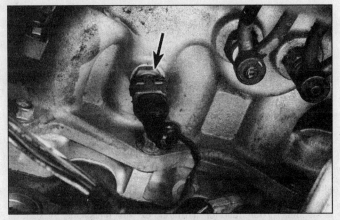

8.2a The coolant temperature sensor (arrow) on feedback carbureted engines is threaded into the underside of the intake manifold

2 To check the sensor, depress the locking tabs, disconnect the electrical connector and measure the resistance across the terminals of the sensor **(see illustrations)**. With the engine completely cold (68-degrees F [20-degrees C]) the resistance should be 2,000 to 3,000 ohms. Next, start the engine and warm it up until it reaches operating

6-10   Chapter 6 Emissions and engine control systems

8.2b To check the coolant temperature sensor, use an ohmmeter to measure the resistance between the two sensor terminals (arrow) (EFI engine shown)

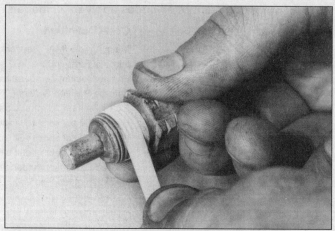

8.5 To prevent leakage, wrap the threads of the coolant temperature sensor with Teflon tape before installing it into the intake manifold

temperature (180-degrees F [80-degrees C]) - the resistance should be 200 to 400 ohms. **Note:** *If necessary, remove the sensor and perform the tests in a pan of heated water to simulate these conditions.*

3   If the resistance values of the coolant temperature sensor are correct, check the circuit for the proper signal voltage. Turn the ignition key ON (engine not running) and check for reference voltage **(see illustrations 8.2a and 8.2b)**. It should be approximately 5 volts.

Replacement
Refer to illustration 8.5

Warning: *Wait until the engine is completely cool before performing this procedure. Also, remove the radiator cap to release any residual pressure in the cooling system, then reinstall the cap.*

4   To remove the sensor, carefully unscrew the sensor. **Caution:** *Handle the coolant sensor with care. Damage to this sensor will affect the operation of the entire fuel injection system.*

5   Before installing the new sensor, wrap the threads with Teflon sealing tape to prevent leakage **(see illustration)**.

6   Installation is the reverse of removal. Check the coolant level and add some, if necessary, to bring it up to the proper level (see Chapter 1).

Oxygen sensor
General description

7   These models are equipped with either a single oxygen sensor system or a dual-stage oxygen sensor system. On dual-stage systems, the main oxygen sensor is mounted ahead of the front catalytic converter and monitors the exhaust gases exiting the engine. The sub oxygen sensor monitors the exhaust gases after they have passed through the front catalytic converter. Each oxygen sensor monitors the oxygen content of the exhaust gas stream. The oxygen content in the exhaust reacts with the oxygen sensor to produce a voltage output which varies from 0.1-volt (high oxygen, lean mixture) to 0.9-volts (low oxygen, rich mixture). The ECM constantly monitors this variable voltage output to determine the ratio of oxygen to fuel in the mixture. The ECM alters the air/fuel mixture ratio by controlling the pulse width (open time) of the fuel injectors. A mixture ratio of 14.7 parts air to 1 part fuel is the ideal mixture ratio for minimizing exhaust emissions, thus allowing the catalytic converter to operate at maximum efficiency. It is this ratio of 14.7 to 1 which the ECM and the oxygen sensor attempt to maintain at all times.

8   The oxygen sensor produces no voltage when the oxygen sensor is below its normal operating temperature of about 600-degrees F. During this initial period before warm-up, the ECM operates in open loop mode.

9   If the engine reaches normal operating temperature and/or has been running for two or more minutes, and if the main oxygen sensor is producing a steady signal voltage below 0.70-volts at 1,500 or more rpm, the ECM will set a Code 15.

10   When there is a problem with the oxygen sensor or its circuit, the ECM operates in the open loop mode - that is, it controls fuel delivery in accordance with a programmed default value instead of feedback information from the oxygen sensor.

11   The proper operation of the oxygen sensor depends on four conditions:

a) **Electrical** - *The low voltages generated by the sensor depend upon good, clean connections which should be checked whenever a malfunction of the sensor is suspected or indicated.*
b) **Outside air supply** - *The sensor is designed to allow air circulation to the internal portion of the sensor. Whenever the sensor is removed and installed or replaced, make sure the air passages are not restricted.*
c) **Proper operating temperature** - *The ECM will not react to the sensor signal until the sensor reaches approximately 600-degrees F. This factor must be taken into consideration when evaluating the performance of the sensor.*
d) **Unleaded fuel** - *The use of unleaded fuel is essential for proper operation of the sensor. Make sure the fuel you are using is of this type.*

12   In addition to observing the above conditions, special care must be taken whenever the sensor is serviced.

a) *The oxygen sensor has a permanently attached pigtail and electrical connector which should not be removed from the sensor. Damage to or removal of the pigtail or electrical connector can adversely affect operation of the sensor.*
b) *Grease, dirt and other contaminants should be kept away from the electrical connector and the louvered end of the sensor.*
c) *Do not use cleaning solvents of any kind on the oxygen sensor.*
d) *Do not drop or roughly handle the sensor.*
e) *The silicone boot must be installed in the correct position to prevent the boot from being melted and to allow the sensor to operate properly.*

Check
Refer to illustration 8.13

13   To check the oxygen sensor use a digital voltmeter to monitor the millivolt signal from the oxygen sensor during actual operating conditions. Locate the oxygen sensor electrical connector and backprobe the sensor wire on the harness side of the oxygen sensor connector **(see illustration)**. To properly backprobe the connector insert a long straight pin (a T-pin is preferred) alongside the wire until the pin contacts the metal wire terminal inside the connector. Connect the positive probe of a voltmeter onto the pin and the negative probe to ground. **Note:** *Refer to the wiring diagrams at the end of Chapter 12 to addi-*

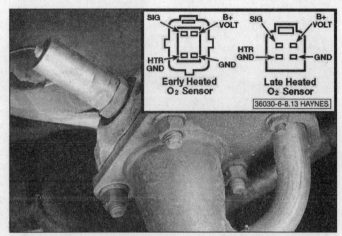

8.13 Insert a pin into the backside of the oxygen sensor connector on the correct terminal and check for a millivolt output signal (SIG) generated by the sensor

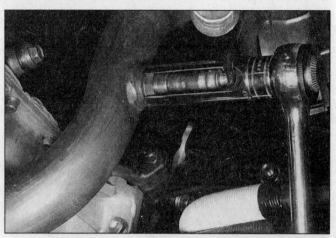

8.21 Slotted sockets are available for easing oxygen sensor removal

tional information on the wire color designations for the oxygen sensor checks.

14 Allow the engine to reach normal operating temperature and check that the oxygen sensor is producing a varying signal voltage between 0.1 and 0.9-volts. **Note:** *Downstream oxygen sensors will produce much slower fluctuating voltage values to reflect the results of the catalyzed exhaust mixture from rich or lean to less presence of CO, HC and NOx molecules. Here the CO_2 and H_2O gaseous forms do not register or react with the oxygen sensors to such a large degree.*

15 Use the OBD system to help diagnose O2 sensor problems by extracting the codes from the computer (see Section 7). On vehicles equipped with the OBD-II system, it will be more difficult to access the trouble codes. The OBD-II system can detect a variety of different oxygen sensor problems and set codes to indicate the specific trouble area. Trouble codes P0131 through P0161 are designated oxygen sensor, circuit and heater control codes. If an OBD-II SCAN tool is not available, have the codes extracted from the PCM by a dealer service department or other qualified automotive repair facility.

16 To check the heated rear oxygen sensor on 1996 and later models, locate the electrical connector at the catalytic converter and check it in the same manner as the main oxygen sensor.

17 Also check to make sure the oxygen sensor heater(s) is supplied with battery voltage **(see illustration 8.13)**. Backprobe the HTR GND and the B+ Volt. terminals with the ignition key ON. Because battery voltage is supplied to the oxygen sensors through a relay, voltage will only be delivered for a very short time (3 seconds) when the ignition key is cycled. Have an assistant turn the ignition key to ON while observing the voltmeter. Refer to Chapter 12 for additional information on the wiring schematics and relays. **Note:** *Not all models are equipped with a heated oxygen sensor. Models with heated oxygen sensors will be equipped with a four-wire electrical connector.*

Replacement

Refer to illustration 8.21

Note: *Because it is installed in the exhaust manifold or pipe, which contracts when cool, the oxygen sensor may be very difficult to loosen when the engine is cold. Rather than risk damage to the sensor (assuming you are planning to reuse it in another manifold or pipe), start and run the engine for a minute or two, then shut it off. Be careful not to burn yourself during the following procedure.*

18 Disconnect the cable from the negative terminal of the battery.
19 Raise the vehicle and place it securely on jackstands.
20 Carefully disconnect the electrical connector from the sensor pigtail lead.
21 Remove the oxygen sensor from the exhaust system **(see illustration)**. **Caution:** *Excessive force may damage the threads.* **Note:** *Some oxygen sensors are threaded directly into the exhaust manifold while others are mounted in the exhaust manifold or pipe with two bolts.*

22 Anti-seize compound must be used on the threads of the sensor to facilitate future removal. The threads of new sensors will already be coated with this compound, but if an old sensor is removed and reinstalled, recoat the threads.
23 Install the sensor and tighten it securely.
24 Reconnect the electrical connector of the pigtail lead to the main engine wiring harness.
25 Lower the vehicle and reconnect the cable to the negative terminal of the battery.

Throttle Position Sensor (TPS)

General description

26 The Throttle Position Sensor (TPS) is located on the end of the throttle shaft on the throttle body (see Chapter 4). By monitoring the output voltage from the TPS, the ECM alters fuel delivery based on throttle valve angle (driver demand). A broken or loose TPS can cause intermittent bursts of fuel from the injector and an unstable idle because the ECM receives a signal that the throttle is moving. All the checks and replacement procedures are covered in Chapter 4.

Airflow sensor/intake air temperature sensor (1990 through 1993 models)

General description

Note: *On 1993 and earlier EFI models, the intake air temperature sensor is incorporated into the airflow sensor.*

27 The airflow sensor (located on top of the air cleaner housing) measures the volume of air entering the intake system using a vane-type potentiometer device. As air enters the air by-pass passage, the measuring plate (vane) swings open and allows an electrical device (potentiometer) to vary its voltage signal according to the position of the measuring plate. This information is relayed to the computer and is used to determine the correct amount of fuel to inject into the combustion chamber for the volume of air (load) that is demanded.

28 The intake air temperature sensor is located inside the airflow sensor. This sensor is a resistor which changes value according to the temperature of the air entering the engine. Low temperatures produce a high resistance value (for example, at 68-degrees F [20-degrees C], the resistance is 2,000 to 3,000 ohms) while high temperatures produce low resistance values (at 176-degrees F [80-degrees C] the resistance is 200 to 400 ohms. The ECM supplies approximately 5-volts (reference voltage) to the air temperature sensor. The IAT sensor alters the voltage according to the temperature of the incoming air. Any problems with the air temperature sensor will usually set a diagnostic code 10.

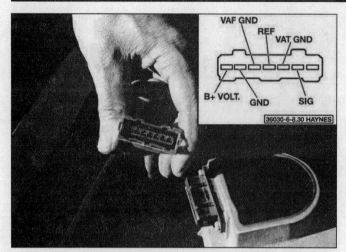

8.30 Check for battery voltage to the airflow sensor with the ignition key ON (engine not running)

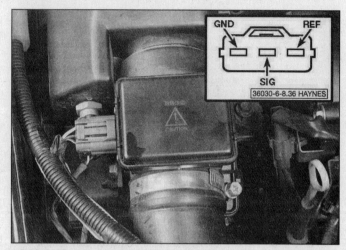

8.36 MAF sensor connector terminal designations on 1994 and later model EFI engines

Check

Refer to illustrations 8.30

29 Remove the airflow sensor and check the body for cracks or damage. with your finger, press the vane in and make sure it moves smoothly and does not bind.

30 Disconnect the airflow sensor harness connector and check for battery voltage to the sensor with the ignition key ON (engine not running) **(see illustration)**. Battery voltage should be present. If not, check the fuel pump relay and power relay (see Chapter 4).

31 Check for reference voltage. Use a digital voltmeter and check for a 5 volt reference signal on the REF terminal. If incorrect, have the ECM diagnosed by a dealer service department. **Note:** *These two checks simply test for power and reference voltage to the airflow sensor. If the checks are correct and the airflow sensor is still suspect, have the airflow meter checked by a dealer service department.*

Replacement

32 Disconnect the electrical connector from the sensor. Remove the intake duct from the sensor and remove the sensor and air cleaner housing top cover from the vehicle.

33 Remove the screws retaining the sensor to the housing and remove the sensor.

34 Installation is the reverse of removal.

Mass Airflow (MAF) sensor (1994 and later models)

General description

35 The airflow sensor is located in the air intake duct. The sensor uses a hot wire sensing element to measure the amount of air entering the intake system. The air passing over the hot wire causes it to cool. Consequently, this change in temperature can be converted into an analog voltage signal to the ECM which in turn, calculates the required fuel injector pulse width. Problems with the MAF sensor or circuit will set a code 8.

Check

Refer to illustration 8.36

36 Disconnect the electrical connector from the MAF sensor and check for reference voltage on the REF terminal with the ignition key ON (engine not running) **(see illustration)**. Check for continuity to ground on the black (GND [-]) wire terminal. Repair the circuits if necessary. **Note:** *These two checks simply test for reference voltage and ground continuity to the MAF sensor. If the checks are correct and the MAF sensor is still suspect, continue with the checks for signal voltage.*

37 Reconnect the connector and using a straight pin, backprobe the SIG wire terminal.

38 With the ignition On, there should be approximately 2.0 volts present. Start the engine and allow it to idle, the voltmeter should now read 1.0 to 2.5 volts. Raise the engine rpm and see if the signal voltage increases slightly (it should). If the voltage is not as specified, check the connectors and the circuit from the ECM to the MAF sensor. If the connectors and circuit are good, replace the MAF sensor.

Replacement

39 Disconnect the electrical connector from the sensor.
40 Loosen the hose clamp and remove the intake duct from the sensor.
41 Remove the screws attaching the sensor to the air cleaner housing and remove the sensor.
42 Installation is the reverse of removal.

Intake Air Temperature (IAT) sensor (1994 and later models)

General description

43 The intake air temperature (IAT) sensor is located inside the air cleaner housing. This sensor is a resistor which changes value according to the temperature of the air entering the engine. Low temperatures produce a high resistance value (for example, at 68 degrees F [20 degrees C] the resistance is 2,000 to 3,000 ohms) while high temperatures produce low resistance values (at 176 degrees F [80 degrees C] the resistance is 200 to 400 ohms. The ECM supplies approximately 5-volts (reference voltage) to the air temperature sensor. The IAT sensor alters the voltage according to the temperature of the incoming air. The signal voltage sent back to the ECM will be high when the air temperature is cold and low when the air temperature is warm. Any problems with the air temperature sensor will usually set a diagnostic code 10.

Check

Refer to illustration 8.45

44 To check the air temperature sensor, disconnect the two terminal electrical connector. Turn the ignition key ON, but do not start the engine.

45 Measure the reference voltage on the REF terminal **(see illustration)**. The meter should read approximately 5-volts.

46 If the reference voltage is not correct, have the ECM diagnosed by a dealer service department or other repair shop.

47 Measure the resistance across the air temperature sensor terminals. The resistance should be HIGH when the air temperature is LOW. Next, start the engine and let it idle. Wait awhile and let the engine reach operating temperature. Turn the ignition OFF, disconnect the air temperature sensor and measure the resistance across the terminals. The resistance should be LOW when the air temperature is HIGH. If the sensor does not exhibit this change in resistance, replace the sensor.

Chapter 6 Emissions and engine control systems

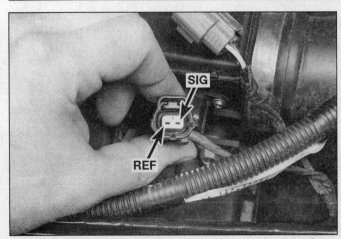

8.45 The IAT sensor is located inside the air cleaner housing

Replacement

48 Disconnect the electrical connector from the sensor.
49 Detach the sensor from the air cleaner housing.
50 Install the sensor and connect the electrical connector.

Crankshaft Position Sensor (1994 and later models)

General description

51 The crankshaft position sensor is located in the distributor. Any diagnostic or repair must be performed by a dealer service department or other qualified automotive repair facility. In the event of crankshaft position sensor problems, the distributor must be replaced as a single unit. Refer to Chapter 5 for additional information. **Note:** *The crankshaft position sensors (CKP and CID) read crankshaft position and engine rpm and relay the signals electronically to the computer. These two sensors are contained within the distributor on 1994 and 1995 models. On 1996 and 1997 models the crankshaft sensor is located near the front oil pump. It reads crankshaft position off the timing pulley. Cylinder detection is detected by the camshaft sensor which is located in the distributor. These sensors are difficult to check - have the ignition system tested by a dealer service department or other qualified automotive repair facility.*

Vehicle Speed Sensor (VSS)

General description

52 The Vehicle Speed Sensor (VSS) is an integral component of the speedometer. As the vehicle drives down the road, the speedometer gear assembly is turned by the speedometer cable. The speedometer cable rotates the speedometer and the VSS simultaneously, giving a speed signal to the ECM. Any problems with the VSS will set a code 6 and must be diagnosed by a dealer service department or other qualified repair shop.

Power steering pressure switch

General description

53 Turning the steering wheel increases power steering fluid pressure and engine load. The pressure switch which is located in the power steering pump will close before the load can cause an idle problem. A pressure switch that will not open or an open circuit from the ECM will cause timing to retard at idle, affecting idle quality. A pressure switch that will not close or an open circuit may cause the engine to die when the power steering system is used heavily.

Check

54 Disconnect the electrical connector to the pressure switch.
55 Connect an ohmmeter across the terminals of the pressure switch. Start the engine and check for continuity as the steering wheel is turned. There should be no continuity when the steering wheel is not being turned.
56 If the switch fails this test, replace the switch.

EGR gas temperature sensor

General description

57 Some models are equipped with an EGR gas temperature sensor mounted near the EGR valve, installed into the EGR tube. This sensor detects the temperature of the exhaust as it flows through the EGR valve. The information is sent to the ECM and in turn the EGR on/off time is regulated precisely and more efficiently. Any malfunction of the EGR gas temperature sensor will set a code 16.

Check

58 Disconnect the electrical connector for the EGR gas temperature sensor and measure the resistance of the sensor. Resistance should decrease as temperature increases.

Removal and installation

59 Disconnect the harness connector for the EGR gas temperature sensor and using and open-end wrench, remove the sensor from the intake manifold.
60 Installation is the reverse of removal.

Crankshaft position sensor (CKP) (1990 through 1993 models)

General information

61 The crankshaft position sensor is mounted in the distributor along with the pick-up coil (see Chapter 5). This type of crankshaft sensor uses a vane with four equally spaced slots which rotates with the distributor shaft. The vane slots pass through a photo optic sensor which senses the slots and converts this interval into an electronic signal that relays crankshaft position to the ECM. Have the crankshaft sensor diagnosed by a dealer service department or other qualified repair shop.

Camshaft position sensor (CMP) (1996 and later models)

General information

62 The camshaft position sensor is located inside the distributor. This Hall Effect switching device relays camshaft position to the ECM. Have the camshaft position sensor diagnosed by a dealer service department. See Chapter 5 for additional information. **Note:** *The crankshaft position sensors (CKP and CID) read crankshaft position and engine rpm and relay the signals electronically to the computer. These two sensors are contained within the distributor on 1994 and 1995 models. On 1996 and 1997 models the crankshaft sensor is located near the front oil pump. It reads crankshaft position off the timing pulley. Cylinder detection is detected by the camshaft sensor which is located in the distributor. These sensors are difficult to check, have the ignition system tested by a dealer service department or other qualified automotive repair facility.*

Clutch Pedal Position switch (manual transaxle)

General description

63 The Clutch Pedal Position switch is located on the clutch release arm (see Chapter 8) under the driver's dash. The clutch pedal position switch acts as a start inhibitor, preventing power from reaching the starter relay until the clutch pedal is depressed. This switch works similar to the Neutral Start switch on automatic transaxle models. These systems work in conjunction with the ignition switch, starter relay (electronic) and starter assembly.

Check

64 Working on the clutch pedal position switch harness connector (ignition switch side), use a voltmeter and with the ignition key ON

Chapter 6 Emissions and engine control systems

9.4 The ECM is located under the dash area - CAREFULLY disconnect the electrical connectors before removal of mounting hardware

10.1a Location of the EVAP system charcoal canister on carbureted engines

10.1b Location of the charcoal canister on fuel-injected engines

(engine not running), check for power to the switch. There should be voltage present.

65 Working on the clutch pedal position switch side of the connector, use an ohmmeter to check the resistance with the pedal up and then check it with the pedal depressed. The switch should register high resistance to zero resistance as the switch is closed. If there is no difference, replace the switch.

66 If both these test results are correct, check the starter system (see Chapter 5).

Adjustment

67 Follow the clutch pedal position switch replacement and adjustment procedures in Chapter 8.

Neutral Start switch (automatic transaxle)

General description

68 The Neutral Start switch, located on the transaxle indicates to the ECM when the transaxle is in Park, Neutral, Drive or Reverse. This information is used for starting, Torque Converter Clutch (TCC), Exhaust Gas Recirculation (EGR) and Idle Air Control (IAC) valve operation. For example, if the signal wire(s) become grounded, it may be difficult to start the engine in Park or Neutral. Because this information sensor requires a SCAN tool to diagnose any problems, have the system tested by a dealer service department or other qualified repair facility.

Adjustment

69 Follow the transaxle shift control cable adjustment in Chapter 7 and check for a distinct "click" when the shift lever selects each gear (Park, Reverse, Neutral, Drive etc.).

70 The shift button should release smoothly and there should not be any cable binding preventing smooth transition between gears.

9 Electronic Control Module (ECM) - removal and installation

Refer to illustration 9.4

1 Disconnect the negative cable from the battery.
2 Remove the trim panels from below the dash area (see Chapter 11).
3 Carefully disconnect the electrical connectors from the ECM. Each connector has a locking tab which must be disengaged before the connector is unplugged.
4 Remove the nuts and bolts from the ECM brackets **(see illustration)**.
5 Lift the ECM from the vehicle.
6 Installation is the reverse of removal.
7 Securely tighten the ECM retaining fasteners during installation.

10 Evaporative Emission Control (EVAP) system

General description

Refer to illustrations 10.1a and 10.1b

1 This system is designed to trap and store fuel that evaporates from the fuel tank, throttle body and intake manifold that would normally enter the atmosphere in the form of hydrocarbon (HC) emissions.

2 The Evaporative Emission Control (EVAP) system consists of a charcoal-filled canister **(see illustrations)**, the lines connecting the canister to the fuel tank, a temperature controlled vacuum valve and a check valve.

3 Fuel vapors are transferred from the fuel tank and throttle body to a canister where they are stored when the engine isn't running. When the engine is running, the fuel vapors are purged from the canister by intake airflow and consumed in the normal combustion process. **Note:** *Early systems use engine vacuum and purge control valves to control the fuel vapors in and out of the charcoal canister. Later systems use electronic purge control solenoids. Have these later systems diagnosed by a dealer service department.*

4 The charcoal canister on early systems is equipped with a check valve that incorporates three check balls. Depending upon the running conditions and the pressure in the fuel tank, the check balls open and close the passageways to the vacuum valve (consequently the throttle body) and fuel tank.

Check

5 Poor idle, stalling and poor driveability can be caused by an inoperative check valve, a damaged canister, split or cracked hoses or hoses connected to the wrong fittings. Check the fuel filler cap for a damaged or deformed gasket (see Chapter 1).

6 Evidence of fuel loss or fuel odor can be caused by liquid fuel leaking from fuel lines, a cracked or damaged canister, an inoperative check valve, disconnected, misrouted, kinked, deteriorated or damaged vapor or control hoses.

7 Inspect each hose attached to the canister for kinks, leaks and cracks along its entire length. Repair or replace as necessary.

8 Look for fuel leaking from the bottom of the canister. If fuel is leaking, replace the canister and check the hoses and hose routing.

9 Inspect the canister. If it's cracked or damaged, replace it.

10 Check for a clogged filter or a stuck check valve. Using low pres-

Chapter 6 Emissions and engine control systems

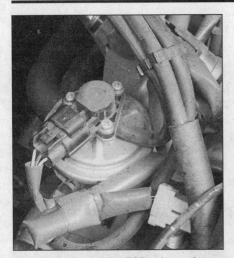

11.1 Location of the EGR valve on later fuel-injected systems

11.4 You can check the EGR valve by trying to move the diaphragm with the tip of your finger to make sure it moves smoothly without binding (this should also cause the engine to stumble when it is running)

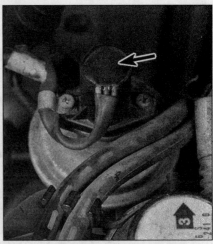

11.7 Location of the EGR valve position sensor (arrow) on an early carbureted engine

sure compressed air, blow into the canister tank pipe. Air should flow freely from the other pipes. If a problem is found, replace the canister.
11   Check the operation of the number 1 purge control valve (large, round valve on top of charcoal canister). With the engine completely cold, use a hand-held pump and apply 4.3 in-Hg to the number 1 purge control valve and observe that air is allowed to pass through port A (small port directly next to the number 1 purge control valve). Air should pass through freely. If not, replace the number 1 purge control valve.

Charcoal canister replacement

12   Clearly label, then detach the vacuum hoses from the canister.
13   Remove the mounting clamp bolts, lower the canister with the bracket, disconnect the hoses from the check valve and remove it from the vehicle.
14   Installation is the reverse of removal.

11 Exhaust Gas Recirculation (EGR) system

General description

Refer to illustration 11.1

1   To reduce oxides of nitrogen emissions, some of the exhaust gases are recirculated through the EGR valve to the intake manifold to lower combustion temperatures **(see illustration)**.
2   The EGR system consists of the EGR valve, the EGR vacuum amplifier, vacuum switching valves (VSV), the Electronic Control Module (ECM) and the EGR gas temperature sensor.

Check

EGR valve

Refer to illustration 11.4
3   Start the engine and allow it to idle.
4   Push up on the EGR valve diaphragm with your finger. The engine should stumble and may even die. **Warning:** *Be careful - the valve may be very hot. If so, wait until the engine has cooled down before performing this test.*
5   Detach the vacuum hose from the EGR valve and attach a hand-held vacuum pump in its place. Apply vacuum to the EGR valve. Vacuum should remain steady and the engine should run poorly.
  a) *If the vacuum does not remain steady and the engine does not run poorly, replace the EGR valve and recheck it.*
  b) *If the vacuum remains steady but the engine does not run poorly, remove the EGR valve and check the valve and the intake mani-*

fold for blockage. Clean or replace parts as necessary and recheck.

EGR valve position sensor (early models)

Refer to illustration 11.7
6   Disconnect the EGR valve position sensor connector.
7   Measure the resistance between terminals at the sensor connector with an ohmmeter **(see illustration)**. The meter readings should fluctuate as the EGR valve is raised and lowered. Use the tip of your finger to raise and lower the EGR diaphragm and be sure the EGR valve is completely cooled down.
8   If the EGR valve position sensor does not indicate any resistance changes, replace it with a new part.
9   Connect the EGR valve position sensor connector.

12 Positive Crankcase Ventilation (PCV) system

General information

Refer to illustration 12.1

1   The Positive Crankcase Ventilation (PCV) system reduces hydrocarbon emissions by scavenging crankcase vapors. It does this by circulating fresh air from the air cleaner through the crankcase, where it mixes with blow-by gases and is then rerouted through a PCV valve to the intake manifold **(see illustration)**.

12.1 Location of the PCV valve on a later fuel-injected engine

6-16 Chapter 6 Emissions and engine control systems

13.1 Typical engine control components on a feedback carburetor system

1. Vacuum switch
2. EGR solenoid number 1
3. High Altitude Compensation (HAC) valve
4. MAP sensor
5. A/C and E/L solenoid valves
6. Air Control Valve
7. O2 sensor
8. HAI vacuum motor
9. Anti-backfire valve

2 The main components of the PCV system are the PCV valve, a fresh air intake and the vacuum hoses connecting these components to the engine.

3 To maintain idle quality, the PCV valve restricts the flow when the intake manifold vacuum is high. If abnormal operating conditions (such as piston ring problems) arise, the system is designed to allow excessive amounts of blow-by gases to flow back through the crankcase vent tube into the air cleaner to be consumed by normal combustion.

4 This system directs the blow-by into the throttle body which, over time, can cause an oily residue build up in the area near the throttle plate. Consequently, it is a good idea to periodically clean this residue from the throttle body. Refer to Chapter 4 for this cleaning procedure.

Check

5 To check the valve, first pull it out of the grommet in the valve cover and shake the valve. It should rattle, indicating that it's not clogged with deposits. If the valve does not rattle, replace it with a new one.

6 Start the engine and allow it to idle, then place your finger over the valve opening. If vacuum is felt, the PCV valve is working properly. If no vacuum is felt, the PCV valve may be bad or the hose may be plugged. Also, check for vacuum leaks at the valve, filler cap and all the hoses.

Replacement

7 Pull straight up on the valve to remove it. Check the rubber grommet for cracks and distortion. If it's damaged, replace it.

8 If the valve is clogged, the hose is also probably plugged.

Remove the hose and clean it with solvent.

9 After cleaning the hose, inspect it for damage, wear and deterioration. Make sure it fits snugly on the fittings.

10 If necessary, install a new PCV valve.

11 Install the clean PCV hose. Make sure that the PCV valve and hose are secure.

13 Feedback carburetor system

Warning: *Gasoline is extremely flammable, so take extra precautions when you work on any part of the fuel system. Don't smoke or allow open flames or bare light bulbs near the work area, and don't work in a garage where a natural gas-type appliance (such as a water heater or a clothes dryer) with a pilot light is present. Since gasoline is carcinogenic, wear latex gloves when there's a possibility of being exposed to fuel, and, if you spill any fuel on your skin, rinse it off immediately with soap and water. Mop up any spills immediately and do not store fuel-soaked rags where they could ignite. The fuel system on fuel-injected models is under constant pressure, so, if any fuel lines are to be disconnected, the fuel pressure in the system must be relieved first (see Section 2). When you perform any kind of work on the fuel system, wear safety glasses and have a Class B type fire extinguisher on hand.*

General information

Refer to illustration 13.1

1 The electronically controlled carburetor emission system, also known as the electronic feedback carburetor system, relies on an elec-

Chapter 6 Emissions and engine control systems

tronic signal which is generated by an exhaust gas oxygen sensor, to control a variety of devices and keep emissions within limits **(see illustration)**. The system works in conjunction with a three-way catalyst to control the levels of carbon monoxide, hydrocarbons and oxides of nitrogen. The feedback carburetor system also works in conjunction with the computer. The two systems share certain sensors and output actuators; therefore, diagnosing the feedback carburetor system will require a thorough check of all the feedback carburetor components (refer to Chapter 6 for additional information).

2 The system operates in two modes: open loop and closed loop. When the engine is cold, the air/fuel mixture is controlled by the computer in accordance with a program designed in at the time of production. The air/fuel mixture during open loop mode will be richer to allow for proper engine warm-up. When the engine is at operating temperature, the system operates in closed loop and the air/fuel mixture is varied depending on the information supplied by the exhaust gas oxygen sensor.

3 Here is a list of the various sensors and output actuators involved with these feedback carburetor systems:

Air bleed control valve
Atmospheric pressure switch (see Section 6)
Bowl vent solenoid valve
Clutch switch (see Chapter 7)
Slow fuel cut solenoid (see Section 4)
Positive Temperature Coefficient (PTC) Heater or Electronic Fuel Evaporator (EFE) (see Section 2)
Engine Coolant Temperature sensor (see Section 8)
Engine Coolant Temperature switch
EGR valve position sensor (California models only) (see Section 8)
Idle switch
Manifold Absolute Pressure sensor
Vacuum solenoid valve
Vacuum switch

4 Have the feedback carburetor system diagnosed by a dealer service department or other qualified automotive repair facility.

14 Catalytic converter

Note: *Because of a Federally mandated extended warranty which covers emissions-related components such as the catalytic converter, check with a dealer service department before replacing the converter at your own expense.*

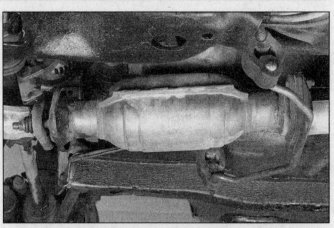

14.2 Let the catalytic converter cool, then spray penetrating lubricant onto the catalytic converter mounting bolts before attempting to unscrew them

General description

Refer to illustration 14.2

1 To reduce hydrocarbon, carbon monoxide and oxides of nitrogen emissions, all vehicles are equipped with a three-way catalyst system which oxidizes and reduces these chemicals, converting them into harmless nitrogen, carbon dioxide and water.

2 The catalytic converter is mounted in the exhaust system much like a muffler **(see illustration)**.

Check

3 Periodically inspect the catalytic converter-to-exhaust pipe mating flanges and bolts. Make sure that there are no loose bolts and no leaks between the flanges.

4 Look for dents in or damage to the catalytic converter protector. If any part of the protector is damaged or dented enough to touch the converter, repair or replace it.

5 Inspect the heat insulator for damage. Make sure that there is adequate clearance between the heat insulator and the catalytic converter.

Replacement

6 To replace the catalytic converter, refer to Chapter 4.

Chapter 6 Emissions and engine control systems

Notes

Chapter 7 Part A
Manual transaxle

Contents

	Section
Back-up light switch - check and replacement	3
General information	1
Lubricant change	See Chapter 1
Lubricant level check	See Chapter 1
Manual transaxle - removal and installation	6
Manual transaxle overhaul - general information	7
Oil seal replacement	4
Shift lever - removal and installation	2
Transaxle mount - check and replacement	5

Specifications

Torque specifications

Ft-lbs (unless otherwise indicated)

Back-up light switch	15 to 21
Speedometer driven gear hold-down bolt	71 to 97 in-lbs
Stabilizer bar and support-to-transaxle nut	28 to 38
Shift rod clevis-to-shift lever nut/bolt	144 to 204 in-lbs
Shift rod-to-selector shift rod adjustment sleeve nut	23 to 34
Transaxle-to-engine bolts	
Flywheel housing reinforcing plate bolts	62 to 86 in-lbs
Upper and lower transaxle bolts	47 to 66
Transaxle-to-engine bracket bolts (all)	27 to 38

1 General information

The vehicles covered by this manual are equipped with a 5-speed manual transaxle or 4-speed automatic transaxle. Information on the manual transaxle is included in this Part of Chapter 7. Service procedures for the automatic transaxle are contained in Chapter 7, Part B.

The manual transaxle is a compact, two-piece, lightweight aluminum alloy housing containing both the transmission and differential assemblies.

Because of the complexity, unavailability of replacement parts and special tools necessary, internal repair procedures for the manual transaxle are beyond the scope of this manual. For readers who wish to tackle a transaxle rebuild, a brief *Manual transaxle overhaul - general information* Section is provided. The bulk of information in this Chapter is devoted to removal and installation procedures.

Chapter 7 Part A Manual transaxle

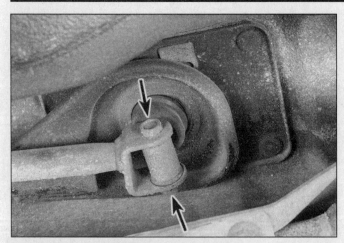

2.4 To disconnect the shift rod from the shift lever, remove this nut and bolt (arrows)

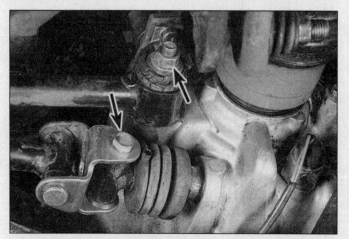

2.5 To disconnect the stabilizer bar and support from the transaxle, remove this nut (upper arrow) and bolt; to disconnect the shift rod from the selector rod, remove this nut and bolt (lower arrow)

2 Shift lever - removal and installation

Refer to illustrations 2.4 and 2.5

1 Remove the center console (see Chapter 11).
2 Remove the shift lever boot retaining screws, the shift lever boot and the boot retainer.
3 Raise the vehicle and support it securely on jackstands.
4 From under the vehicle, disconnect the shift rod from the shift lever (see illustration).
5 Disconnect the shift lever stabilizer bar and support from the transaxle (see illustration).
6 Remove the shift lever and the stabilizer bar and support as a single assembly.
7 To disengage the shift lever from the stabilizer bar and support, remove the shift lever seal, pry out the shift lever support shaft spring, then remove the shift lever support spring, the lower boot, the bushing and the bushing boot from the stabilizer bar and support. Note the order in which the spring, lower boot, bushing and bushing boot are removed, then slide them off the shift lever.
8 Inspect the bushing in the forward end of the stabilizer bar and support, the bushing at the base of the shift lever, and all rubber boots, for wear and damage. Replace bushings and boots as necessary.
9 Installation is the reverse of removal. When reassembling the shift lever to the stabilizer bar and support, make sure that you install the bushing boot, the bushing, the lower boot and the support spring, in that order. Be sure to tighten all fasteners to the torque listed in this Chapter's Specifications.

3 Back-up light switch - check and replacement

Refer to illustration 3.1

1 The back-up light switch (see illustration) is located on the front of the transaxle case.

Check

2 Turn the ignition key to the On position and move the shift lever to the Reverse position. The switch should close the back-up light circuit and turn on the back-up lights.
3 If it doesn't, check the back-up light fuse (see Chapter 12).
4 If the fuse is okay, verify that there's voltage available on the battery side of the switch (with the ignition turned to On).
5 If there's no voltage on the battery side of the switch, check the wire between the fuse and the switch; if there is voltage, put the shift lever in reverse and see if there's voltage on the ground side of the switch.
6 If there's no voltage on the ground side of the switch, replace the switch (see below); if there is voltage, note whether one or both back-up lights are out.
7 If only one bulb is out, replace it; if they're both out, the bulbs could be the problem, but it's more likely that the wire between the switch and the bulbs has an open somewhere.

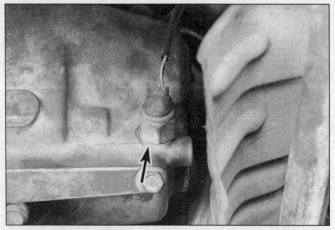

3.1 The back-up light switch (arrow) is located on the front side of the transaxle case

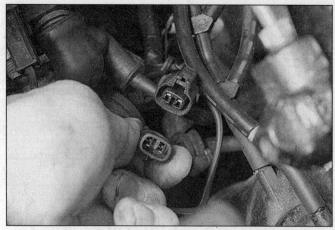

3.8 Before unscrewing the back-up light switch from the transaxle, unplug this electrical connector

4.4 Carefully pry out the side-gear seal with a seal removal tool or a large screwdriver; make sure you don't damage the seal bore or the new seal may leak

4.6 Use a seal installer, a large socket or a piece of pipe to install the new seal

Replacement

Refer to illustration 3.8

8 Unplug the electrical connectors for the back-up light switch **(see illustration)**.
9 Unscrew and remove the old switch.
10 To test the new switch before installation, simply check continuity across the switch terminals: with the plunger depressed, there should be continuity; with the plunger free, there should be no continuity.
11 Using a new O-ring, screw in the new switch and tighten it to the torque listed in this Chapter's Specifications.
12 Plug in the electrical connector.
13 Check the switch to ensure that the circuit is working properly.

4 Oil seal - replacement

1 Oil leaks frequently occur because of worn differential side gear oil seals and/or the speedometer driven gear O-ring and seal. Replacement of the side-gear seals or the speedometer driven gear O-ring and seal is relatively easy, because repairs can be made without removing the transaxle from the vehicle.

Driveaxle oil seals

Refer to illustrations 4.4 and 4.6

2 The driveaxle oil seals are located on the sides of the transaxle, where the inner ends of the driveaxles are splined into the driveaxle side gears. If you suspect that a driveaxle oil seal is leaking, raise the vehicle and support it securely on jackstands. If the seal is leaking, you'll see lubricant on the side of the transaxle, below the seal.
3 Remove the driveaxle (see Chapter 8).
4 Using a screwdriver or pry bar, carefully pry the oil seal out of the transaxle bore **(see illustration)**.
5 If the seal cannot be removed with a screwdriver or pry bar, a special oil seal removal tool (available at auto parts stores) will be required.
6 Using a seal installer, a large section of pipe or a large deep socket as a drift, install the new oil seal. Drive it into the bore squarely and make sure it's completely seated **(see illustration)**. A fully-seated seal should be flush with the surface of the transaxle housing.
7 Lubricate the lip of the new seal with multi-purpose grease, then install the driveaxle (see Chapter 8). Be careful not to damage the lip of the new seal.

Speedometer driven gear O-ring and seal

Refer to illustrations 4.9 and 4.11

8 Raise the vehicle and place it securely on jackstands.
9 Locate the speedometer cable retainer **(see illustration)**. Use a pair of pliers to loosen the retainer, then unscrew it from the driven gear assembly and detach the speedometer cable from the transaxle.
10 Remove the driven-gear hold-down bolt and pull the driven gear assembly straight up to remove it from the transaxle.
11 Remove the old O-ring **(see illustration)** and install a new one.

4.9 The speedometer cable retainer and driven gear assembly are located on the right rear part of the transaxle, right above the shift lever stabilizer bar and support

4.11 Replace this O-ring (arrow) on the speedometer driven gear assembly

5.2 Insert a large screwdriver or prybar between each of the transaxle mounting brackets and the insulator (the rubber part) of each mount, then try to lever the transaxle upward, off the mount.

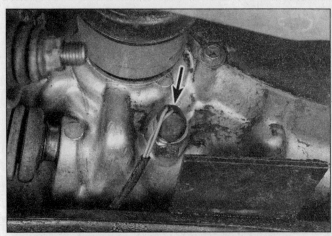

6.3 The Park/Neutral position switch (arrow) is located on the right side of the transaxle, just below the right inner CV joint; trace the leads back to the connector and unplug it

12   To replace the old seal, dig it out with a hooked scribe or a small screwdriver, then tap the new seal into place with a hammer and a small deep socket.

13   Installation is the reverse of removal. Be sure to tighten the driven-gear hold-down bolt to the torque listed in this Chapter's Specifications.

5   Transaxle mount - check and replacement

Check

Refer to illustration 5.2

1   Raise the front of the vehicle and place it securely on jackstands.
2   Insert a large screwdriver or prybar between each of the transaxle mounting brackets and the insulator (the rubber part) of each mount **(see illustration)**, then try to lever the transaxle upward, off the mount.
3   The transaxle should not move excessively away from the mount. If it does, or if the rubber insulator is torn or badly cracked, replace it (see below).
4   Remove the jackstands and lower the vehicle.

Replacement

5   Raise the vehicle and place it securely on jackstands.
6   Support the transaxle with a floor jack.
7   Remove the nut from the insulator-to-bracket bolt, raise the transaxle slightly if necessary, pull out the bolt from the mounting bracket and remove the mount insulator. **Warning:** *Never place your hands between the transaxle and the frame as the jack could slip and serious injury could result. Use a long screwdriver or other tool to remove the mount.*
8   Installation is the reverse of removal. Tighten the insulator-to-bracket nut and bolt to securely.
9   Remove the jackstands and lower the vehicle.

6   Manual transaxle - removal and installation

Removal

Refer to illustrations 6.3, 6.5, 6.14a, 6.14b, 6.15 and 6.17

1   Disconnect the cables from the battery. **Warning:** *When removing the battery cables always detach the negative cable first and hook it up last.*
2   Unplug the electrical connectors for the back-up light switch **(see illustration 3.8)**.

6.5 To detach the ground strap from the transaxle, remove this bolt (arrow)

3   Unplug the electrical connector for the Park/Neutral position switch **(see illustration)**. The switch is on the back side of the transaxle, right below the right inner CV joint, but the connector is near the two connectors for the back-up light switch.
4   Remove the clutch cable adjusting nut, disengage the clutch cable from the release lever and from the cable bracket (see Chapter 8).
5   Remove the ground strap screw and detach the ground strap from the transaxle **(see illustration)**.
6   Remove the starter (see Chapter 5).
7   Unscrew the speedometer cable retainer **(see illustration 4.9)** and disconnect the speedometer cable from the transaxle. If the retainer is difficult to loosen, use a pair of pliers to break it loose.
8   Remove the two upper transaxle-to-engine bolts.
9   Support the engine with an engine hoist. The engine must be securely supported at all times while the transaxle is out of the vehicle.
10   Loosen the front wheel lug nuts, raise the vehicle and support it securely on jackstands. Remove the wheels.
11   Drain the transaxle lubricant into a suitable container (see Chapter 1).
12   Separate the driveaxles from the transaxle (see Chapter 8). Support the inner ends of the driveaxles with wire or rope. **Note:** *It's not necessary to remove the driveaxles from the hubs.*
13   Remove the adjustment sleeve for the transaxle selector shift rod from the input shift shaft and disconnect the shift rod from the transaxle **(see illustration 2.5)**. Remove the shift lever stabilizer bar

Chapter 7 Part A Manual transaxle   7A-5

6.14a To remove the engine/transaxle support member, remove the front transaxle mounting bracket nuts (upper arrows) and the front support retaining bolts (lower arrows) . . .

6.14b . . . then remove the rear engine mounting bracket nuts (upper arrows) and the rear support retaining bolts (lower arrows)

6.15 To detach the rear transaxle-to-engine bracket, remove these three bolts (arrows) (one bolt not visible in this photo)

6.17 To remove the flywheel housing reinforcement plates, remove these bolts (arrows) (two of the three bolts for the front plate not visible in this photo)

nut, lock washer and flat washer, and remove the shift lever stabilizer bar and support **(see illustration 2.5)**.

14 The engine/transaxle support is the large longitudinal crossmember that supports the engine and transaxle. To remove the support, remove the front and rear transaxle mounting bracket-to-support nuts and the four support retaining bolts **(see illustrations)**. Remove the through-bolts and nuts from each mount and remove the front and rear mounts. Unbolt the front and rear mount brackets and remove the brackets.

15 Remove the three bolts from the rear transaxle-to-engine bracket **(see illustration)** and remove the bracket. Remove the three bolts from the front transaxle-to-engine bracket and remove the bracket.

16 Support the transaxle with a jack (preferably a special jack made for this purpose). If you're using a floor jack, be sure to place a wood block between the lifting pad and the transaxle to protect the cast aluminum housing. Safety chains will help steady the transaxle on the jack.

17 Remove the four flywheel housing reinforcing plate bolts **(see illustration)**.

18 Remove the two remaining transaxle bolts.

19 Separate the transaxle from the engine by carefully rolling the jack to the left. Once the input shaft is clear of the splines in the clutch hub, lower the transaxle and remove it from under the vehicle. Try to keep the transaxle as level as possible.

20 The clutch components can now be inspected (see Chapter 8). In most cases, new clutch components should be routinely installed whenever the transaxle is removed. This is also a good time to inspect the condition of the driveaxle boots and make any necessary repairs (see Chapter 8).

Installation

21 Install any clutch components removed (see Chapter 8).

22 With the transaxle secured to the jack as on removal, raise it into position and then carefully slide it toward the engine, engaging the input shaft with the splines in the clutch hub. Do not use excessive force to install the transaxle - if the input shaft does not slide into place, readjust the angle of the transaxle so it is level and/or turn the input shaft so the splines engage properly with the clutch.

23 Install the four flywheel housing reinforcing plate bolts and tighten them to the torque listed in this Chapter's Specifications.

24 Install the two lower transaxle-to-engine bolts and tighten them to the torque listed in this Chapter's Specifications.

25 Remove the transaxle support jack.

26 Install the front and rear transaxle mounting brackets and the mounts and tighten all nuts and bolts securely. Install the engine/transaxle support and tighten the front and rear transaxle mount-to-support nuts and the four support retaining bolts securely.

27 Install the front transaxle-to-engine bracket and tighten the transaxle-to-engine bolts to the torque listed in this Chapter's Specifications. Install the rear transaxle bracket and tighten the three transaxle-to-engine bolts to the torque listed in this Chapter's Specifications.

28 Install the washer on the control rod-to-support bar stud. Install the shift lever stabilizer bar and support on the control rod-to-support bar stud. Install the washer, lock washer and shift lever stabilizer bar and support nut. Tighten the shift lever stabilizer bar and support nut to the torque listed in this Chapter's Specifications.

29 Position the transaxle shift rod and clevis on the input shift shaft and install the selector shift rod adjustment sleeve. Tighten the transaxle selector shift rod adjustment sleeve nut to the torque listed in this Chapter's Specifications.

30 Route the Park/Neutral position switch wiring harness over the top of the rear engine support.

31 Reattach the driveaxles to the transaxle (see Chapter 8).

32 Install the wheels. Remove the jackstands, lower the vehicle and tighten the wheel lug nuts to the torque listed in Chapter 1 Specifications.

33 Install the two upper transaxle-to-engine bolts and tighten them to the torque listed in this Chapter's Specifications. One of these bolts is installed through the heater pipe bracket.

34 Attach the ground strap to the transaxle.

35 Reconnect the speedometer cable to the threaded sleeve on the transaxle and *hand tighten* the retainer.

36 Install the starter (see Chapter 5).

37 Route the clutch cable correctly, so that it has no kinks or sharp bends in it, then thread it through the cable bracket, reattach it to the release lever, install the adjusting nut and adjust the cable (see Chapter 8).

38 Plug in the electrical connectors for the Park/Neutral position switch and for the backup light switch.

39 Fill the transaxle with lubricant (see Chapter 1).

40 Connect the battery cables. **Warning:** *When connecting the battery cables always attach the positive cable first.*

41 Road test the vehicle to check for proper transaxle operation and check for leakage.

7 Manual transaxle overhaul - general information

1 Overhauling a manual transaxle is a difficult job for the do-it-yourselfer. It involves the disassembly and reassembly of many small parts. Numerous clearances must be precisely measured and, if necessary, changed with select-fit spacers and snap-rings. As a result, if transaxle problems arise, it can be removed and installed by a competent do-it-yourselfer, but overhaul should be left to a transmission repair shop. Rebuilt transaxles may be available - check with your dealer parts department and auto parts stores. At any rate, the time and money involved in an overhaul is almost sure to exceed the cost of a rebuilt unit.

2 Nevertheless, it's not impossible for an inexperienced mechanic to rebuild a transaxle if the manufactures shop manuals and special tools are available and the job is done in a deliberate step-by-step manner so nothing is overlooked.

3 The tools necessary for an overhaul include internal and external snap-ring pliers, a bearing puller, a slide hammer, a set of pin punches, a dial indicator and possibly a hydraulic press. In addition, a large, sturdy workbench and a vise or transaxle stand will be required.

4 During disassembly of the transaxle, make careful notes of how each piece comes off, where it fits in relation to other pieces and what holds it in place. Your notes plus the manufacturer's shop manual, which contains exploded views, will make it much easier to get the transaxle back together.

5 Before taking the transaxle apart for repair, it will help if you have some idea what area of the transaxle is malfunctioning. Certain problems can be closely tied to specific areas in the transaxle, which can make component examination and replacement easier. Refer to the *Troubleshooting* Section at the front of this manual for information regarding possible sources of trouble.

Chapter 7 Part B
Automatic transaxle

Contents

	Section		Section
Automatic transaxle - removal and installation	8	Intermediate band - adjustment	7
Automatic transaxle fluid and filter change	See Chapter 1	Park/neutral start switch - check and replacement	5
Automatic transaxle fluid level check	See Chapter 1	Shift cable - removal, installation and adjustment	4
Diagnosis - general	2	Shift lever - removal and installation	3
General information	1	Shift-lock system - description, check and cable replacement	6

Specifications

General
Fluid type and capacity .... See Chapter 1

Torque specifications
Ft-lbs (unless otherwise indicated)

Intermediate band
- Adjustment bolt .... 108 to 132 in-lbs
- Locknut .... 41 to 59

Park/neutral start switch .... 14 to 19
Torque converter-to-driveplate bolts .... 26 to 36
Engine-to-transaxle bolts
- Festiva models .... 41 to 59
- Aspire models .... 47 to 66
Gusset-to-transaxle bolts .... 27 to 38

1 General information

All vehicles covered in this manual come equipped with either a 4- or 5-speed manual transaxle or a 3-speed automatic transaxle. All information on the automatic transaxle is included in this Part of Chapter 7. Information for the manual transaxle can be found in Part A of this Chapter.

Due to the complexity of the automatic transaxle covered in this manual and to the specialized equipment necessary to perform most service operations, this Chapter contains only those procedures related to general diagnosis, routine maintenance, adjustment and removal and installation.

If the transaxle requires major repair work, it should be left to a dealer service department or an automotive or transmission repair shop. You can, however, remove and install the transaxle yourself and save the expense, even if the repair work is done by a transmission shop.

2 Diagnosis - general

Note: *Automatic transaxle malfunctions may be caused by five general conditions: poor engine performance, improper adjustments, hydraulic malfunctions, mechanical malfunctions or malfunctions in the computer or its signal network. Diagnosis of these problems should always begin with a check of the easily repaired items: fluid level and condition (see Chapter 1), shift cable adjustment and shift lever installation. Next, perform a road test to determine if the problem has been corrected or if more diagnosis is necessary. If the problem persists after the preliminary tests and corrections are completed, additional diagnosis should be done by a dealer service department or transmission repair shop. Refer to the Troubleshooting Section at the front of this manual for information on symptoms of transaxle problems.*

Preliminary checks

1 Drive the vehicle to warm the transaxle to normal operating temperature.
2 Check the fluid level as described in Chapter 1:
   a) If the fluid level is unusually low, add enough fluid to bring the level within the designated area of the dipstick, then check for external leaks (see following).
   b) If the fluid level is abnormally high, drain off the excess, then check the drained fluid for contamination by coolant. The presence of engine coolant in the automatic transmission fluid indicates that a failure has occurred in the internal radiator walls that separate the coolant from the transmission fluid (see Chapter 3).
   c) If the fluid is foaming, drain it and refill the transaxle, then check for coolant in the fluid, or a high fluid level.
3 Check the engine idle speed. **Note:** *If the engine is malfunctioning, do not proceed with the preliminary checks until it has been repaired and runs normally.*
4 Check and, if necessary, adjust the shift cable (see Section 4).
5 Inspect the shift lever-to-shift cable connection under the console (see Section 3) and the manual lever on the transaxle (see Section 5). Make sure that both are operating properly and smoothly.

Fluid leak diagnosis

6 Most fluid leaks are easy to locate visually. Repair usually consists of replacing a seal or gasket. If a leak is difficult to find, the following procedure may help.
7 Identify the fluid. Make sure it's transmission fluid and not engine oil or brake fluid (automatic transmission fluid is a deep red color).
8 Try to pinpoint the source of the leak. Drive the vehicle several miles, then park it over a large sheet of cardboard. After a minute or two, you should be able to locate the leak by determining the source of the fluid dripping onto the cardboard.
9 Make a careful visual inspection of the suspected component and the area immediately around it. Pay particular attention to gasket mating surfaces. A mirror is often helpful for finding leaks in areas that are hard to see.
10 If the leak still cannot be found, clean the suspected area thoroughly with a degreaser or solvent, then dry it.
11 Drive the vehicle for several miles at normal operating temperature and varying speeds. After driving the vehicle, visually inspect the suspected component again.
12 Once the leak has been located, the cause must be determined before it can be properly repaired. If a gasket is replaced but the sealing flange is bent, the new gasket will not stop the leak. The bent flange must be straightened.
13 Before attempting to repair a leak, check to make sure that the following conditions are corrected or they may cause another leak. **Note:** *Some of the following conditions cannot be fixed without highly specialized tools and expertise. Such problems must be referred to a transmission shop or a dealer service department.*

Gasket leaks

14 Check the pan periodically. Make sure the bolts are tight, no bolts

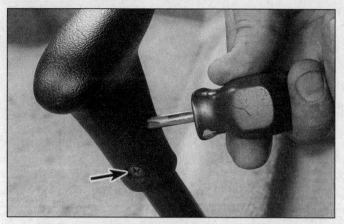

3.2 To remove the shift lever knob, remove these two screws (arrow)

are missing, the gasket is in good condition and the pan is flat (dents in the pan may indicate damage to the valve body inside).
15 If the pan gasket is leaking, the fluid level or the fluid pressure may be too high, the vent may be plugged, the pan bolts may be too tight, the pan sealing flange may be warped, the sealing surface of the transaxle housing may be damaged, the gasket may be damaged or the transaxle casting may be cracked or porous. If sealant instead of gasket material has been used to form a seal between the pan and the transaxle housing, it may be the wrong sealant.

Seal leaks

16 If a transaxle seal is leaking, the fluid level or pressure may be too high, the vent may be plugged, the seal bore may be damaged, the seal itself may be damaged or improperly installed, the surface of the shaft protruding through the seal may be damaged or a loose bearing may be causing excessive shaft movement.
17 Make sure the dipstick tube seal is in good condition and the tube is properly seated. Periodically check the area around the speedometer sensor for leakage. If transmission fluid is evident, check the O-ring for damage.

Case leaks

18 If the case itself appears to be leaking, the casting is porous and will have to be repaired or replaced.
19 Make sure the oil cooler hose fittings are tight and in good condition.

Fluid comes out vent pipe or fill tube

20 If this condition occurs the possible causes are, the transaxle is overfilled, there is coolant in the fluid, the case is porous, the dipstick is incorrect, the vent is plugged or the drain-back holes are plugged.

3 Shift lever - removal and installation

Removal

Refer to illustration 3.2, 3.3a, 3.3b, 3.4, 3.6a and 3.6b

Warning: *If the vehicle is equipped with an airbag, disconnect the negative battery cable and wait two minutes before working in the vicinity of the impact sensors, steering column or instrument panel to avoid the possibility of accidental deployment of the airbag(s), which could cause personal injury (see Chapter 12).*

Note: *The illustrations accompanying this Section depict a shift lever assembly used in an Aspire model; aside from a slightly different shift cable/shift lever connection, the shift lever setup in older Festiva models is similar to the unit shown here.*

1 Remove the center console (see Chapter 11).
2 On Festiva models, back off the locknut and unscrew the shift lever knob. On Aspire models, loosen the shift knob set screws **(see**

Chapter 7 Part B Automatic transaxle

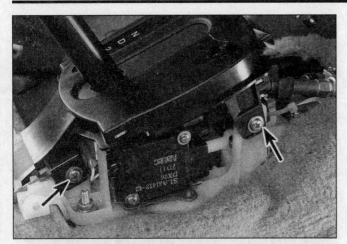

3.3a To detach the shift lever trim bezel, remove the two retaining screws (arrows) from the right side and the other from the left side (not shown) . . .

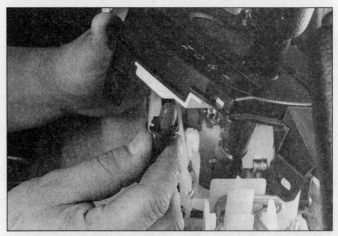

3.3b . . . lift up the bezel and pull out the shift indicator bulb

illustration) and remove the knob from the shifter.
3   Remove the gear position indicator trim bezel screws, lift up the bezel, remove the gearshift indicator bulb from the bezel and remove the bezel **(see illustrations)**.
4   Disconnect the shift cable from the shift lever and from the shift lever base **(see illustration)**.
5   Disconnect the shift/interlock cable from the shift lever assembly **(see illustration 3.4)**.
6   Unplug the electrical connectors behind the shift lever base, remove the shift lever base retaining nuts **(see illustrations)**.
7   On Aspire models, the shift lever assembly can now be removed.
8   On Festiva models, the shift lever is removed from under the vehicle. First, bend back the three retaining clips on the shift lever base far enough to clear the floorpan. Then raise the vehicle and place it securely on jackstands. Disconnect the exhaust pipe hangers to allow enough clearance to remove the shift lever assembly. Remove the two shift cable bracket bolts. Remove the shift lever assembly.
9   If you're replacing the shift lever assembly, remove the shift lock actuator retaining screws, remove the actuator and install it on the new shift lever assembly.

Installation

Festiva models

Refer to illustrations 3.16 and 3.23

10 Insert the shift lever assembly up through the hole in the floorpan and, working from inside the vehicle, bend over the three tangs so they

3.4 To detach the shift cable from the shift lever assembly, remove these two bolts (left arrows) and remove the nut (right arrow) that connects the cable end to the pin on the shift lever; to detach the shift-lock cable from the shift lever assembly, remove this locknut (upper arrow), then disengage the cable from the bracket

hook on to the floorpan.
11 Underneath the vehicle, insert the shift cable into the shift lever assembly; make sure the end of the cable passes through the T-joint. Tighten the exhaust pipe hangers. Remove the jackstands and lower

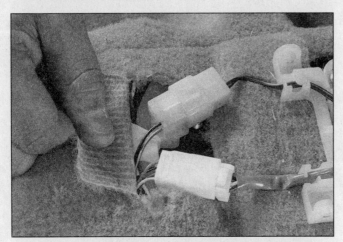

3.6a To remove the shift lever assembly, unplug these two electrical connectors . . .

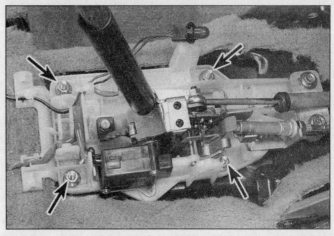

3.6b . . . then remove the four shift lever base retaining nuts

7B-4   Chapter 7 Part B   Automatic transaxle

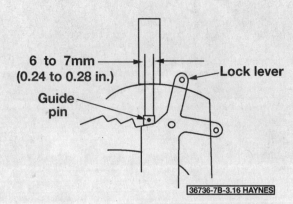

3.16 On Festiva models, lightly press the shift lever pushrod and verify that the overlap between the guide pin and the lock lever is within 0.24 to 0.28 inch

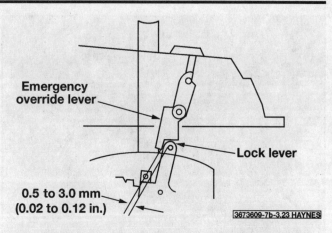

3.23 Verify that the clearance between the lock lever and the emergency override lever are within the indicated clearance

the vehicle.

12   Inside the vehicle, align the holes in the lock lever and guide plate, then install an alignment pin to hold the lock lever in this position **(see illustration)**. Make sure that the shift lock actuator arm engages the lock lever.

13   Attach the cable to the shift lever and install the adjustment nut.

14   Install the four shift lever assembly retaining nuts and tighten them securely.

15   Remove the alignment pin.

16   Lightly press the shift lever pushrod and verify that the overlap between the guide pin and the lock lever is within 0.24 to 0.28 inch **(see illustration)**.

17   Adjust the shift cable (see Section 4).

18   Put the shift lever in the Park position.

19   Align the holes in the slider and the shift indicator bezel, then insert an alignment pin to hole the slider in this position.

20   Route the shift indicator bulb wires through the gear position indicator trim bezel routing clips and insert the bulb in its housing, then turn it counterclockwise to lock it into position.

21   Place the shift indicator trim bezel in position, install the retaining screws and tighten them securely.

22   Remove the alignment pin.

23   Verify that the clearance between the lock lever and the emergency override lever are within the indicated clearance **(see illustration)**.

24   Install the locknut and shift lever knob. Tighten the locknut securely.

25   Install the center console (see Chapter 11).

Aspire models

26   Installation is the reverse of removal.

27   Be sure to adjust the shift cable when you're finished (see Section 4).

4   Shift cable - removal, installation and adjustment

Warning: *If the vehicle is equipped with an airbag, disconnect the negative battery cable, then the positive battery cable and wait two minutes before working in the vicinity of the impact sensors, steering column or instrument panel to avoid the possibility of accidental deployment of the airbag(s), which could cause personal injury (see Chapter 12).*

1   If the shift lever is hard to move from one position to another, disconnect the cable at the transaxle then operate the shift lever with the cable disconnected. If the shift lever now moves smoothly through all positions, adjust the cable.

Removal and installation

Refer to illustrations 4.4 and 4.5

2   Remove the center console and the dash (see Chapter 11).

3   Remove the heater core (see Chapter 3).

4   Remove the cotter pin **(see illustration)** and disengage the shift cable from the shift lever on the transaxle.

5   Remove the locking C-clip **(see illustration)** from the shift cable bracket and disengage the cable from the bracket.

4.4 To detach the shift cable from the manual lever, remove this cotter pin (arrow)

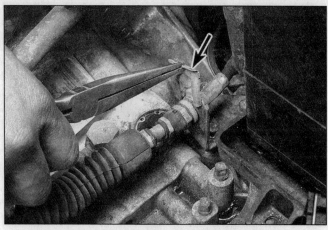

4.5 To detach the shift cable from the cable bracket, remove this locking clip (arrow)

Chapter 7 Part B Automatic transaxle

5.1 The park/neutral start switch (lower arrow) is located on the front of the transaxle; the switch connector (upper arrow) is attached to the bracket for the manual lever selector shaft

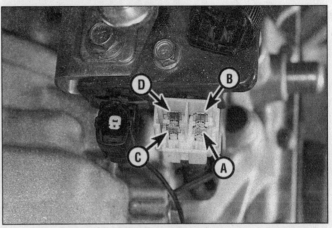

5.2 Terminal guide for the Park/neutral start switch (Aspire models)

A   Ignition switch feed (park/neutral switch)
B   To starter motor
C   Ignition switch feed (back-up light switch)
D   To back-up lights

6   Remove the center console (see Chapter 11).
7   On Festiva models, back off the locknut and unscrew the shift lever knob. On Aspire models, remove the shift knob retaining screws **(see illustration 3.2)** and remove the knob from the shift lever.
8   Remove the shift lever trim bezel and unplug the shift indicator bulb from the bezel **(see illustrations 3.3a and 3.3b)**.
9   Disconnect the shift cable from the shift lever base and from the shift lever **(see illustration 3.4)**.
10   Pry the shift cable grommet from the firewall, pull the cable through the firewall from the passenger compartment side and remove the cable.
11   Installation is the reverse of removal.
12   When you're done, adjust the shift cable as follows.

Adjustment

Festiva models

13   Remove the console (see Chapter 11).
14   Unscrew the shift indicator trim bezel and pull it up the shift lever (to remove it, you'll have to remove the shift lever knob).
15   Loosen the shift cable adjusting nuts.
16   Put the shift lever in the Neutral position.
17   On the transaxle, put the manual lever in the Neutral position by moving the manual lever all the way to Park, then bringing it back two detents (clicks).
18   Tighten the shift cable adjusting nuts evenly and securely.
19   Move the shift lever from Park to 1st gear and verify that all detents can be felt and the shift lever moves freely and smoothly. If it doesn't, readjust it.
20   Install the shift indicator trim bezel and the console.

Aspire models

21   Place the shift lever in the Park (P) position.
22   Remove the shift lever knob and the shift lever trim panel (see Steps 6 through 8).
23   Make sure that the detent spring roller is in the Park (P) detent. Roll the vehicle back and forth to ensure that the detent spring roller is fully engaged with the Park detent.
24   *Loosen - don't remove -* the two shift cable-to-shift lever base retaining bolts **(see illustration 3.4)**.
25   At the transaxle, put the manual lever in the Park position.
26   Make sure that the shift lever inside the vehicle is still in Park.
27   Lightly press the selector cam (the little pointed piece on the upper end of the shift lever) and verify that the guide plate (the "stepped" piece at the lower end of the shift lever) and guide pin clearances are within the indicated dimensions. Put the shift lever in Neutral and Drive and verify that the guide plate and guide clearances are still within these same dimensions. If they're not, readjust the shift cable.
28   When the guide plate and guide pin clearance are within specification for Park, Neutral and Drive, tighten the shift cable-to-shift lever base retaining bolts to the torque listed in this Chapter's Specifications.
29   Verify that the shift lever can be shifted easily and smoothly in all gears. The engine should start only in the Park or Neutral positions.
30   Install the shift lever trim panel and shift knob (see Section 3).

5   Park/neutral start switch - check and replacement

Check

Refer to illustrations 5.1 and 5.2

Note: *Before performing this procedure, be sure the shift cable is correctly adjusted.*

1   Locate the park/neutral start switch electrical connector(s) on top of the transaxle **(see illustration)**. On Festiva models, there are separate connectors for the park/neutral start switch and the back-up light switch; the park/neutral start switch connector has a black wire/blue stripe and a black wire/red stripe; the back-up light switch has a black wire/yellow stripe and a red wire/yellow stripe. Aspire models have a single four-terminal connector, a white wire and a black wire/white stripe for the park/neutral start circuit, and a black wire/yellow stripe and a red wire/green stripe for the back-up light circuit.
2   On Festiva models, backprobe the terminal for the black wire/blue stripe with the positive probe of a voltmeter; on Aspire models, backprobe terminal A of the connector **(see illustration)**. Ground the voltmeter's negative probe, turn the ignition key to On and verify that the terminal is getting power. If it is, turn off the ignition switch and proceed to the next Step. If it isn't, troubleshoot the wire between this terminal and the ignition switch before proceeding (see Wiring Diagrams at the end of Chapter 12).
3   Unplug the switch connector. Place the shift lever in the Park position. Using an ohmmeter, verify that there's continuity between the two terminals of the park/neutral switch on Festiva models, or between terminal A and terminal B on Aspire models. Place the shift lever in the Neutral position and verify that there's continuity between the same two terminals. Put the shift lever in all other gear positions and verify that there is no continuity between these two terminals. If the switch fails any of these continuity checks, replace it.
4   Put the shift lever in the Reverse position and verify that there is continuity between the two terminals for the back-up light switch on Festiva models, or between terminal C and terminal D on Aspire models. If there isn't, replace the switch.

6.2 The shift-lock actuator is housed inside this small plastic box, which is mounted on the shift lever base; to detach the actuator from the shift lever assembly, simply remove the actuator retaining screws and, on Aspire models (shown), separate the actuator from its mounting bracket once it's removed from the shift lever base

Replacement

5 Unplug the switch electrical connector **(see illustration 5.1)**.
6 Place a drain pan under the transaxle; some fluid will leak out when the switch is unscrewed.
7 Using a long extension and a crow's foot wrench, unscrew and remove the switch.
8 Coat the threads of the new switch with thread sealant and install the switch. Tighten the switch to the torque listed in this Chapter's Specifications.
9 Test the switch (see Steps 1 through 4).
10 Plug in the switch electrical connector.
11 Check and, if necessary, add transaxle fluid (see Chapter 1).

6 Shift-lock system - description, check and cable replacement

Warning: *If the vehicle is equipped with an air bag, disconnect the negative battery cable, then the positive battery cable and wait two minutes before working in the vicinity of the impact sensors, steering column or instrument panel to avoid the possibility of accidental deployment of the airbag(s), which could cause personal injury (see Chapter 12).*

6.10 To disconnect the shift-lock cable from the ignition key lock cylinder, remove this bracket bolt (arrow) and disengage the cable eye from the pin on the key lock cylinder

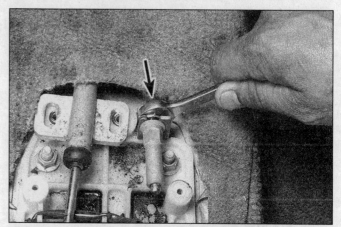

6.7 To detach the shift-lock cable from the shift lever base, loosen the locknut on the front of the cable bracket; unless you plan to discard the cable, don't loosen the adjustment nut behind the bracket - it's preset at the factory

Description

Refer to illustration 6.2

1 When the ignition switch is in the On position, the shift-lock system locks the shift lever; it cannot be moved to any other gear position unless the brake pedal is depressed (and the shift release button is depressed). Conversely, the ignition key cannot be turned to the Lock position when the shift lever is in any position other than Park. And once the ignition key is removed, the shift lever cannot be shifted out of Park.
2 The shift-lock system consists of a cable that connects the shift lever to the ignition key lock cylinder, and an electrically operated shift lock actuator that locks and unlocks the shift lever **(see illustration)**. When the ignition key is turned to the On position, current flows from the ignition switch to the shift lock actuator; when the brake pedal is depressed, current flows to ground through the actuator, moving the lock lever inside the actuator to its "unlock" position, allowing the shift lever to be moved out of the Park position.

Check

3 Place the shift lever in the Park position. The ignition switch should rotate freely from Off to the Lock position. Move the shift lever to the Drive position. The ignition switch should not be able to rotate from Off to the Lock position.
4 With the ignition switch in the Off position, you should not be able to move the shift lever out of the Park position.
5 If the shift-lock system doesn't operate as described, it should be serviced.

Component replacement

Shift-lock cable

Refer to illustrations 6.7 and 6.10

6 Remove the center console (see Chapter 11).
7 Loosen the shift-lock cable locknut in front of the bracket on the shift lever base **(see illustration)**. (If you're planning to install the same shift-lock cable, *i.e.* you're not actually replacing the old cable, do NOT loosen the nut behind the bracket; it's preset at the factory and loosening it will put the cable out of adjustment).
8 Using a flashlight, trace the cable forward up into the dash and remove the cable bracket(s); there are two on Festiva models, one on Aspire models.
9 Remove the steering column covers (see Chapter 11).
10 Remove the shift-lock cable bracket bolt and disengage the shift-lock cable from the pin on the ignition key lock cylinder **(see illustration)**.
11 Connect a three or four-foot long section of mechanic's wire to

Chapter 7 Part B Automatic transaxle

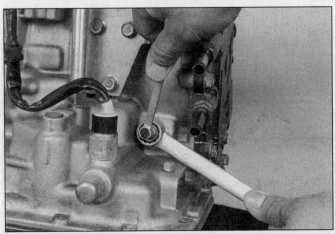

7.2 To adjust the intermediate band, hold the adjustment bolt and loosen the locknut as shown, then tighten the adjustment bolt to 108 to 132 in-lbs, back off the bolt exactly three turns, then hold the bolt again and tighten the locknut to the torque listed in this Chapter's Specifications (transaxle removed from vehicle for clarity

8.12 To remove the driveplate access cover, remove these bolts (arrows)

the shift-lock cable and pull the cable out of the dash. **Caution:** *If you don't use mechanic's wire, you might route the new cable incorrectly.*
12 Disconnect the mechanic's wire from the old shift-lock cable, reattach it to the new cable and route the new cable through the dash.
13 Installation is the reverse of removal.

Shift-lock actuator

14 Remove the center console (see Chapter 11).
15 Remove the gear position indicator trim bezel (see Section 3).
16 Unplug the electrical connector for the shift-lock actuator.
17 On Festiva models, remove the shift-lock actuator retaining screws, disengage the actuator from the shift-lock cable and remove the actuator.
18 On Aspire models, remove the shift lever assembly (see Section 3), remove the shift-lock actuator retaining screws from the underside of the shift lever base, remove the shift-lock actuator and mounting bracket as a single assembly, and unscrew the actuator from the mounting bracket.
19 Installation is the reverse of removal. Make sure that the shift lever assembly is correctly adjusted (see Section 3).

7 Intermediate band - adjustment

Refer to illustration 7.2

1 Raise the front of the vehicle and place it securely on jackstands.
2 Loosen the intermediate band adjustment bolt locknut **(see illustration)**, tighten the adjustment bolt to the torque listed in this Chapter's Specifications, then back it out exactly three turns.
3 Hold the adjuster bolt to prevent it from turning and tighten the locknut to the torque listed in this Chapter's Specifications.
4 Remove the jackstands and lower the vehicle.

8 Automatic transaxle - removal and installation

Removal

Refer to illustrations 8.12, 8.13 and 8.14

1 Disconnect the negative battery cable.
2 Loosen the front wheel lug nuts. Raise the vehicle and place it securely on jackstands. Remove both front wheels.
3 Drain the transaxle fluid (see Chapter 1).
4 Unscrew the speedometer cable retainer and disconnect the

speedometer cable from the driven gear assembly **(see illustration 4.9 in Chapter 7A)**.
5 Unplug all electrical connectors and detach any wiring harness brackets from the transaxle. Detach any ground wires from the transaxle. Detach any vacuum hoses from the transaxle.
6 Disconnect the shift cable from the transaxle manual lever and detach it from the cable bracket (see Section 4). Set the cable aside so that it's out of the way.
7 Support the engine from above with a hoist (see Chapter 2).
8 Separate the driveaxles from the transaxle (see Chapter 8). Support the inner ends of the driveaxles with wire or rope. **Note:** *It's not necessary to remove the driveaxles from the hubs.*
9 Disconnect the oil cooler hoses from the metal oil cooler lines. Plug the ends to avoid contamination.
10 The engine/transaxle support is the large longitudinal crossmember that supports the engine and transaxle. To remove the support, remove the front and rear transaxle mounting bracket-to-support nuts and the four support retaining bolts **(see illustrations 6.14a and 6.14b in Chapter 7A)**. Remove the through-bolts and nuts from each mount and remove the front and rear mounts. Unbolt the front and rear mount brackets and remove the brackets.
11 Remove the gusset plate-to-transaxle bolts.
12 Remove the driveplate access cover **(see illustration)**.
13 Make an alignment mark on the torque converter and the driveplate **(see illustration)**. This will ensure proper alignment during installation.

8.13 Be sure to mark the relationship of the driveplate and torque converter to ensure that they're still dynamically balanced when they're reattached to one another

14 Remove the four driveplate-to-torque converter bolts (see illustration). Rotate the engine crankshaft by turning the damper pulley clockwise to expose the bolts. Remove all four bolts.
15 Remove the starter motor (see Chapter 5).
16 Support the transaxle with a transmission jack, if available, or with a floor jack. Safety chains will help steady the transaxle on the jack.
17 Remove the engine-to-transaxle bolts.
18 Make a final check that all wires and hoses have been disconnected from the transaxle, then move the transaxle and jack toward the side of the vehicle until the transaxle is clear of the engine locating dowels. Make sure you keep the transaxle level as you do this. **Note:** *As soon as the transaxle clears the engine, attach a small C-clamp to the transaxle housing to hold the torque converter to the transaxle during the remainder of the removal procedure.*
19 Lower the transaxle and remove it from under the vehicle.

Installation

Note: *To ensure the torque converter is fully seated, rotate it a couple of turns while pushing in on it. If it isn't seated, it will "clunk" into place (it may even "clunk" more than once). Also lubricate the hub of the torque converter with multi-purpose grease.*
20 With the transaxle secured to the jack, raise it into position and carefully slide it forward. Remove the C-clamp and rotate the torque converter to align the torque converter bolt holes with the driveplate bolt holes, aligning the marks you made in Step 13. Do not use excessive force to install the transaxle - if so something binds and the transaxle won't mate with the engine, alter the angle of the transaxle slightly until it does mate. **Caution:** *Do NOT use the transaxle-to-engine bolts to force the transaxle to the engine. Doing so could crack or damage major components. If you experience difficulties, have an assistant help line up the dowel pins on the engine block with the transaxle. Some wiggling of the engine and/or transaxle will probably be necessary to secure proper alignment of the two components.*
21 Install the engine-to-transaxle bolts and tighten them to the torque listed in this Chapter's Specifications.
22 Remove the transmission jack.
23 Install the starter motor (see Chapter 5).
24 Install the four driveplate-to-torque converter bolts and tighten them to the torque listed in this Chapter's Specifications.
25 Install the driveplate access cover and tighten the bolts securely.
26 Install the gusset plate-to-transaxle bolts and tighten them to the torque listed in this Chapter's Specifications.
27 Install the front and rear transaxle mounting brackets and the

8.14 Jam a screwdriver between the driveplate ring gear and the transaxle housing to immobilize the driveplate so you can loosen the four driveplate-to-torque converter bolts

mounts and tighten all nuts and bolts securely. Install the engine/transaxle support and tighten the front and rear transaxle mount-to-support nuts and the four support retaining bolts securely.
28 Install both driveaxle assemblies (see Chapter 8).
29 Connect the oil cooler hoses to the metal oil cooler lines. Make sure that the hoses overlap the metal lines by at least 1-inch. Tighten the hose clamp screws securely.
30 Install the wheels and lug nuts. Remove the jackstands and lower the vehicle. Tighten the lug nuts to the torque listed in the Chapter 1 Specifications.
31 Remove the engine hoist.
32 Attach the shift cable to the cable bracket and to the transaxle manual lever (see Section 4).
33 Connect any disconnected vacuum hoses. Connect all electrical connectors. Connect the ground wire to the transaxle.
34 Connect the speedometer cable to the driven gear assembly (see Chapter 7A).
35 Connect the negative battery cable.
36 Fill the transaxle with the specified lubricant (see Chapter 1).
37 Road test the vehicle and check for proper transaxle operation and check for fluid leaks. Shut off the engine and recheck the fluids.

Chapter 8
Clutch and driveaxles

Contents

	Section		Section
Clutch - description and check	2	Driveaxle - removal and installation	9
Clutch cable - replacement	3	Driveaxle boot check	See Chapter 1
Clutch components - removal, inspection and installation	4	Driveaxle boot - replacement	10
Clutch pedal height and freeplay check and adjustment	See Chapter 1	Driveaxles - general information and inspection	8
		Flywheel - removal and installation	See Chapter 2
Clutch release bearing and lever - removal, inspection and installation	5	General information	1
		Oil seal - replacement	See Chapter 7B
Clutch start switch - check and replacement	7	Pilot bearing - inspection and replacement	6

Specifications

Clutch
Fluid type	See Chapter 1
Pedal freeplay	See Chapter 1

Driveaxles
Inner CV joint length (measured from the center of each boot clamp)
- Festiva .... 3-1/2 inches
- Aspire .... 4-7/64 inches

Torque specifications
Ft-lbs

Clutch
- Pressure plate-to-flywheel bolts .... 13 to 20
- Release fork-to-pivot shaft bolt .... 26 to 30

Driveaxle/hub locknut .... 116 to 174

Wheel lug nuts .... See Chapter 1

Chapter 8 Clutch and driveaxles

1 General information

The information in this Chapter deals with the components from the rear of the engine to the front wheels, except for the transaxle, which is dealt with in Chapters 7A and 7B. For the purposes of this Chapter, these components are grouped into two categories: clutch and driveaxles. Separate Sections within this Chapter offer general descriptions and checking procedures for both groups.

Since nearly all the procedures covered in this Chapter involve working under the vehicle, make sure it's securely supported on sturdy jackstands or a hoist where the vehicle can be easily raised and lowered.

2 Clutch - description and check

1 All models with a manual transaxle use a single dry plate, diaphragm spring type clutch. The clutch disc has a splined hub which allows it to slide along the splines of the transmission input shaft. The clutch and pressure plate are held in contact by spring pressure exerted by the diaphragm in the pressure plate.
2 The clutch release system is cable operated. The release system includes the clutch pedal, the clutch cable, the release lever and the release bearing.
3 When the clutch pedal is depressed, it pulls the clutch cable, which pulls the release lever. As the lever moved, it slides the release bearing along the input shaft, and the release bearing pushes against the fingers of the diaphragm spring in the pressure plate assembly, which releases the clutch plate.
4 Terminology can be a problem regarding the clutch components because common names have in some cases changed from that used by the manufacturer. For example, the driven plate is also called the clutch plate or disc and the clutch release bearing is sometimes called a throwout bearing.
5 Unless you're replacing components with obvious damage, perform some preliminary checks to diagnose a clutch system malfunction.

 a) *To check "clutch spin down time," run the engine at normal idle speed with the transmission in Neutral (clutch pedal up - engaged). Disengage the clutch (pedal down), wait nine seconds and shift the transmission into Reverse. No grinding noise should be heard. A grinding noise would most likely indicate a problem in the pressure plate or the clutch disc.*
 b) *To check for complete clutch release, run the engine (with the parking brake applied to prevent movement) and hold the clutch pedal approximately 1/4-inch from the floor. Shift the transmission between first gear and Reverse several times. If the shift is not smooth, component failure is indicated.*
 c) *Visually inspect the clutch pedal bushing at the top of the clutch pedal to make sure there is no sticking or excessive wear.*
 d) *A clutch pedal that is difficult to operate is most likely caused by a faulty clutch cable. Check the cable where it enters the casing for fraying, rust or other signs of corrosion. If it looks good, lubricate the cable with penetrating oil. If pedal operation improves, the cable is worn out and should be replaced.*

3 Clutch cable - replacement

Refer to illustration 3.1

1 In the engine compartment, remove the clutch cable adjuster nut **(see illustration)** and disconnect the cable from the release lever.
2 Disengage the clutch cable from the cable bracket.
3 Working inside the vehicle, use a flashlight to locate the rear end of the clutch cable under the dash at the top of the clutch pedal.
4 Remove the clutch cable C-clip and disengage the cable from the bracket.
5 Pull up on the clutch pedal (this lowers the upper end of the pedal and moves it toward the clutch cable) and disengage the cable clevis from the upper end of the pedal.

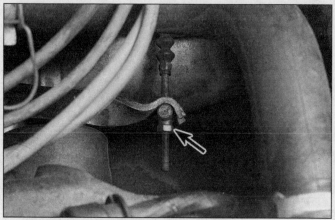

3.1 Remove the adjuster nut (arrow) and disengage the clutch cable from the clutch release lever

6 Pry out the cable grommet at the firewall and pull the cable through the firewall and into the engine compartment.
7 Installation is the reverse of removal.
8 Adjust the clutch pedal height and freeplay when you're done (see Chapter 1).

4 Clutch components - removal, inspection and installation

Warning: *Dust produced by clutch wear and deposited on clutch components may contain asbestos, which is hazardous to your health. DO NOT blow it out with compressed air and DO NOT inhale it. DO NOT use gasoline or petroleum-based solvents to remove the dust. Brake system cleaner should be used to flush the dust into a drain pan. After the clutch components are wiped clean with a rag, dispose of the contaminated rags and cleaner in a labeled, covered container.*

Removal

Refer to illustrations 4.5 and 4.6

1 Access to the clutch components is normally accomplished by removing the transaxle, leaving the engine in the vehicle. If, of course, the engine is being removed for major overhaul, then the opportunity should always be taken to check the clutch for wear and replace worn components as necessary. However, the relatively low cost of the clutch components compared to the time and labor involved in gaining access to them warrants their replacement any time the engine or transaxle is removed, unless they are new or in near-perfect condition. The following procedures assume that the engine will stay in place.
2 Remove the transaxle from the vehicle (see Chapter 7A). Support the engine while the transaxle is out. Preferably, an engine hoist should be used to support it from above. However, if a jack is used underneath the engine, make sure a piece of wood is used between the jack and oil pan to spread the load. **Caution:** *The pick-up for the oil pump is very close to the bottom of the oil pan. If the pan is bent or distorted in any way, engine oil starvation could occur.*
3 The release lever and release bearing can remain attached to the transaxle for the time being.
4 To support the clutch disc during removal, install a clutch alignment tool through the clutch disc hub.
5 Carefully inspect the flywheel and pressure plate for indexing marks. The marks are usually an X, an O or a white letter. If they cannot be found, scribe marks yourself so the pressure plate and the flywheel will be in the same alignment during installation **(see illustration)**.
6 Slowly loosen the pressure plate-to-flywheel bolts **(see illustration)**. Work in a diagonal pattern and loosen each bolt a little at a time until all spring pressure is relieved. Then hold the pressure plate securely and completely remove the bolts, followed by the pressure plate and clutch disc.

Chapter 8 Clutch and driveaxles

4.5 Mark the relationship of the pressure plate to the flywheel (in case you're going to reuse the same pressure plate)

4.6 To detach the pressure plate from the flywheel, remove these bolts (arrows) gradually and evenly; tighten them the same way, until they're all snug, before torquing them

Inspection

Refer to illustrations 4.9, 4.11a and 4.11b

7 Ordinarily, when a problem occurs in the clutch, it can be attributed to wear of the clutch driven plate assembly (clutch disc). However, all components should be inspected at this time.

8 Inspect the flywheel for cracks, heat checking, score marks and other damage. If the imperfections are slight, a machine shop can resurface it to make it flat and smooth. Refer to Chapter 2 for the flywheel removal procedure.

9 Inspect the lining on the clutch disc. There should be at least 1/16-inch of lining above the rivet heads. Check for loose rivets, distortion, cracks, broken springs and other obvious damage **(see illustration)**. As mentioned above, ordinarily the clutch disc is replaced as a matter of course, so if in doubt about the condition, replace it with a new one.

10 The release bearing and pilot bearing should be replaced along with the clutch disc (see Sections 5 and 6).

11 Check the machined surface and the diaphragm spring fingers of the pressure plate **(see illustrations)**. If the surface is grooved or otherwise damaged, replace the pressure plate assembly. Also check for obvious damage, distortion, cracking, etc. Light glazing can be removed with emery cloth or sandpaper. If a new pressure plate is indicated, new or factory rebuilt units are available.

4.9 Examine the clutch disc for evidence of excessive wear, such as smeared friction material, chewed-up rivets, worn hub splines and distorted damper cushions or springs

4.11a Examine the pressure plate friction surface for score marks, cracks and evidence of overheating (blue spots)

NORMAL FINGER WEAR

EXCESSIVE FINGER WEAR

BROKEN OR BENT FINGERS

4.11b Replace the pressure plate if any of these conditions are noted

4.13 Center the clutch disc in the pressure plate with a clutch alignment tool

5.2 To remove the release bearing, simply slide it off the input shaft

Installation

Refer to illustration 4.13

12 Before installation, carefully wipe the flywheel and pressure plate machined surfaces clean. It's important that no oil or grease is on these surfaces or the lining of the clutch disc. Handle these parts only with clean hands.

13 Position the clutch disc and pressure plate with the clutch held in place with an alignment tool **(see illustration)**. Make sure it's installed properly (most replacement clutch plates will be marked "flywheel side" or something similar - if not marked, install the clutch disc with the damper springs or cushion toward the transaxle).

14 Tighten the pressure plate-to-flywheel bolts only finger tight, working around the pressure plate.

15 Center the clutch disc by ensuring the alignment tool is through the splined hub and into the pilot bearing. Wiggle the tool up, down or side-to-side as needed to bottom the tool. Tighten the pressure plate-to-flywheel bolts a little at a time, working in a crisscross pattern to prevent distortion of the cover. After all of the bolts are snug, tighten them to the torque listed in this Chapter's Specifications. Remove the alignment tool.

16 Using high-temperature grease, lubricate the inner groove of the release bearing (see Section 5). Also apply a light film of grease on the release lever contact areas, the pilot bearing and the transaxle input shaft bearing retainer.

17 Install the clutch release bearing (see Section 5).

18 Install the transaxle and all components removed previously, tightening all fasteners to the proper torque specifications.

5 Clutch release bearing and lever - removal, inspection and installation

Warning: *Dust produced by clutch wear and deposited on clutch components may contain asbestos, which is hazardous to your health. DO NOT blow it out with compressed air and DO NOT inhale it. DO NOT use gasoline or petroleum-based solvents to remove the dust. Brake system cleaner should be used to flush it into a drain pan. After the clutch components are wiped clean with a rag, dispose of the contaminated rags and cleaner in a labeled, covered container.*

Removal

Refer to illustrations 5.2 and 5.3

1 Remove the transaxle (see Chapter 7).

2 Remove the clutch release bearing from the input shaft **(see illustration)**.

3 To remove the release lever and pivot shaft, remove the bolt that attaches the release fork to the shaft **(see illustration)**, then pull out the lever and shaft. (It's not necessary to remove the fork and shaft unless the shaft is bent or damaged, the release fork fingers are bent or damaged, or the pivot shaft bushings or grommet are excessively worn.)

Inspection

Refer to illustration 5.4

4 Hold the bearing by the outer race and rotate the inner race while applying pressure **(see illustration)**. If the bearing doesn't turn smoothly or if it's noisy, replace the bearing/hub assembly with a new one. Wipe the bearing with a clean rag and inspect it for damage, wear and cracks. Don't immerse the bearing in solvent - it's a sealed bearing; soaking or dipping it in solvent will ruin it. Also check the release lever pivot shaft bushings and grommet for cracks and excessive wear.

Installation

Refer to illustrations 5.5, 5.6 and 5.7

5 Fill the inner groove of the release bearing with high-temperature grease and lubricate the friction points for the release fork fingers on the back side of the bearing **(see illustration)**. Also apply a light coat of the same grease to the transaxle input shaft splines and the front bearing retainer.

6 If you removed the release lever pivot shaft and release fork, lubricate the pivot shaft bushings, insert the pivot shaft through the first bushing, then through the release fork, then through the other bushing. Make sure the threaded hole in the fork and in the shaft are aligned **(see illustration)**, then install the fork-to-pivot shaft bolt and tighten it to the torque listed in this Chapter's Specifications.

7 Lubricate the friction pads on the ends of the release fork fingers with high-temperature grease **(see illustration)**.

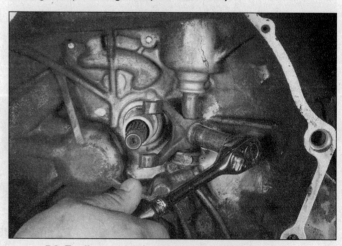

5.3 To disconnect the release fork from the release lever pivot shaft, remove this bolt

Chapter 8 Clutch and driveaxles

5.4 To check the operation of the bearing, hold it by the outer race and rotate the inner race while applying pressure - the bearing should turn smoothly - if it doesn't, replace it

5.5 Fill the release bearing inner groove with high-temperature grease and lubricate the friction surfaces (arrows) for the release fork fingers on the back side of the bearing

5.6 Align the threaded hole in the release fork with the threaded hole in the release lever pivot shaft

8 Slide the release bearing onto the transaxle input shaft. Make sure the release bearing and release fork fingers are properly engaged.
9 Apply a light coat of high-temperature grease to the face of the release bearing where it contacts the pressure plate diaphragm fingers.
10 Install the transaxle (see Chapter 7A).

6 Pilot bearing - inspection and replacement

Refer to illustrations 6.5a, 6.5b and 6.6

1 A pilot bearing, pressed into the rear of the crankshaft, supports the front of the transmission input shaft. The needle roller bearing is greased at the factory and does not require additional lubrication. The pilot bearing should be inspected whenever the clutch components are removed. Due to its inaccessibility, if you are in doubt as to its condition, replace it with a new one.
2 Remove the transmission (see Chapter 7, Part A).
3 Remove the clutch components (see Section 4).
4 Inspect the bearing for excessive wear, scoring, lack of grease, dryness or obvious damage. If any of these conditions are noted, the bearing should be replaced. A flashlight will be helpful to direct light into the recess.
5 The bearing must be pulled from its hole in the crankshaft, gripping the bearing at the rear. Special tools are available, but you may be able to get by with a an alternative tool or fabricated tool. One method that works well is to use a blind-hole puller/slide-hammer; this setup

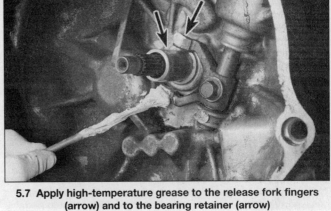

5.7 Apply high-temperature grease to the release fork fingers (arrow) and to the bearing retainer (arrow)

uses a small tip with two adjustable hooks **(see illustration)**. Such tips are commonly available for slide-hammers, and are often included with better-quality slide-hammer kits. If a slide-hammer with the correct hooked tip is not available, try to find a threaded blind-hole puller that will fit into the bearing hole and hook behind the bearing **(see illustration)**.

6.5a Removing the pilot bearing with a slide-hammer and blind-hole puller

6.5b A threaded blind-hole puller is another tool that's handy for removing the pilot bearing

6.6 To install the new pilot bearing, line it up square to the bore, then carefully tap it into place with a seal driver, a large socket or even a clutch alignment tool

7.4 The clutch start switch (left arrow) is attached to a bracket at the upper end of the clutch pedal; that's the brake light switch (right arrow) above the brake pedal

6 To install the new bearing, lightly lubricate the outside surface with multi-purpose grease, then drive it into the recess with a hammer and a bearing driver or a clutch alignment tool **(see illustration)**. Make sure that the bearing seal faces toward the transmission. Don't allow the pilot bearing to become cocked in the bore. Tap it into place until it's flush with the edge of the bearing bore.
7 Lubricate the pilot bearing with high-temperature grease.
8 Install the clutch components (see Section 4).
9 Install the transmission (see Chapter 7A).

7 Clutch start switch - check and replacement

Check

Refer to illustration 7.4
1 Check the clutch pedal height and freeplay (see Chapter 1).
2 Verify that the engine will not start when the clutch pedal is released. Verify that the engine will start when the clutch pedal is depressed all the way.
3 If the clutch start switch doesn't perform as described, adjust or, if necessary, replace it.
4 Locate the switch **(see illustration)** at the upper end of the clutch pedal assembly and unplug the electrical connector.
5 Verify that there is continuity between the clutch start switch terminals when the switch is On (pedal depressed).
6 Verify that no continuity exists between the switch terminals when the switch is Off (pedal released).
7 If the switch fails either of the tests, replace it.

Replacement

Refer to illustration 7.8
8 Unplug the electrical connector. Loosen the locknuts **(see illustration)** and remove the switch.
9 Installation is the reverse of removal.
10 Adjust the clutch pedal height and freeplay (see Chapter 1).
11 Verify again that the engine doesn't start when the clutch pedal is released, and does start when the pedal is depressed.

8 Driveaxles - general information and inspection

1 Power is transmitted from the transaxle to the wheels through a pair of driveaxles. The inner end of each driveaxle is splined into the differential side gears. The outer ends of the driveaxles are splined to the axle hubs and locked in place by a large locknut.

2 The inner ends of the driveaxles are equipped with sliding constant velocity joints, which are capable of both angular and axial motion. Most inner joint assemblies consist of a tripod bearing and a joint housing (outer race) in which the joint is free to slide in and out as the driveaxle moves up and down with the wheel. Early inner joints are the "ball-and-cage" type, which consist of an inner hub/race, six ball bearings, an outer cage, and a outer race/housing. The inner joints can be disassembled and cleaned in the event of a boot failure (see Section 10), but if any parts are damaged, the joints must be replaced as a unit.
3 The outer CV joints are the "ball-and-cage" type, which have ball bearings running between an inner race and an outer cage, allowing angular but not axial movement. The outer joints should be cleaned, inspected and repacked when replacing the boot, but they cannot be disassembled. If an outer joint is damaged, it must be replaced along with the axleshaft (the outer joint and axleshaft are sold as a single component).
4 The boots should be inspected periodically for damage and leaking lubricant. Torn CV joint boots must be replaced immediately or the joints can be damaged. Boot replacement involves removal of the driveaxle (see Section 9). **Note:** *Some auto parts stores carry "split" type replacement boots, which can be installed without removing the driveaxle from the vehicle. This is a convenient alternative, but it should only be considered a temporary fix. At any rate, the driveaxle still must*

7.8 To remove the clutch start switch, unplug the electrical connector (not visible in this photo), remove the lower locknut (arrow) and pull the switch out of the bracket

Chapter 8 Clutch and driveaxles

9.4 Use a large prybar to immobilize the hub while loosening the driveaxle hub nut

9.5 Using a soft-faced hammer, strike the end of the driveaxle sharply with a hammer; when it breaks free, it will move noticeably

be removed and the CV joint disassembled and cleaned to ensure the joint is free from contaminants such as moisture and dirt which will accelerate CV joint wear. The most common symptom of worn or damaged CV joints, besides lubricant leaks, is a clicking noise in turns, a clunk when accelerating after coasting, and vibration at highway speeds. To check for wear in the CV joints and driveaxle shafts, grasp each axle (one at a time) and rotate it in both directions while holding the CV joint housings, feeling for play indicating worn splines or sloppy CV joints. Also check the driveaxle shafts for cracks, dents and distortion.

9 Driveaxle - removal and installation

Removal

Refer to illustrations 9.4, 9.5, 9.9 and 9.10

1 Disconnect the cable from the negative terminal of the battery.
2 Apply the parking brake.
3 Loosen the front wheel lug nuts. Using a hammer and punch, unstake the driveaxle hub locknut and loosen it. Raise the vehicle and support it securely on jackstands. Remove the wheels.
4 Remove the driveaxle hub locknut. To prevent the hub from turning, wedge a prybar between two of the wheel studs and allow the prybar to rest against the ground **(see illustration)**.

5 To loosen the driveaxle from the hub splines, tap the end of the driveaxle with a soft-faced hammer **(see illustration)** or a hammer and a brass punch. **Note:** *Don't attempt to push the end of the driveaxle through the hub yet. Applying force to the end of the driveaxle, beyond just breaking it loose from the hub, can damage the driveaxle or transaxle. If the driveaxle is stuck in the hub splines and won't move, it may be necessary to remove the brake disc (see Chapter 9) and push it from the hub with a two-jaw puller after Step 8 is performed.*
6 Remove any engine splash shields that would interfere with driveaxle removal.
7 Remove the nuts from the stabilizer bar brackets (see Chapter 10).
8 Separate the lower ball joint from the control arm (see Chapter 10).
9 Pull out on the steering knuckle and detach the driveaxle from the hub **(see illustration)**. Don't let the driveaxle hang by the inner CV joint after the outer end has been detached from the steering knuckle, as the inner joint could become damaged. Support the outer end of the driveaxle with a piece of wire, if necessary.
10 Place a drain pan underneath the transaxle to catch the lubricant that will spill out when the driveaxles are removed. Gently pry the inner CV joint out of the transaxle, being careful not to damage the dust cover or oil seal **(see illustration)**.
11 Refer to Chapter 7 for the driveaxle oil seal replacement procedure.

9.9 Pull the steering knuckle out and slide the end of the driveaxle out of the hub

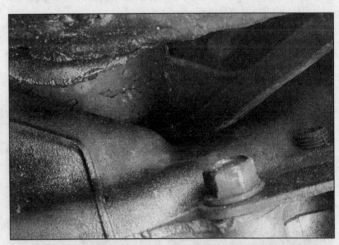

9.10 To separate the inner end of the driveaxle from the transaxle, pry on the CV joint housing like this with a large screwdriver or prybar - you may need to give the prybar a sharp rap with a brass hammer

Chapter 8 Clutch and driveaxles

10.2a Pry up the retaining tabs on the boot clamps . . .

10.2b . . . then open the clamps and remove them from the boot

10.3 Once the boot is detached from the inner CV joint housing, the housing can be removed

10.4 Use a center punch to place marks (arrows) on the tripod and the driveaxle to ensure that they're properly reassembled

10.5 Remove the snap-ring from the groove in the end of the axleshaft

Installation

12 Installation is the reverse of the removal procedure, but with the following additional points:

a) *Install a new circlip on the end of the driveaxle inner CV joint, apply molybdenum based grease to the splines and wipe the transaxle oil seal with transaxle oil.*
b) *With the end gap of the circlip facing up, push the driveaxle sharply in to seat the circlip on the inner CV joint in the groove of the differential side gear.*
c) *Install a new driveaxle hub locknut, tighten it to the torque listed in this Chapter's Specifications and stake the locknut with a punch.*
d) *Tighten the balljoint pinch bolt to the torque listed in the Chapter 10 Specifications.*
e) *Install the wheel and lug nuts, lower the vehicle and tighten the lug nuts to the torque listed in the Chapter 1 Specifications.*
f) *Check the transaxle lubricant and add, if necessary, to bring it to the proper level (see Chapter 1).*

10 Driveaxle boot - replacement

Note: *If the CV joints must be overhauled (usually due to torn boots), explore all options before beginning the job. Complete rebuilt driveaxles are available on an exchange basis, which eliminates much time and work. Whichever route you choose to take, check on the cost and availability of parts before disassembling the vehicle.*

All units

1 Remove the driveaxle (see Section 9).

Inner CV joint

Tripod type

Disassembly

Refer to illustrations 10.2a, 10.2b, 10.3, 10.4, 10.5 and 10.6

2 Remove the boot clamps **(see illustrations)**.
3 Pull the boot back from the inner CV joint, remove the retainer ring **(see illustration)** and slide the joint housing off.
4 Use a center punch to mark the tripod and axleshaft to ensure that they are reassembled properly **(see illustration)**.
5 Remove the snap-ring from the end of the axleshaft with a pair of snap-ring pliers **(see illustration)**.

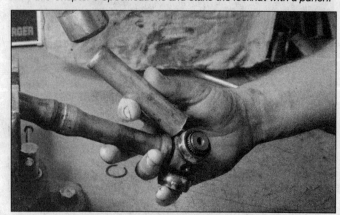

10.6 Drive the tripod joint from the axleshaft with a brass punch and hammer - make sure you don't damage the bearing surfaces or the splines on the shaft

Chapter 8 Clutch and driveaxles

6   Use a hammer and a brass punch to drive the tripod joint from the driveaxle **(see illustration)**.

Check

7   Clean all components with solvent to remove the grease, and check for cracks, pitting, scoring and other signs of wear.

Reassembly

Refer to illustrations 10.8a, 10.8b, 10.10a, 10.10b and 10.10c

8   Slide the clamps and boot onto the axleshaft. It's a good idea to wrap the axleshaft splines with tape to prevent damaging the boot **(see illustration)**. Place the tripod on the shaft and install the snap-ring **(see illustration)**. Apply grease to the tripod assembly, the inside of the joint housing and the inside of the boot.

9   Slide the boot into place, making sure both ends seat in their grooves.

10   Adjust the length of the joint to the length listed in this Chapter's Specifications, equalize the pressure in the boot **(see illustrations)**, then tighten and secure the boot clamps. **Note:** *Measure the length of the joint from the center of one boot clamp to the center of the other boot clamp.* Proceed to Step 30.

Double-offset (ball-and-cage) type

Disassembly

Refer to illustrations 10.11, 10.12, 10.13, 10.14 and 10.15

11   Remove both boot clamps **(see illustrations 10.2a and 10.2b)** and discard them. Slide the boot out of the way **(see illustration)**.

10.8a Wrap the splined area of the axleshaft with tape to prevent damage to the boots when installing them

12   Mark the shaft, the inner race, the cage and the outer race (housing) so they can be reassembled in the same way **(see illustration)**.

13   Pry the wire ring bearing retainer from the housing **(see illustration)**.

10.8b Install the tripod with the chamfered (tapered) ends of the splines facing toward the axleshaft

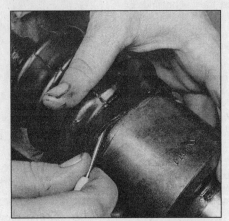

10.10a Adjust the length of the joint then equalize the pressure inside the boot by inserting a small, dull screwdriver between the boot and the CV joint housing

10.10b To install the new clamps, bend the tang down . . .

10.10c . . . and tap the tabs down to hold it in place

10.11 Remove both boot clamps, then slide the boot down the axleshaft so it's out of the way

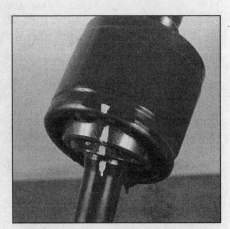

10.12 Mark the shaft, inner race, cage and outer race (housing) so they can be reassembled in the same relationship to each other

8-10 Chapter 8 Clutch and driveaxles

10.13 Pry the retainer from the housing with a small screwdriver

10.14 Slide the housing off the bearing assembly - some of the ball bearings may fall out when the race is removed, so be ready to catch them

14 Pull the housing off the inner bearing assembly **(see illustration)**.
15 Remove the snap-ring from the groove in the axleshaft with a pair of snap-ring pliers **(see illustration)**.
16 Slide the inner race off the axleshaft.
17 Using a screwdriver or piece of wood, pry the ball bearings from the cage. Be careful not to scratch the inner race, the ball bearings or the cage.
18 Remove the cage.

Inspection

Refer to illustrations 10.19a and 10.19b
19 Clean the components with solvent to remove all traces of grease. Inspect the cage and races for pitting, score marks, cracks and other signs of wear and damage **(see illustrations)**. Shiny, polished spots are normal and will not adversely affect CV joint performance.

Reassembly

Refer to illustration 10.25
20 Wrap the axleshaft splines with tape to avoid damaging the boot. Slide the small boot clamp and boot onto the axleshaft, then remove the tape. Slide the large boot clamp over the boot.
21 Install the cage on the axleshaft with the smaller diameter side of the cage facing toward the boot.
22 Install the inner race onto the axleshaft with the matchmark on the race (or the larger diameter side) aligned with the mark on the end of the axleshaft.

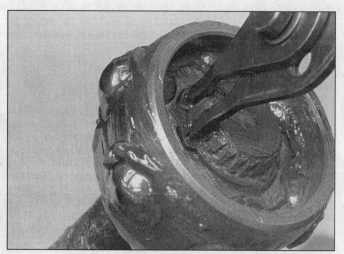

10.15 Remove the snap-ring from the groove in the axleshaft with a pair of snap-ring pliers

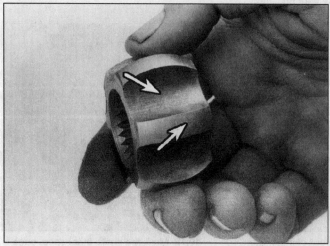

10.19a Inspect the inner race lands and grooves for pitting, score marks, cracks and other signs of wear and damage

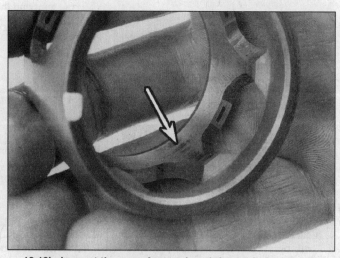

10.19b Inspect the cage for cracks, pitting and score marks (shiny, polished spots are normal and will not adversely affect CV joint performance)

Chapter 8 Clutch and driveaxles

10.25 Pack the inner race and cage assembly with grease, by hand, until grease is worked completely into the assembly (also note that the larger diameter side, or "bulge," is facing out)

10.34 After the old grease has been rinsed away and the cleaning solvent has been blown out with compressed air, rotate the outer joint through its full range of motion and inspect the bearing surfaces for wear or damage - if any of the balls, the race or the cage look damaged, replace the outer joint and shaft assembly

23 Install the snap-ring in the groove. Make sure it's completely seated by pushing on the inner race.
24 Move the cage up over the inner race, aligning the match marks. Press the ball bearings into the cage windows with your thumbs. If they won't stay in place, apply CV joint grease to hold them.
25 Fill the outer race and boot with CV joint grease (normally included with the new boot kit). Pack the inner race and cage assembly with grease, by hand, until grease is worked completely into the assembly **(see illustration)**.
26 Slide the inner race, balls and cage into the CV joint housing and install the wire ring bearing retainer.
27 Wipe any excess grease from the axle boot groove on the outer race. Seat the small diameter of the boot in the recessed area on the axleshaft. Push the other end of the boot onto the CV joint housing and move the race in or out until the length of the joint is as listed in this Chapter's Specifications.
28 Equalize the pressure in the boot by inserting a dull screwdriver between the boot and the outer race **(see illustration 10.10a)**. Don't damage the boot with the tool.
29 Tighten the boot clamps **(see illustrations 10.10a and 10.10b)**.

All inner CV joints

30 Install a new circlip on the inner CV joint stub axle.
31 Install the driveaxle (see Section 9).

Outer CV joint

Refer to illustration 10.34

32 Remove the boot clamps **(see illustrations 10.2a and 10.2b)** and slide the boot back far enough to inspect the joint. If the axleshaft is rough and you're planning to re-use the same boot, wrap the axleshaft with tape to protect the boot from damage.
33 Thoroughly wash the outer CV joint in clean solvent and blow dry it with compressed air, if available. The outer joint can't be disassembled, so it's difficult to wash away all the old grease and to rid the bearing of solvent once it's clean. But it's imperative that the job be done thoroughly, so take your time and do it right.
34 Bend the outer CV joint housing at an angle to the driveaxle to expose the bearings, inner race and cage. Inspect the bearing surfaces for signs of wear. If the joint is worn, replace it (see step 38) **(see illustration)**.
35 If the boot is damaged but the joint is OK, remove the inner CV joint and boot (see Steps 2 through 6 for tripod and Steps 11 through 18 for double-offset). If the shaft is equipped with a dynamic damper, mark or measure its position on the shaft, then remove the clamp and slide it off.
36 Slide the new outer boot onto the driveaxle. It's a good idea to wrap vinyl tape around the shaft splines to prevent damage to the boot **(see illustration 10.8a)**. When the boot is in position, add the specified amount of grease (included in the boot replacement kit) to the outer joint and the boot (pack the joint with as much grease as it will hold and put the rest into the boot). Slide the boot on the rest of the way, equalize the pressure inside the boot and install the new clamps **(see illustrations 10.10a, 10.10b and 10.10c)**. Slide the dynamic damper (if equipped) and a new clamp onto the shaft, aligning it with the mark made in Step 35. Tighten the clamp.
37 Slide on the small clamp and inner boot and install the inner CV joint (see Steps 8 through 10 for tripod joints and 20 through 29 for double-offset). Be sure to adjust the length of the inner joint before tightening the clamps.

All units

39 Install the driveaxle (see Section 9).

Notes

Chapter 9 Brakes

Contents

	Section		Section
Anti-lock Brake System (ABS) - general information	2	Disc brake pads (front) - replacement	3
Brake check	See Chapter 1	Drum brake shoes - replacement	6
Brake disc - inspection, removal and installation	5	General information	1
Brake fluid level check	See Chapter 1	Master cylinder - removal, overhaul and installation	8
Brake hoses and lines - inspection and replacement	9	Parking brake - adjustment	12
Brake hydraulic system - bleeding	10	Parking brake cables - replacement	13
Brake light switch - check and replacement	15	Power brake booster - check, removal and installation	11
Brake pedal - adjustment	14	Wheel cylinder - removal, overhaul and installation	7
Disc brake caliper - removal, overhaul and installation	4		

Specifications

General
Brake fluid type .... See Chapter 1
Brake pedal adjustment
    Pedal height (pedal-to-floor) .... 8.03 to 8.23 inches
    Pedal freeplay .... 0.16 to 0.28 inch
Power brake booster pushrod clearance
    Festiva
        No vacuum .... 0.016 to 0.024 inch
        20 in-Hg vacuum applied .... 0 to 0.012 inch
    Aspire
        No vacuum .... 0 inch
        20 in-Hg vacuum applied .... 0.004 to 0.016 inch

Disc brakes
Minimum brake pad thickness .... See Chapter 1
Minimum disc thickness *
    Festiva .... 0.43 inch
    Aspire
        Automatic .... 0.78 inch
        Manual .... 0.63 inch

Drum brakes
Drum inside diameter
Standard
    Festiva .... 6.69 inches
    Aspire .... 7.86 inches
Maximum *
    Festiva .... 6.75 inches
    Aspire .... 7.93 inches

* **Note:** *If different specifications are cast into the disc or drum, they supersede information printed here.*

Parking brake
Parking brake lever travel .... 11 to 16 clicks

Torque specifications
	Ft-lbs (unless otherwise indicated)
Brake hose-to-caliper banjo bolts	16 to 22
Caliper mounting bolts	29 to 36
Hub-to-disc bolts (Festiva)	33 to 40
Master cylinder-to-brake booster nuts	84 to 144 in-lbs
Power brake booster mounting nuts	14 to 19
Wheel cylinder mounting bolts	84 to 108 in-lbs
Wheel lug nuts	See Chapter 1

1 General information

The vehicles covered by this manual are equipped with hydraulically operated front and rear brake systems. The front brakes are disc type and the rear brakes are drum type. Both the front and rear brakes are self-adjusting. The disc brakes automatically compensate for pad wear, while the drum brakes incorporate an adjustment mechanism which is activated as the parking brake is applied. Some Aspire models are equipped with an optional anti-lock brake system (see Section 2).

Hydraulic system

The hydraulic system consists of two separate circuits. The master cylinder has separate reservoirs for the two circuits, and, in the event of a leak or failure in one hydraulic circuit, the other circuit will remain operative. A dual proportioning valve on the firewall provides brake balance between the front and rear brakes.

Power brake booster

The power brake booster, utilizing engine manifold vacuum and atmospheric pressure to provide assistance to the hydraulically operated brakes, is mounted on the firewall in the engine compartment.

Parking brake

The parking brake operates the rear brakes only, through cable actuation. It's activated by a lever mounted in the center console.

Service

After completing any operation involving disassembly of any part of the brake system, always test drive the vehicle to check for proper braking performance before resuming normal driving. When testing the brakes, perform the tests on a clean, dry, flat surface. Conditions other than these can lead to inaccurate test results.

Test the brakes at various speeds with both light and heavy pedal pressure. The vehicle should stop evenly without pulling to one side or the other. Avoid locking the brakes, because this slides the tires and diminishes braking efficiency and control of the vehicle.

Tires, vehicle load and wheel alignment are factors which also affect braking performance.

2 Anti-lock Brake System (ABS) - general information

1 The Anti-lock Brake System (ABS) is designed to maintain vehicle steerability, directional stability and optimum deceleration under severe braking conditions and on most road surfaces. It does this by monitoring the rotational speed of each wheel and controlling the brake line pressure to each wheel during braking. This prevents the wheel from locking up.

Components
Actuator assembly
2 The ABS hydraulic unit consists of an electric hydraulic pump, solenoid valves, flow control valves, buffer and damper chamber, and is located in the engine compartment. The electric pump provides hydraulic pressure to the brakes, modulating brake line pressure during ABS operation, by turning on/off the solenoid valves and opening/closing the flow control valves in the ABS hydraulic unit. The buffer chamber stores hydraulic fluid from the brakes for smooth decrease of pressure. The damper chamber decreases pump noise and vibration. The hydraulic unit is controlled by the ABS control module, located under the dash at the driver's side. A fail-safe mode, operated by the fail-safe relay, mounted in the engine compartment near the left side headlight, returns the brake system to conventional operation if there is a malfunction in the ABS, and the ABS warning light comes ON.

Speed sensors
3 The speed sensors, which are located at each wheel, generate small electrical pulsation's when the toothed sensor rotors are turning, sending a variable voltage signal to the ABS control module indicating wheel rotational speed.
4 The front speed sensors are mounted at the front wheel hubs in close relationship to the toothed sensor rotors, which are integral with the outer constant velocity (CV) joints.
5 The rear wheel sensors are bolted to the brake backing plates or axle carriers. The sensor rotors are integral with the rear brake hub.

ABS computer
6 The ABS control module, mounted behind the driver's side kick panel, is the "brain" of the ABS system. The function of the control module is to accept and process information received from the wheel speed sensors to control the hydraulic line pressure, avoiding wheel lock up. The module also constantly monitors the system, even under normal driving conditions, to find faults within the system.
7 If a problem develops within the system, an "ABS" light will glow on the dashboard. A diagnostic code will also be stored in the ECU, which, when retrieved by a service technician, will indicate the problem area or component.

Diagnosis and repair
8 If a dashboard warning light comes on and stays on while the vehicle is in operation, the ABS system requires attention. Although a special electronic ABS diagnostic tester is necessary to properly diagnose the system, the home mechanic can perform a few preliminary

3.1 Always wash the brakes with brake cleaner before caliper removal

3.4a Depress the piston into the caliper as far as it will go with a C-clamp to make room for the new brake pads

3.4b If you don't have a C-clamp that's big enough to fit over the caliper, use a pair of adjustable pliers instead

checks before taking the vehicle to a dealer service department or other repair shop which is equipped with a tester.
a) Check the brake fluid level in the reservoir.
b) Check that all electrical connectors are securely connected.
c) Check the fuses.

Caution: *Do not attempt to use the vehicle until repairs are accomplished.*

9 Additional checks that can be performed to assist you before dealer diagnosis and/or repair, that may help you discuss the problem with the dealer service representatives, are as follows:

10 With a fully charged vehicle battery, turn the ignition switch ON, and verify that the ABS warning light goes out after a few seconds.

11 If the light stays on, the ABS control module is detecting a failure and will not activate the ABS hydraulic unit. See the dealer for servicing. Turn the ignition switch to OFF.

12 Carefully jack up the vehicle on a level surface and support it securely with jackstands. Shift the transaxle to neutral or N range.

13 Release the parking brake. Rotate each wheel by hand, making sure that excessive brake drag does not exist.

14 Locate the Data Link Connector under the hood at the firewall, near the battery, and place a jumper wire between terminals TBS and GND.

15 Starting with the right front wheel, have an assistant depress the brake pedal while you check that the right front wheel will not rotate by hand.

16 With the brake pedal still depressed, turn the ignition switch ON and verify that the brake is released momentarily (approximately 1/2 second) and the wheel turns when pressure reduction from the ABS automatically cycles on and off. Perform the same check on the other wheels in order, left front, right rear, and left rear.

17 If the system tests are satisfactory for momentary brake release (the previous two steps above), then the lines to the ABS hydraulic unit are okay, the braking system including the hydraulic unit is okay, the electrical system (solenoid, ABS motor) in the ABS hydraulic unit is okay, and the ABS control module and its output system including relay, solenoid, wiring harness are okay. However, the above test does not check the ABS input system/harness, intermittent failures or fluid leakage, so any further diagnosis and repair should be handled by a dealer service department. **Caution:** *Do not drive the vehicle until repairs are completed.*

3 Disc brake pads (front) - replacement

Refer to illustrations 3.1, 3.4a, 3.4b, 3.5, 3.6, 3.7, 3.8 and 3.10
Warning: *Dust created by the brake system may contain asbestos, which is harmful to your health. Never blow it out with compressed air and don't inhale any of it. Clean the brake assembly with brake cleaner before any brake work* **(see illustration 3.1)**. *An approved filtering mask should be worn when working on the brakes. Do not, under any circumstances, use petroleum-based solvents to clean brake parts. Use brake system cleaner.*
Note: *If an overhaul is indicated (usually because of fluid leakage), explore all options before beginning the job. New and factory rebuilt calipers are available on an exchange basis, which makes this job quite easy. If you decide to rebuild the calipers, make sure a rebuild kit is available before proceeding. Always rebuild the calipers in pairs (front pair and/or rear pair) - never rebuild just one of them. Be careful when handling the new brake pads - do not touch the lining surface with your fingers to eliminate any oil contamination, which will affect braking efficiency.*

1 Loosen the front wheel lug nuts, raise the front of the vehicle and place it securely on jackstands. Remove the wheel. **Note:** *Work on one brake assembly at a time, using the opposite side brake assembly for reference if necessary.* Wash the brake assembly with brake system cleaner before beginning work **(see illustration)**.

2 If you are checking the brake pads for wear, see Chapter 1. Inspect the brake disc (see Section 5). If machining is necessary, remove the brake disc as described in Section 5.

3 Open the hood and remove the cap from the brake fluid reservoir.

4 With the old pads in place, press against the pad on the piston side of the caliper, displacing the caliper piston inward fully **(see illustrations)**. As the piston is pressed inward, watch the fluid level in the brake fluid reservoir rise, being careful to remove any excess so that fluid will not spill over.

3.5 Disengage the M-shaped anti-rattle spring from the brake pads

3.6 Disengage the W-shaped pin retaining spring from the pad retaining pins

3.7 Remove the pad retaining pins and the anti-rattle spring

3.8 Remove the brake pads - be sure to swap any shims on the old pads to the new pads

3.10 Apply anti-squeal compound to the brake pad backing plate shims

5   Disengage the M-shaped anti-rattle spring from the brake pads **(see illustration)**.
6   Remove the W-shaped pin retaining spring from the upper and lower pins **(see illustration)**.
7   Remove the brake pad pins **(see illustration)**.
8   Remove the brake pads **(see illustration)**.
9   There are two shims on the inner brake pad and one shim on the outer pad. Remove these shims, clean them up, and install them on the new pads.
10   Apply a bead of anti-squeal compound to the brake pad backing plate shims **(see illustration)**. Be careful to not get any on or near the brake pad friction surfaces.
11   With the caliper piston fully depressed, install the new pads and shims.
12   Install the pad pins, the W-shaped pin retaining spring and the M-shaped anti-rattle spring. If you have difficulty remembering how the correct installation looks, refer to the other front caliper assembly.
13   Repeat Steps 4 through 12 for the opposite wheel brake pad replacement.
14   Check the brake fluid level and remove or add brake fluid as necessary. Reinstall the reservoir cap. **Warning:** *Press the brake pedal several times and recheck the brake fluid level in the reservoir before driving the vehicle.*

4 Disc brake caliper - removal, overhaul and installation

Warning: *Dust created by the brake system may contain asbestos, which is harmful to your health. Never blow it out with compressed air and don't inhale any of it. An approved filtering mask should be worn when working on the brakes. Do not, under any circumstances, use petroleum-based solvents to clean brake parts. Use brake system cleaner.*
Note: *If an overhaul is indicated (usually because of fluid leakage), explore all options before beginning the job. New and factory rebuilt calipers are available on an exchange basis, which makes this job quite easy. If you decide to rebuild the calipers, make sure a rebuild kit is available before proceeding. Always rebuild the calipers in pairs - never rebuild just one of them.*

Removal

Refer to illustrations 4.2a, 4.2b, 4.7 and 4.8

1   Loosen the front wheel lug nuts, raise the front of the vehicle and place it securely on jackstands. Remove the wheel and prepare for disassembly.
2   Remove the brake hose bolt (banjo bolt) and disconnect the brake

Chapter 9 Brakes

4.2a Remove the brake hose banjo bolt

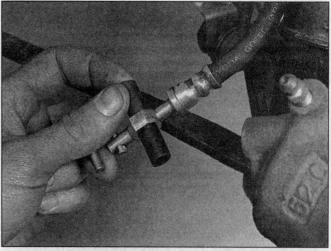

4.2b Using a piece of rubber hose of the appropriate size, plug the brake line; this will prevent brake fluid from leaking out and dirt and moisture from contaminating the fluid in the hose

hose from the caliper. Plug the brake hose to keep contaminants out of the brake system and to prevent losing any more brake fluid than is necessary **(see illustrations)**. **Note:** *Don't disconnect the hose if you are only removing the caliper for access to other components.*

3   Remove the W-shaped clip from the caliper **(see illustration 3.5)**.
4   Remove the M-shaped spring from the caliper **(see illustration 3.6)**.
5   Remove the pad pins from the caliper **(see illustration 3.7)**.
6   Push the caliper piston inward if necessary **(see illustration 3.4a)** to remove the pads, the anti-squeak shim, the outer shim, and the inner shim **(see illustration 3.8)**.
7   Remove the caliper mounting bolts **(see illustration)**.
8   Remove the caliper **(see illustration)** and the two anti-squeak caps, if equipped, from the projections on the knuckle.

Overhaul

Refer to illustrations 4.9, 4.11, 4.13 and 4.15

9   To overhaul the caliper, remove the bolt sleeves and the boots **(see illustration)**.
10 Remove the rubber cap and bleeder screw.
11 Remove the caliper piston retaining ring and dust boot **(see illustration)**.
12 Before you remove the piston, place a wood block or some rags between the piston and caliper to prevent damage as it is removed.

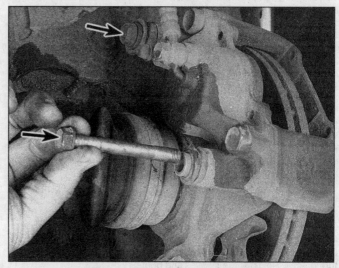

4.7 Remove the caliper mounting bolts

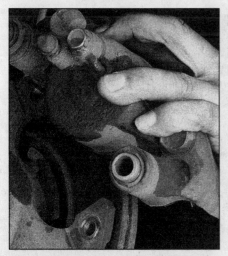

4.8 Remove the caliper and the two anti-squeak caps, if equipped from the steering knuckle (not shown)

4.9 Remove the caliper bolt sleeves and boots

4.11 Using a small screwdriver or hooked tool, remove the dust boot retaining ring

4.13 With the caliper padded to catch the piston, use compressed air to force the piston out of its bore - make sure your hands or fingers are not between the piston and caliper

13 To remove the piston from the caliper, apply compressed air to the brake fluid hose connection on the caliper body **(see illustration)**. Use only enough pressure to ease the piston out of its bore. **Warning:** *Be careful not to place your fingers between the piston and the caliper, as the piston may come out with some force.*

14 Inspect the mating surfaces of the piston and caliper bore wall. If there is any scoring, rust, pitting or bright areas, replace the complete caliper unit with a new one. Crocus cloth can be used to remove light corrosion and stains.

15 If these components are in good condition, remove the piston seal from the caliper bore **(see illustration)**.

16 Wash all the components with brake system cleaner and allow them to dry.

17 To reassemble the caliper, you should already have the correct rebuild kit for your vehicle.

18 Submerge the new piston seal and the piston in brake fluid and install the piston seal in its groove in the caliper bore.

19 Install the piston into the caliper bore. Do not force the piston into the bore, but make sure it is squarely in place, then apply firm (but not excessive) force by hand to install it.

20 Install the new piston dust boot and retaining ring.

21 Reinstall the bleeder screw and rubber cap.

22 Clean the bolt sleeves and lightly coat them with high-temperature grease. Reinstall the sleeves and boots.

23 At this time, inspect the brake disc to be sure that it is reusable (see Section 5).

4.15 Remove the seal from the piston using a plastic or wooden tool, such as a pencil

Installation

24 Install the caliper by reversing the removal procedure. Remember to replace the copper sealing washers (gaskets) at the brake hose-to-caliper connection (new washers normally come with the rebuild kit).

25 Bleed the brake circuit according to the procedure in Section 10. Make sure there are no leaks from the hose connections. Test the brakes carefully before returning the vehicle to normal service.

---

5 Brake disc - inspection, removal and installation

Refer to illustration 5.2

1 Loosen the wheel lug nuts, raise the vehicle and support it securely on jackstands. Remove the wheel.

2 Remove the brake pads (see Section 3) and caliper (see Section 4). It's not necessary to disconnect the brake hose. After removing the caliper, suspend it out of the way with a piece of wire **(see illustration)**. On Aspire models, the brake disc can be slid off the hub after the caliper has been removed. If the vehicle you are working on is of this design, reinstall the lug nuts and tighten them securely before performing the following runout check.

Inspection

Refer to illustrations 5.3, 5.4a, 5.4b, 5.5a and 5.5b

3 Visually inspect the disc surface for scoring and other damage **(see illustration)**. Light scratches and shallow grooves are normal

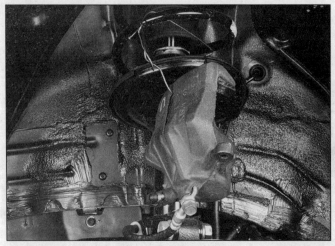

5.2 Hang the caliper with a piece of wire - do not allow the caliper to hang by the brake hose

5.3 The brake pads on this vehicle were obviously neglected, as they wore down to the rivets and cut deep grooves into the disc - wear this severe means the disc must be replaced

Chapter 9 Brakes

5.4a To check disc runout, mount a dial indicator as shown and rotate the disc

5.4b Using a swirling motion, remove the glaze from the disc surface with sandpaper or emery cloth

5.5a The minimum wear dimension is cast into the back side of the disc

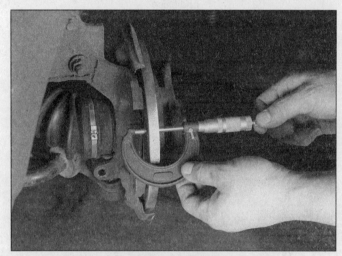

5.5b Use a micrometer to measure disc thickness

after use and may not always be detrimental to brake operation, but deep scoring - over 0.015-inch - requires disc removal and refinishing by an automotive machine shop. Be sure to check both sides of the disc. If pulsating has been noticed during application of the brakes, suspect disc runout.

4 To check disc runout, place a dial indicator at a point about 1/2-inch from the outer edge of the disc **(see illustration)**. Set the indicator to zero and turn the disc. The difference between the highest and lowest indicator reading should not exceed the specified allowable runout maximum. If it does, the disc should be refinished by an automotive machine shop. **Note:** *Professionals recommend resurfacing of brake discs regardless of the dial indicator reading (to produce a smooth, flat surface that will eliminate brake pedal pulsation's and other undesirable symptoms related to questionable discs). At the very least, if you elect not to have the discs resurfaced, deglaze them with medium-grit emery cloth (use a swirling motion to ensure a non-directional finish)* **(see illustration).**

5 It is absolutely critical that the disc not be machined to a thickness under the specified minimum allowable thickness. This thickness is stamped onto the inner diameter of the disc **(see illustration)**. The disc thickness can be checked with a micrometer **(see illustration)**.

Removal and installation

6 On Aspire models, the brake disc can be slid off the hub after the caliper has been removed. If the vehicle you are working on is of this design, remove the lug nuts that were installed to perform the above checks and slide the disc off the hub.

7 On Festiva models, remove the steering knuckle and hub assembly from the vehicle (see Chapter 10).

8 Because of the special tools needed to remove and install the hub/disc assembly from the steering knuckle on Festiva models, and because the hub bearing preload must be adjusted properly on these models, the steering knuckle and hub assembly must be removed and installed by a dealer service department or automotive machine shop.

9 On Festiva models, install the steering knuckle and hub assembly, referring to the procedure described in Chapter 10.

10 Install the caliper (see Section 4) and brake pads (see Section 3). Tighten the caliper bolts to the torque listed in this Chapter's Specifications.

11 Install the wheel, then lower the vehicle to the ground. Depress the brake pedal a few times to bring the brake pads into contact with the disc. Bleeding of the system will not be necessary unless the brake hose was disconnected from the caliper. Check the operation of the brakes carefully before placing the vehicle into normal service.

6 Drum brake shoes - replacement

Refer to illustrations 6.4a through 6.4aa and 6.5

Warning: *Drum brake shoes must be replaced on both wheels (rear) at the same time - never replace the shoes on only one wheel. Also, the dust created by the brake system may contain asbestos, which is harmful to your health. Never blow it out with compressed air and don't*

Chapter 9 Brakes

6.4a Remove the grease cap from the drum with a hammer and chisel

6.4b Remove the wheel bearing locknut. Caution: *On some models, the right-side wheel bearing locknut may be left-hand thread - turn clockwise to loosen*

6.4c Remove the washer

6.4d Remove the outer wheel bearing, then remove the brake drum

inhale any of brake dust. An approved filtering mask should be worn when working on the brakes. Do not, under any circumstances, use petroleum-based solvents to clean brake parts. Use brake system cleaner only!

6.4e Before removing anything, place a drain pan under the brake assembly, clean the brake assembly with brake cleaner and allow it to dry; DO NOT USE COMPRESSED AIR TO BLOW OFF BRAKE DUST!

Caution: *Whenever the brake shoes are replaced, the return and hold-down springs should also be replaced. Due to the continuous heating/cooling cycle to which the springs are subjected, spring tension decreases over a period of time and may allow the shoes to drag on the drum and wear at a much faster rate than normal.*

1   Loosen the wheel lug nuts, raise the rear of the vehicle and support it securely on jackstands. Block the front wheels to keep the vehicle from rolling.
2   Release the parking brake.
3   Remove the wheel. If checking brake shoe linings for wear, see Chapter 1. Also, check the brake wheel cylinder for any signs of fluid leakage. If fluid leakage is found, repair the wheel cylinder (see Section 7). **Note:** *All four rear brake shoes at both rear wheels must be replaced at the same time, but to avoid mixing up parts, work on only one brake assembly, and when complete, repair the opposite wheel before driving the vehicle.*
4   Follow the accompanying illustrations for the brake shoe replacement procedure **(see illustrations 6.4a through 6.4aa)**. Be sure to stay in order and read the caption under each illustration. **Note:** *If the brake drum cannot be easily pulled off the axle and shoe backing plate assembly, make sure the parking brake is completely released. If the drum still cannot be pulled off, the brake shoes will have to be retracted. This is done by loosening the parking brake cable nut until the parking brake lever at the brake backing plate returns to its stop. The drum should now come off.*

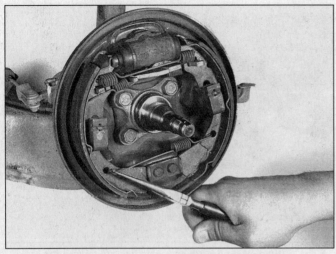

6.4f Unhook the lower return spring from the shoes

6.4g Rotate the rear shoe hold-down clip pin 90-degrees . . .

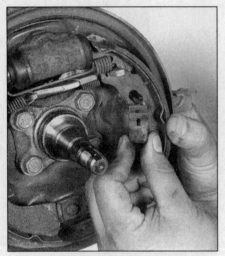

6.4h . . . then remove the rear shoe hold-down clip

6.4i Disengage the rear shoe from the adjuster assembly . . .

6.4j . . . then unhook the upper return spring and remove the rear shoe; unhook the other end of the upper return spring from the other shoe and set the spring aside

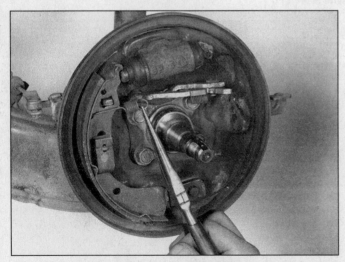

6.4k Unhook this small spring from the adjuster assembly and from the front shoe and set the spring aside

6.4l Remove the front shoe hold-down clip and remove the front shoe

6.4m To reset the self-adjuster mechanism, pull the adjuster cam to the rear . . .

6.4n . . . then swing the adjuster cam inward, toward the brake backing plate

6.4o Apply high-temperature grease to the shoe contact points on the backing plate

6.4p Place the front shoe in position, making sure it's properly engaged with the notch in the wheel cylinder piston and with the hooked forward end of the self-adjuster, insert the hold-down clip pin through the backside of the brake backing plate . . .

6.4q . . . install the hold-down clip and rotate the pin 90-degrees to lock the clip into place

6.4r Hook the upper return spring into the front shoe . . .

6.4s . . . hook the other end of the spring into the rear shoe . . .

Chapter 9 Brakes

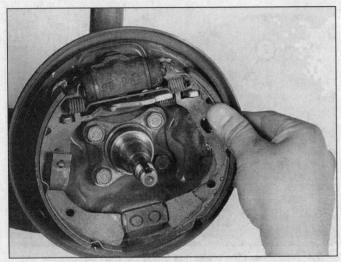

6.4t ... install the rear shoe, making sure it's properly engaged with the wheel cylinder piston and the hooked end of the self-adjuster ...

6.4u ... install the pin and hold-down clip ...

6.4v ... and lock the clip into place

6.4w Hook the lower return spring into the rear shoe ...

6.4x ... and into the front shoe

6.4y Hook the self-adjuster spring into the front shoe ...

6.4z ... and hook the other end of the spring onto the self-adjuster

6.4aa After the drum, outer wheel bearing, washer and locknut are installed, adjust the wheel bearing preload (see Chapter 1) and stake the nut or install the cotter pin

6.5 The maximum drum diameter is cast into the drum

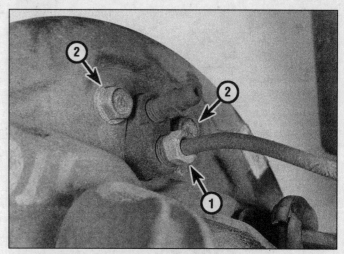

7.4 Disconnect the brake line fitting (1), then remove the two wheel cylinder bolts (2)

5 Before reinstalling the brake drum, check it for cracks, score marks, deep scratches and hard spots, which will appear as small discolored areas. If the hard spots cannot be removed with fine emery cloth or if any of the other conditions listed above exist, the drum must be taken to an automotive machine shop to have it resurfaced. **Note:** *Professionals recommend resurfacing the drums each time a brake job is done. Resurfacing will eliminate the possibility of out-of-round drums.* If the drums are worn so much that they cannot be resurfaced without exceeding the maximum allowable diameter, which is stamped into the drum **(see illustration)**, then new ones will be required. At the very least, if you elect not to have the drums resurfaced, remove the glaze from the surface with emery cloth or sandpaper, using a swirling motion.

6 Before reinstalling the brake drum, have an assistant depress the brake pedal while you check for operation of the automatic adjuster. Install the brake drum on the axle flange.

7 Mount the wheel and install the wheel lug nuts. Make sure the new brake shoes are adjusted so there is no brake drag with the brake pedal released.

8 Lower the vehicle and tighten the lug nuts to the torque listed in the Chapter 1 Specifications.

9 Make a number of forward and reverse stops and operate the parking brake and check for satisfactory pedal action. Check the operation of the brakes carefully before driving the vehicle.

7 Wheel cylinder - removal, overhaul and installation

Note: *If an overhaul is indicated (usually because of fluid leaks or sticky operation), explore all options before beginning the job. New wheel cylinders are available, which makes this job quite easy. If you decide to rebuild the wheel cylinder, make sure a rebuild kit is available before proceeding. Never overhaul only one wheel cylinder - always rebuild both of them. If the wheel cylinder shows evidence of brake fluid leakage, all brake components must be cleaned and the brake shoes must be replaced if contaminated with brake fluid (see Section 6).*

Removal

Refer to illustration 7.4

1 Raise the rear of the vehicle and support it securely on jackstands. Block the front wheels to keep the vehicle from rolling.

2 Remove the brake shoe assembly (see Section 6).

3 Remove all dirt and foreign material from around the wheel cylinder.

4 Disconnect the brake line **(see illustration)** with a flare-nut wrench, if available. Don't pull the brake line away from the wheel cylinder.

5 Remove the wheel cylinder mounting bolts.

6 Detach the wheel cylinder from the brake backing plate and place it on a clean workbench. Immediately plug the brake line to prevent fluid loss and contamination.

Overhaul

Refer to illustration 7.7

7 Remove the bleeder screw, dust boots, piston cups, pistons, spring, and spring caps from the wheel cylinder body **(see illustration)**.

8 Clean the wheel cylinder with brake fluid, denatured alcohol or brake system cleaner. **Warning:** *Do not, under any circumstances, use petroleum-based solvents to clean brake parts!*

9 Use filtered, unlubricated compressed air to dry the wheel cylinder and blow out the passages.

10 Check the bore for corrosion and score marks. Crocus cloth (see your local automotive parts supplier) can be used to remove light corrosion and stains, but the cylinder must be replaced with a new one if the defects cannot be removed easily, or if the bore is scored.

11 Lubricate the wheel cylinder bore, new piston cups and pistons with brake fluid. **Warning:** *Always use fresh brake fluid from a new, previously unopened container.*

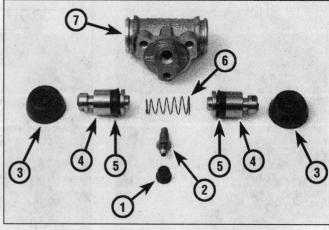

7.7 Typical wheel cylinder assembly

1	Bleeder screw cap	5	Piston cup
2	Bleeder screw	6	Spring
3	Dust boot	7	Wheel cylinder housing
4	Piston		

Chapter 9 Brakes

8.1 Unplug the electrical connector (right arrow) for the fluid level warning switch; to detach the master cylinder from the power brake booster, remove the two mounting nuts (lower arrows)

12 Assemble the wheel cylinder components. Make sure the piston cup lips face inward.

Installation

13 Place the wheel cylinder in position and install the mounting bolts finger tight. Connect the brake line to the cylinder, being careful not to cross-thread the fitting, but do not tighten the brake line at this time.
14 Tighten the wheel cylinder mounting bolts to the torque listed in this Chapter's Specifications.
15 Tighten the brake line securely and install the brake shoe assembly (see Section 6).
16 Bleed the brakes (see Section 10).
17 Check the operation of the brakes carefully before driving the vehicle.

8 Master cylinder - removal, overhaul and installation

Note: *Before deciding to overhaul the master cylinder, check on the availability and cost of a new or factory rebuilt unit and also the availability of a rebuild kit. If you decide to rebuild the cylinder, inspect the bore as described in Step 12 before purchasing parts.*

Removal

Refer to illustrations 8.1 and 8.3

1 Unplug the electrical connector for the fluid level warning switch **(see illustration)**. Check continuity at the level sensor terminals; no continuity should be measured when the fluid level is above MIN.
2 Carefully remove the brake fluid reservoir cap and remove as much fluid as possible from the reservoir with a syringe. Check continuity of the level sensor; continuity should be measured when fluid level is below the MIN level.
3 Place rags under the fittings and prepare caps or plastic bags to cover the ends of the lines once they are disconnected. **Caution:** *Brake fluid will damage paint. Cover all body parts and be careful not to spill fluid during this procedure.* Loosen the fittings at the ends of the brake lines where they enter the master cylinder **(see illustration)**. To prevent rounding off the flats, use a flare-nut wrench, which wraps around the fitting hex.
4 Pull the brake lines away from the master cylinder and plug the ends to prevent contamination.
5 Remove the nuts and washers attaching the master cylinder to the power brake booster.
6 Pull the master cylinder off the studs to remove it. Again, be careful not to spill the fluid as this is done.

8.3 Loosen the brake line fittings with a flare-nut wrench

Overhaul

Refer to illustrations 8.8a, 8.8b, 8.8c, 8.9, 8.10, 8.11a, 8.11b, 8.14 and 8.15

7 Before attempting the overhaul of the master cylinder, obtain the proper rebuild kit, which will contain the necessary replacement parts and also any instructions which are specific to your model.
8 Remove the reservoir retaining screw, pull off the reservoir and

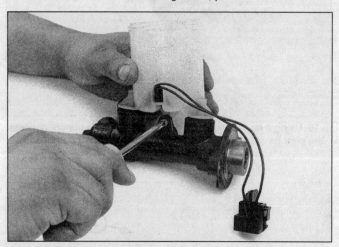

8.8a To detach the brake fluid reservoir from the master cylinder, remove this screw . . .

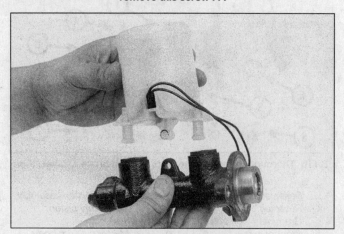

8.8b . . . and firmly pull the reservoir straight up

8.8c After the reservoir has been removed, pull the grommets from the master cylinder body; if they're hard, cracked, damaged or have been leaking, replace them

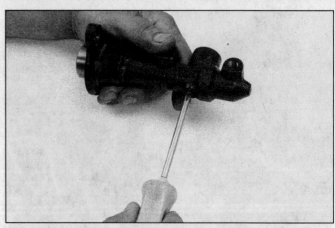

8.9 Using a Phillips screwdriver, remove the stopper screw; you may have to depress the pistons slightly to loosen the bolt (be sure to replace the stopper screw sealing washer)

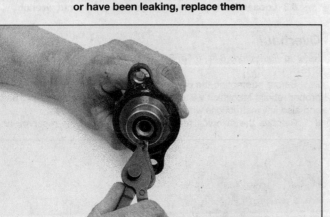

8.10 Using a large Phillips screwdriver or a brass punch, depress the pistons again and remove the snap-ring with a pair of snap-ring pliers (punch not shown for clarity)

8.11a After the stopper screw and snap-ring have been removed, remove the pistons; if either piston is difficult to remove, tap the master cylinder on a block of wood as shown

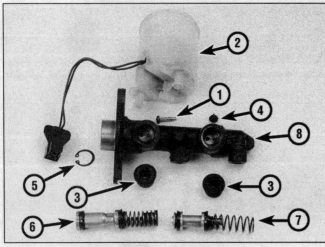

8.11b Thoroughly wash all the parts in fresh brake cleaner, then lay them out for inspection

1. Reservoir retaining screw
2. Reservoir
3. Grommets
4. Stopper screw
5. Snap ring
6. Primary piston assembly
7. Secondary piston assembly
8. Master cylinder body

remove the grommets (**see illustrations**).

9 Remove the stopper screw from the master cylinder (**see illustration**).

10 Place the master cylinder in a bench vise and use a punch or Phillips screwdriver to depress the pistons until they bottom against the other end of the master cylinder. Hold the pistons in this position and carefully remove the snap-ring at the end of the master cylinder (**see illustration**).

11 The internal components can now be removed from the bore (**see illustrations**). Make a note of the proper order of the components so they can be returned to their original locations. **Note:** *The two springs are different, so pay particular attention to their installed order.*

12 Carefully inspect the bore of the master cylinder. Any deep score marks or other damage mean that a new master cylinder is required. DO NOT attempt to hone the bore.

13 Replace all parts included in the rebuild kit, following any instructions in the kit. Clean all re-used parts with brake system cleaner. **Warning:** *Do not use any petroleum-based solvents. During reassembly, lubricate all parts liberally with clean, fresh brake fluid.*

14 Depress the pistons into the bore and install the new snap-ring (**see illustration**), making sure it's seated properly in the groove.

15 Depress the pistons again, until the secondary piston bottoms in the bore, and install the stopper screw (**see illustration**).

16 Install the reservoir grommets, reservoir and reservoir mounting screw.

17 Before installing the master cylinder, it should be bench bled. Since you'll have to apply pressure to the master cylinder piston and,

Chapter 9 Brakes

8.14 Depress the pistons and install a new snap ring

8.15 Depress the pistons until the secondary piston bottoms against the end of the bore and install the stopper screw

9.3 Unscrew the brake line threaded fitting with a flare-nut wrench to protect the fitting corners from being rounded off

at the same time, control flow from the brake line outlets, the master cylinder should be mounted in a vise, with the jaws of the vise clamping on the mounting flange.

18 Insert threaded plugs into the brake line outlet holes and snug them down so no air will leak past them, but not so tight that they cannot be easily loosened.

19 Fill the reservoir with brake fluid of the recommended type (see Chapter 1).

20 Remove one plug and push the piston assembly into the bore to expel the air from the master cylinder. A large Phillips screwdriver can be used to push on the piston assembly.

21 To prevent air from being drawn back into the master cylinder, the plug must be replaced and snugged down before releasing the pressure on the piston.

22 Repeat the procedure until only brake fluid is expelled from the brake line outlet hole. When only brake fluid is expelled, repeat the procedure at the other outlet hole and plug. Be sure to keep the master cylinder reservoir filled with brake fluid to prevent the introduction of any additional air into the master cylinder system while bleeding it.

23 Since high pressure is not involved in the bench bleeding procedure, an alternative to the removal and replacement of the plugs with each stroke of the piston assembly is available. Before pushing in on the piston assembly, remove the plug as described in Step 20. Before releasing the piston, however, instead of replacing the plug, simply put your finger tightly over the hole to keep air from being drawn back into the master cylinder. Wait several seconds for brake fluid to be drawn from the reservoir into the bore, then depress the piston again, removing your finger as brake fluid is expelled. Be sure to put your finger back over the hole each time before releasing the piston, and when the bleeding procedure is complete for that outlet, replace the plug and tighten it before going on to the other port.

Installation

24 Install the master cylinder over the studs on the power brake booster, install the washers, and tighten the nuts only finger-tight at this time. Don't forget to use a new gasket.

25 Thread the brake line fittings into the master cylinder. Since the master cylinder is still loose, it can be moved slightly so the fittings thread in easily by hand. Be careful not to strip the threads as the fittings are tightened.

26 Tighten the master cylinder mounting nuts to the torque listed in this Chapter's Specifications. Tighten the brake line fittings securely using a flare-nut wrench.

27 Fill the master cylinder reservoir with fluid, then bleed the master cylinder and the brake system (see Section 11). To bleed the master cylinder on the vehicle, have an assistant depress the brake pedal and hold it down while you loosen the fitting to allow air and fluid to escape. Tighten the fitting, then allow your assistant to return the pedal

to its rest position. Repeat this procedure on all fittings until the fluid is free of air bubbles. Check the operation of the brake system carefully before driving the vehicle.

9 Brake hoses and lines - inspection and replacement

Inspection

1 About every six months, with the vehicle raised and supported securely on jackstands, the rubber hoses which connect the steel brake lines with the front and rear brake assemblies should be inspected for cracks, chafing of the outer cover, leaks, blisters and other damage. These are important and vulnerable parts of the brake system and inspection should be complete. A light and mirror will be helpful for a thorough check. If a hose exhibits any of the above conditions, replace it with a new one.

Replacement

Front brake hose

Refer to illustrations 9.3 and 9.4

2 Loosen the wheel lug nuts, raise the vehicle and support it securely on jackstands. Remove the wheel.

3 At the frame bracket, unscrew the brake line fitting from the hose **(see illustration)**. Use a flare-nut wrench to prevent rounding off the corners and hold the hose fitting with an open-end wrench.

4 Remove the U-clip from the female fitting at the bracket with a pair of pliers **(see illustration)**, then pass the hose through the bracket.

9.4 Pull off the U-clip with a pair of pliers

5 At the caliper end of the hose, remove the banjo fitting bolt, then separate the hose from the caliper. Note that there are two copper sealing washers on either side of the fitting - these sealing washers should be replaced with new ones during installation.
6 To install the hose, pass the caliper fitting end through any bracket, as necessary, then connect the fitting to the caliper with the banjo bolt and new copper sealing washers. Make sure the locating lug on the fitting is engaged with the hole in the caliper, then tighten the bolt to the torque listed in this Chapter's Specifications.
7 Push the hose support into the strut bracket and install the U-clip, as necessary. Make sure the hose is not twisted between the caliper and any brackets.
8 Route the hose into the frame bracket, again making sure it is not twisted, then connect the brake line fitting, starting the threads by hand. Install the U-clip, then tighten the fitting securely.
9 Bleed the caliper (see Section 10).
10 Install the wheel and lug nuts, lower the vehicle and tighten the lug nuts to the torque listed in the Chapter 1 Specifications.

Rear brake hose

11 The rear brake hose serves as the flexible connection between two rigid metal lines, one on the body and the other on the suspension. Both ends of the hose are attached to these metal lines with threaded fittings and U-clips. Refer to Steps 2, 3 and 4 above. Be sure to bleed the wheel cylinder when you're done (see Section 10).

Metal brake lines

12 When replacing brake lines, be sure to use the correct parts. Don't use copper tubing for any brake system components. Purchase steel brake lines from a dealer or auto parts store.
13 Prefabricated brake line, with the tube ends already flared and fittings installed, is preferred and may be available at auto parts stores and dealer parts departments. These lines are also bent to the proper shapes. If pre-bent lines are not available, remove the defective line and purchase a straight section of brake line (with flared ends and fittings attached) that measures as close as possible to the original. Using the proper tubing bending tools, bend the new line to resemble the original.
14 When installing the new line, make sure it is securely supported in the bracket(s) and has plenty of clearance between moving or hot components.
15 After installation, check the master cylinder fluid level and add fluid as necessary. Bleed the brake system (see Section 10) and test the brakes carefully before driving the vehicle in traffic.

10 Brake hydraulic system - bleeding

Refer to illustration 10.8
Warning: *Wear eye protection when bleeding the brake system. If the fluid comes in contact with your eyes, immediately rinse them with water and seek medical attention. Do not get any brake fluid on the brake pads.*
Note: *Bleeding the hydraulic system is necessary to remove any air that manages to find its way into the system when it has been opened during removal and installation of a hose, line, caliper or master cylinder.*

1 If a brake line is disconnected at the brake master cylinder, or if air has entered it due to low fluid level in the brake master cylinder reservoir, start bleeding the brakes at the wheel cylinder or caliper farthest from the brake master cylinder, and move to the next closest wheel cylinder or caliper until all four wheels are bled.
2 If a brake line was disconnected only at a wheel, then only that caliper or wheel cylinder must be bled.
3 If a brake line is disconnected at a fitting located between the master cylinder and any of the brakes, that part of the system served by the disconnected brake line must be bled.
4 Remove any residual vacuum from the brake power booster by applying the brake several times with the engine off.
5 Remove the master cylinder reservoir cover and fill the reservoir

10.8 When bleeding the brakes, a hose is connected to the bleed screw at the caliper or wheel cylinder and then submerged in brake fluid - air will be seen as bubbles in the tube and container (all air must be expelled before moving to the next wheel)

with brake fluid. Reinstall the cover. **Note:** *Check the fluid level often during the bleeding operation and add fluid as necessary to prevent the fluid level from falling low enough to allow air bubbles into the master cylinder.*
6 Have an assistant on hand, as well as a supply of new brake fluid, a clear plastic container partially filled with clean brake fluid, a length of plastic, rubber or vinyl tubing to fit over the bleeder valve and a wrench to open and close the bleeder valve.
7 For a brake system bleeding of all four wheels, begin at the right rear wheel; remove the bleeder cap and attach the vinyl hose to the bleeder valve. Loosen the bleeder valve slightly, then tighten it to a point where it is snug but can still be loosened quickly and easily. If bleeding only specific wheels or portions of the brake system, follow the appropriate Steps below.
8 Place one end of the tubing over the bleeder valve and submerge the other end in brake fluid in the container **(see illustration)**.
9 Have the assistant pump the brakes slowly a few times to get pressure in the system, then hold the pedal down firmly.
10 While the pedal is held down, open the bleeder valve just enough to allow a flow of fluid to leave the valve. Watch for air bubbles to exit the submerged end of the tubing. When the fluid flow slows after a couple of seconds, close the valve and have your assistant release the pedal.
11 Repeat Steps 9 and 10 until no more air is seen leaving the tubing, then tighten the bleeder valve securely and proceed to the left front wheel, the left rear wheel and the right front wheel, in that order, and perform the same procedure. Be sure to check the fluid in the master cylinder reservoir frequently, keeping it about 3/4 full during bleeding. **Note:** *Always use new, fresh brake fluid. Old fluid or fluid from an opened container contains moisture which will deteriorate the brake system components.*
12 At the end of the operation, refill the master cylinder with fluid to the MAX mark on the reservoir.
13 Check the operation of the brakes. The pedal should feel solid when depressed, with no sponginess. If necessary, repeat the entire brake system bleeding. **Warning:** *Do not operate the vehicle if you are in doubt about the effectiveness of the brake system.*

11 Power brake booster - check, removal and installation

Operating check

1 Depress the brake pedal several times with the engine off and make sure there's no change in the pedal reserve distance (distance

Chapter 9 Brakes

11.10 To disconnect the power brake booster pushrod from the brake pedal, remove the retaining clip and clevis pin (arrows)

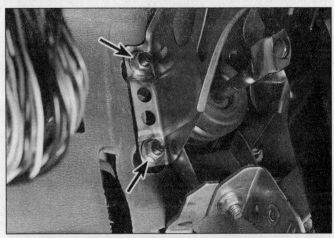

11.11 To detach the booster from the firewall, remove the four mounting nuts (arrows) (right nuts not visible in this photo)

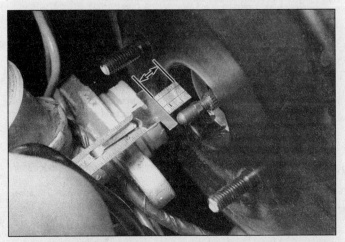

11.15a Measure the distance that the pushrod protrudes from the master cylinder mounting surface on the power brake booster

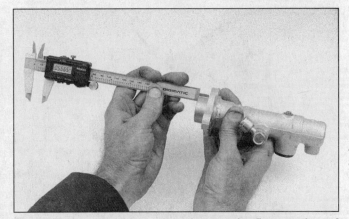

11.15b Measure the distance from the mounting flange to the end of the master cylinder

from the pedal to the floor).

2 Depress the pedal and start the engine. If the pedal goes down slightly, operation is normal.

Air tightness check

3 Start the engine and turn it off after one or two minutes. Depress the brake pedal slowly several times. If the pedal depresses less each time, the booster is airtight.

4 Depress the brake pedal while the engine is running, then stop the engine with the pedal held depressed. If there is no change in the pedal reserve travel after holding the pedal for 30 seconds, the booster is airtight. **Note:** *If the air tightness check fails in either Step above, first try checking and/or replacing the power brake booster vacuum hose/check valve and repeat the air tightness check.*

Removal

Refer to illustrations 11.10 and 11.11

5 Power brake booster units shouldn't be disassembled. They require special tools not normally found in most automotive repair stations or shops. Because of its critical relationship to brake performance, the booster should be replaced with a new or rebuilt one.

6 Disconnect the vacuum hose/check valve leading from the engine to the booster. Be careful not to damage the hose when removing it from the booster fitting.

7 Remove the brake master cylinder (see Section 10).

8 Remove the steering column lower finish panel (see Chapter 11).

9 Remove the pedal return spring.

10 Locate the pushrod clevis connecting the booster to the brake pedal **(see illustration)**. Remove the cotter pin from the clevis pin with pliers and pull out the clevis pin.

11 Using a flashlight, locate the four mounting nuts **(see illustration)** holding the brake booster to the firewall and remove them.

12 Slide the booster straight out from the firewall until the studs clear the holes. Be careful not to tear or damage the brake booster gasket between the firewall and the booster.

Installation

Refer to illustrations 11.15a, 11.15b, 11.15c and 11.16

13 Installation is basically the reverse of removal. Tighten the booster mounting nuts to the torque listed in this Chapter's Specifications. Be sure to use a new clevis retaining clip if the old clip is loose.

14 When installing the power brake booster vacuum hose/check valve, be sure to install the vacuum hose/check valve with the arrows on the vacuum hose toward the engine.

15 If the power brake booster unit is being replaced, check the pushrod clearance as follows:

a) *Measure the distance that the pushrod protrudes from the master cylinder mounting surface on the front of the power brake booster* **(see illustration)**. *Jot down this measurement - this will be called "dimension A."*

b) *Measure the distance from the mounting flange (with the gasket in place, if equipped) to the end of the master cylinder as shown* **(see illustration)**. *Jot down this measurement - this will be called "dimension B."*

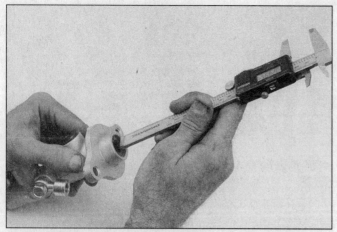

11.15c Measure the distance from the piston pocket to the end of the master cylinder

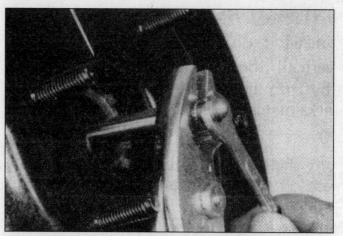

11.16 To adjust the length of the booster pushrod, loosen the locknut and turn the rod in or out as necessary

c) Using a vernier caliper, measure the distance from the end of the master cylinder to the bottom of the pocket in the piston **(see illustration)**. Jot down this measurement - this will be called "dimension C."

d) Subtract dimension B from dimension C, then subtract dimension A from the difference between B and C. This is the actual pushrod clearance.

e) Compare your calculation to the pushrod clearance listed in this Chapter's Specifications. If necessary, adjust the pushrod length to achieve the correct clearance.

16 To adjust the pushrod length, loosen the pushrod locknut and turn the knurled part of the pushrod with a pair of pliers **(see illustration)**. Recheck the clearance. Repeat this step as often as necessary until the clearance is correct.

17 After the master cylinder and brake lines have been installed, adjust the brake pedal height and freeplay (see Section 14) and bleed the brakes (see Section 10).

12 Parking brake - adjustment

Refer to illustration 12.3

1 If the parking brake cable is correctly adjusted, the parking brake lever, when applied, should travel 11 to 16 clicks to lock up the rear wheels. If the lever requires less than the specified minimum number of clicks, the parking brake may not be releasing completely and could cause the rear brakes to drag. If the lever can be pulled up more than the specified maximum number of clicks, the parking brake may not hold adequately on an incline, allowing the car to roll.

2 To access the parking brake cable adjuster, remove the rear console (see Chapter 11).

3 Remove the locking clip from the adjuster nut. Turn the adjuster nut **(see illustration)** until the lever applies the parking brake system within the specified number of clicks. Tighten the locknut. Install the locking clip.

4 Install the rear console.

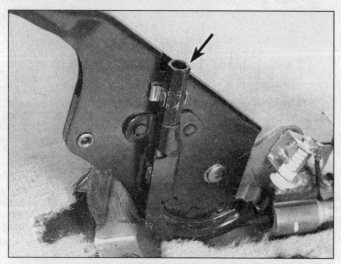

12.3 If the parking brake lever doesn't apply the rear brakes with the specified number of clicks, turn this adjuster nut (arrow) until it does

13 Parking brake cables - replacement

Front cable

Refer to illustrations 13.3, 13.9 and 13.10

1 Remove the rear console (see Chapter 11).

2 Remove the locking clip, then remove the adjuster nut **(see illustration 12.3)**.

3 Unplug the electrical connector for the parking brake indicator light switch **(see illustration)**.

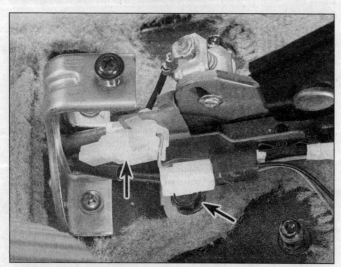

13.3 Unplug the electrical connector (left arrow) for the parking brake light indicator switch; to detach the parking brake lever from the floorpan, remove the two large retaining bolts (arrow indicates right bolt; left bolt, not visible in this photo, on other side of lever)

Chapter 9 Brakes

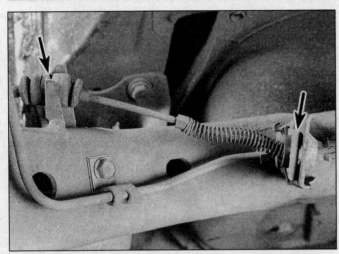

13.9 On the rear axle, remove this lock clip (right arrow) to disengage the front parking brake cable from the inner bracket; to disengage the cable from the outer bracket, simply pull out the grommet (left arrow) (the rear cable uses the same setup)

13.10 To disconnect either parking brake cable from the rear drum brake, remove the cotter pin and pull out this clevis pin (arrow)

4 Remove the two parking brake lever mounting bolts **(see illustration 13.3)**, lift up the parking brake lever and disengage the parking brake cable from the lever.
5 Remove the rear seat (see Chapter 11).
6 Remove the luggage compartment carpeting hold-down pins and peel back the carpeting the expose the parking brake cable cover.
7 Remove the two parking brake cable cover retaining screws.
8 Loosen the rear wheel lug nuts, raise the rear of the vehicle and support it securely on jackstands. Block the front wheels. Remove the wheels.
9 The left cable is attached to two cable brackets on the rear axle. Remove the locking clip that attaches the front cable to the bracket that's closer to the center **(see illustration)**.
10 Remove the cotter pin and clevis pin attaching the front cable to the lever at the left rear drum brake **(see illustration)**.
11 Disengage the rubber grommet from the outer left cable bracket on the axle **(see illustration 13.9)**.
12 Remove the bolt that attaches the cable bracket to the fuel tank.
13 Remove the two equalizer retaining bolts.
14 To remove the front cable, pull it through the floorpan from underneath the vehicle.
15 Installation is the reverse of removal. Apply a light coat of grease to the cable clevis pin.
16 Adjust the parking brake cable when you're done (see Section 12).

Rear cable

17 The rear cable, which extends from the equalizer to the right rear drum brake, is removed in a similar fashion to the front cable. Refer to Steps 8 through 13.
18 Adjust the parking brake cable when you're done (see Section 12).

14 Brake pedal - adjustment

Pedal height

Refer to illustrations 14.1 and 14.5

1 Measure the pedal height **(see illustration)** and compare your measurement to the pedal height listed in this Chapter's Specifications.
2 If the pedal height is incorrect, inspect the pedal pivot (at the top of the pedal) for missing, damaged or excessively worn bushings. Replace damaged bushings as necessary, then measure pedal height again.
3 If the pedal height is still incorrect, adjust it as follows:
4 Unplug the electrical connector for the brake light switch.
5 Loosen the brake light switch jam nuts **(see illustration)** and move the switch to the rear until it no longer contacts the brake pedal.
6 Loosen the power brake booster-to-brake pedal pushrod locknut.

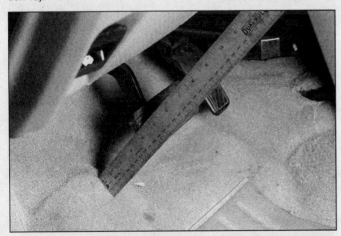

14.1 Brake pedal height is the distance between the pedal and the floorboard when the pedal is released

14.5 Before the brake pedal height can be adjusted, the brake light switch must be moved back so that the switch plunger doesn't contact the pedal; to do so, back off these jam nuts (arrows) and pull the switch back

9-20   Chapter 9   Brakes

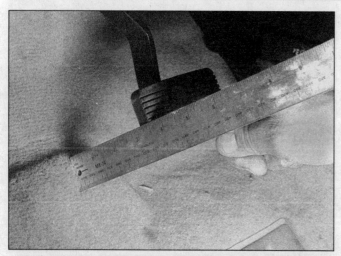

14.13 Brake pedal freeplay is the distance between the pedal when released and the point at which some resistance is first felt when the pedal is depressed

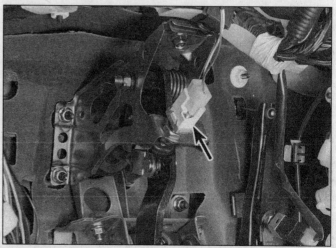

15.1 The brake light switch (arrow) is mounted on a bracket at the top of the brake pedal; to remove it, simply unplug the electrical connector, loosen the upper locknut, unscrew the lower locknut and pull out the switch

7   Adjust the pedal height by turning the pushrod in or out.
8   Tighten the pushrod locknut.
9   Adjust the brake light switch (see Section 15).
10 Verify that brake lights come on when the brake pedal is depressed, and go off when the brake pedal is released.
11 Check and, if necessary, adjust the brake pedal freeplay (see below).

Pedal freeplay

Refer to illustration 14.13

12 Stop the engine, if it's running, and depress the brake pedal several times until there's no more vacuum left in the booster.
13 Gently press the pedal by hand until you feel some resistance, then measure the distance between the fully released pedal and the point at which you feel resistance **(see illustration)**. Compare your measurement with the pedal freeplay listed in this Chapter's Specifications. If the pedal freeplay is incorrect, adjust it as follows:
14 Loosen the brake booster-to-brake pedal pushrod locknut.
15 Adjust pedal freeplay by turning the pushrod in or out. (If the pedal freeplay cannot be adjusted properly, troubleshoot the brake system.)
16 Tighten the pushrod locknut.
17 Adjust the brake light switch (see Section 15).

15   Brake light switch - check, adjustment and replacement

Check

Refer to illustration 15.1

1   The brake light switch **(see illustration)** is located on a bracket at the top of the brake pedal. The switch activates the brake lights at the rear of the vehicle when the pedal is depressed.

2   To check the brake light switch, simply note whether the brake lights come on when the pedal is depressed and go off when the pedal is released. If they do not function correctly, adjust the switch (see below).
3   If the lights still do not come on even after the switch is adjusted, either voltage is not reaching the switch, the switch is defective, or the circuit between the switch and the lights is defective. There is always the remote possibility that all of the brake light bulbs are burned out, but this is not very likely.
4   Use a voltmeter or test light to verify that there's voltage present at one side of the switch connector. If no voltage is present, troubleshoot the circuit from the switch to the fuse box. If there is voltage present, check for voltage on the other terminal when the brake pedal is depressed. If no voltage is present, replace the switch. If there is voltage present, troubleshoot the circuit from the switch to the brake lights (see the *wiring diagrams* at the end of Chapter 12).

Adjustment

5   Unplug the electrical connector from the switch.
6   Loosen the switch locknuts **(see illustration 14.5)**.
7   Hook up an ohmmeter across the switch connector terminals.
8   Move the switch until it produces continuity when the brake pedal is depressed, and produces no continuity when the pedal is released.
9   Tighten the locknuts.
10 Plug in the connector.

Replacement

11 Unplug the electrical connector from the switch.
12 Loosen the upper switch locknut, unscrew the lower locknut and remove the switch from the pedal bracket.
13 Installation is the reverse of removal.
14 Adjust the switch (see above).
15 Adjust the brake pedal (see Section 14).

Chapter 10
Suspension and steering systems

Contents

	Section
Axle beam - removal, inspection and installation	10
Balljoints - check and replacement	6
Control arm - removal, inspection and installation	5
General information	1
Hub and bearing assembly (front) - removal and installation	8
Power steering fluid level check	See Chapter 1
Power steering pump - removal and installation	15
Power steering system -bleeding	16
Shock absorber/coil spring (rear) - removal, inspection and installation	9
Stabilizer bar and bushings - removal and installation	4

	Section
Steering and suspension check	See Chapter 1
Steering gear - removal and installation	14
Steering gear boots - replacement	13
Steering knuckle and hub - removal and installation	7
Steering wheel - removal and installation	11
Strut assembly (front) - removal, inspection and installation	2
Strut/shock absorber or coil spring - replacement	3
Tie-rod ends - removal and installation	12
Wheel alignment - general information	18
Wheels and tires - general information	17

Specifications

Torque specifications
Ft-lbs

Front suspension
Balljoint-to-steering knuckle pinch bolt/nut	32 to 40
Control arm pivot bolt	32 to 40
Stabilizer bar	
Bracket nuts	40 to 50
Stabilizer-to-control arm nuts	47 to 57
Strut/coil spring assembly	
Strut-to-steering knuckle bolts/nuts	69 to 86
Strut upper mounting nuts	34 to 46
Piston rod nut	40 to 50

Rear suspension
Axle beam pivot bolt nuts	69 to 86
Shock absorber	
Upper piston rod nut	144 to 216
Lower shock-to-axle beam bolt	50 to 60

Steering
Steering gear bracket bolts or nuts	
Festiva	23 to 34
Aspire	27 to 38
Steering wheel	
Horn pad screws (airbag-equipped models)	36 to 53 in-lbs
Steering wheel nut	29 to 36
Tie-rod end-to-steering knuckle nuts	
Festiva	26 to 30
Aspire	31 to 42
Universal joint-to-input shaft pinch bolt	13 to 20

Chapter 10 Suspension and steering systems

1.1 Typical front suspension and steering components

1. Stabilizer bar
2. Stabilizer bar bushing brackets
3. Control arms
4. Balljoints
5. Driveaxles
6. Steering gear boot
7. Steering gear assembly
8. Tie-rod ends
9. Strut/coil spring assembly

1.2 Typical rear suspension components

1. Shock absorber and coil spring assembly
2. Axle beam

Chapter 10 Suspension and steering systems

1 General information

Refer to illustrations 1.1 and 1.2

The front suspension **(see illustration)** is a MacPherson strut design. The upper end of each strut is attached to the vehicle's strut towers. The lower end of the strut is bolted to the upper end of the steering knuckle. The steering knuckle is attached to a balljoint mounted on the outer end of the suspension control arm. A stabilizer bar, which reduces vehicle roll during cornering, is bolted to the vehicle and the ends are connected to the control arms.

The rear suspension **(see illustration)** utilizes integral shock absorber/coil spring assemblies. The upper end of each shock is attached to the vehicle body. The lower end of each shock is attached to the axle beam. The axle beam is attached to the vehicle by a pair of arms welded to each end of the beam.

A rack-and-pinion type steering gear, which is located behind the engine/transaxle assembly on the lower firewall, actuates the tie-rods, which are attached to the steering knuckles. The inner ends of the tie-rods are protected by rubber dust boots which should be inspected periodically for secure attachment, tears and leaking lubricant.

The power assist system consists of a belt-driven pump and a series of metal lines and flexible hoses connecting the pump to the steering gear. The fluid level in the power steering pump reservoir should be checked periodically (see Chapter 1).

The steering wheel operates the steering shaft, which actuates the steering gear through universal joints. Looseness in the steering can be caused by wear in the steering shaft universal joints, the steering gear, the tie-rod ends and loose retaining bolts.

Frequently, when working on the suspension or steering system components, you may come across fasteners which seem impossible to loosen. These fasteners on the underside of the vehicle are continually subjected to water, road grime, mud, etc., and can become rusted or "frozen," making them extremely difficult to remove. In order to unscrew these stubborn fasteners without damaging them (or other components), be sure to use lots of penetrating oil and allow it to soak in for a while. Using a wire brush to clean exposed threads will also ease removal of the nut or bolt and prevent damage to the threads. Sometimes a sharp blow with a hammer and punch will break the bond between a nut and bolt threads, but care must be taken to prevent the punch from slipping off the fastener and ruining the threads. Heating the stuck fastener and surrounding area with a torch sometimes helps too, but isn't recommended because of the obvious dangers associated with fire. Long breaker bars and extension, or "cheater," pipes will increase leverage, but never use an extension pipe on a ratchet - the ratcheting mechanism could be damaged. Sometimes tightening the nut or bolt first will help to break it loose. Fasteners that require drastic measures to remove should always be replaced with new ones.

Since most of the procedures dealt with in this Chapter involve jacking up the vehicle and working underneath it, a good pair of jackstands will be needed. A hydraulic floor jack is the preferred type of jack to lift the vehicle, and it can also be used to support certain components during various operations. **Warning:** *Never, under any circumstances, rely on a jack to support the vehicle while working on it. Whenever any of the suspension or steering fasteners are loosened or removed they must be inspected and, if necessary, replaced with new ones of the same part number or of original equipment quality and design. Torque specifications must be followed for proper reassembly and component retention. Never attempt to heat or straighten any suspension or steering components. Instead, replace any bent or damaged part with a new one.*

2 Strut assembly (front) - removal, inspection and installation

Removal

Refer to illustrations 2.3 and 2.5

1 Loosen the wheel lug nuts, raise the vehicle and support it securely on jackstands. Remove the wheel.
2 Unclip the brake hose bracket from the strut. If the vehicle is equipped with ABS, detach the speed sensor wiring harness from the strut by removing the clamp bracket bolt.
3 Remove the strut-to-knuckle nuts **(see illustration)** and knock the bolts out with a hammer and punch.
4 Separate the strut from the steering knuckle. Be careful not to overextend the inner CV joint. Also, do not let the steering knuckle fall outward and strain the brake hose.
5 Mark the relationship of one of the upper mounting studs to the body. This will ensure that the strut upper mount is reinstalled in the same position, preserving the camber adjustment. Support the strut and spring assembly with one hand and remove the strut-to-body nuts **(see illustration)**. Remove the assembly out from the fenderwell.

Inspection

6 Check the strut body for leaking fluid, dents, cracks and other obvious damage which would warrant repair or replacement.
7 Check the coil spring for chips or cracks in the spring coating (this will cause premature spring failure due to corrosion). Inspect the spring seat for cuts and general deterioration.
8 If any undesirable conditions exist, proceed to the strut disassembly procedure (see Section 3).

Installation

9 Guide the strut assembly up into the fenderwell and insert the upper mounting studs through the holes in the body. Once the studs protrude, install the nuts to support the strut from falling back down-

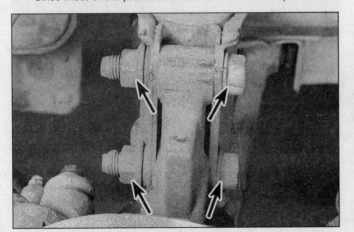

2.3 To detach the strut assembly from the steering knuckle, remove the two nuts (arrows), then knock out the bolts with a hammer and punch

2.5 To detach the upper end of the strut assembly from the body, remove the upper mounting nuts (arrows)

Chapter 10 Suspension and steering systems

3.3 Install the spring compressor according to the tool manufacturer's instructions and compress the spring until all pressure is relieved from the upper spring seat

3.4 Remove the piston rod nut

ward. The help of an assistant is recommended, as the strut is heavy and awkward.

10 Slide the steering knuckle into the strut flange and insert the two bolts. Install the nuts and tighten them to the torque listed in this Chapter's Specifications.

11 Connect the brake hose to the clip on the strut. If the vehicle is equipped with ABS, install the speed sensor wiring harness bracket.

12 Install the wheel and lug nuts, then lower the vehicle and tighten the lug nuts to the torque listed in the Chapter 1 Specifications.

13 Tighten the upper mounting nuts to the torque listed in this Chapter's Specifications.

14 Drive the vehicle to an alignment shop to have the front end alignment checked, and if necessary, adjusted. **Note:** *When disposing of used struts, return them to your automotive parts store or dealer - some struts are gas-charged and require special disposal procedures.*

3   Strut/shock absorber or spring - replacement

1   If the struts or coil springs exhibit the telltale signs of wear (leaking fluid, loss of damping capability, chipped, sagging or cracked coil springs) explore all options before beginning any work. The strut/shock absorber assemblies are not serviceable and must be replaced if a problem develops. However, strut assemblies complete with springs may be available on an exchange basis, which eliminates much time and work. Whichever repair/replacement method you choose, check on the cost and availability of parts before disassembling your vehicle. **Warning:** *Disassembling a strut is potentially dangerous and utmost attention must be directed to the job, or serious injury may result. Use only a high-quality spring compressor and carefully follow the manu-*

facturer's instructions furnished with the tool. After removing the coil spring from the strut assembly, set it aside in a safe, isolated area.

Disassembly

Refer to illustrations 3.3, 3.4, 3.5, 3.6 and 3.7

2   Remove the strut assembly following the procedure described in the previous Section. Mount the strut assembly in a vise. Line the vise jaws with wood or rags to prevent damage to the unit and do not tighten the vise excessively.

3   Following the tool manufacturer's instructions, install the spring compressor (which can be obtained at most auto parts stores or equipment yards on a daily rental basis) on the spring and compress it sufficiently to relieve all pressure from the upper spring seat **(see illustration).** This can be verified by wiggling the spring.

4   Loosen the piston rod nut **(see illustration).** It may be necessary to hold the rod from turning while loosening the nut.

5   Remove the nut and mounting bracket **(see illustration).** There should already be a white factory alignment spot on the bracket; this spot must be on the same side of the strut as the steering knuckle mounting bracket. If there's no factory mark, make your own mark on the mounting bracket so that it will be reinstalled with the same orientation. Inspect the bearing in the mounting bracket for smooth operation. If it does not turn smoothly, replace the mounting bracket. Check the rubber portion of the mounting bracket for cracking and general deterioration. If there is any separation of the rubber, replace it.

6   Lift the spring seat and upper insulator from the piston rod **(see illustration).** Check the rubber spring seat for cracking and hardness, replacing it if necessary.

7   Carefully lift the compressed spring from the assembly **(see illustration)** and set it in a safe place. **Warning:** *Keep the ends of the spring pointed away from your body.*

3.5 Lift the suspension mounting bracket off the piston rod

3.6 Remove the spring seat from the piston rod

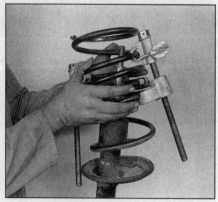

3.7 Remove the compressed spring assembly - keep the ends of the spring pointed away from your body

Chapter 10 Suspension and steering systems

3.11 When installing the spring, make sure the end fits into the recessed portion of the lower seat (arrow)

4.2 To detach the stabilizer bar brackets from the vehicle, remove these nuts or bolts (arrows)

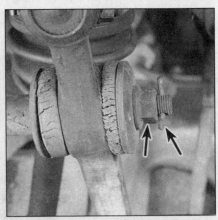

4.3 To disconnect the stabilizer bar from the control arms, remove the cotter pin and nut (arrows) from each end

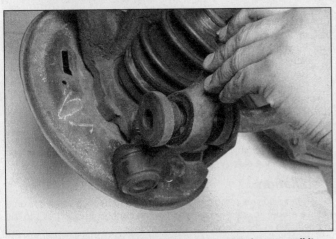

4.4 To separate the stabilizer from the control arms, pull it forward until it's disengaged from the bushings in the control arms

8 Slide the rubber bumper off the piston rod.
9 Check the lower insulator (if equipped) for wear, cracking and hardness and replace it if necessary.

Reassembly

Refer to illustration 3.11

10 If the lower insulator is being replaced, set it into position with the step (dropped portion) seated in the lowest part of the seat. Extend the piston rod to its full length and install the rubber bumper. Apply rubber lubricant to the rubber bumper contact surfaces.
11 Place the coil spring onto the lower insulator, with the end of the spring resting in the step (lowest part of the insulator) **(see illustration)**.
12 Install the upper insulator and upper spring seat. Apply rubber lubricant to the upper insulator contact surfaces.
13 Install the mounting bracket on the piston rod. Make sure the white factory alignment spot, or the mark you made prior to disassembly, faces out (same direction as the strut mounting bracket.
14 Install the piston rod nut and partially tighten it.
15 Remove the spring compressor tool.
16 Tighten the piston rod nut to the torque listed in this Chapter's Specifications.
17 Install the strut assembly (see Section 2).

4 Stabilizer bar and bushings - removal and installation

Refer to illustrations 4.2, 4.3 and 4.4

1 Loosen the front wheel lug nuts. Raise the front of the vehicle and support it securely on jackstands. Apply the parking brake and block the rear wheels to keep the vehicle from rolling off the stands. Remove the front wheels.
2 Remove the stabilizer bar bushing bracket nuts **(see illustration)** and remove the brackets.
3 Remove the cotter pins and nuts from the ends of the stabilizer bar **(see illustration)**.
4 Pull the stabilizer forward and disengage the ends from the control arms **(see illustration)**.
5 Remove and inspect the stabilizer bar bushings. If they're warped, cracked, torn or otherwise damaged, replace them.
6 Installation is the reverse of removal. Make sure that the concave washers are installed with the dished side facing out, away from the bushings. Be sure to tighten all fasteners to the torque listed in this Chapter's Specifications.

5 Control arm - removal, inspection and installation

Refer to illustrations 5.3, 5.4a and 5.4b

1 Loosen the wheel lug nuts on the side to be dismantled, raise the front of the vehicle, support it securely on jackstands and remove the wheel.
2 Remove the stabilizer bar-to-control arm nut and concave washer **(see illustration 4.3)**.
3 Remove the bolt and nut from the inner control arm pivot **(see illustration)**.

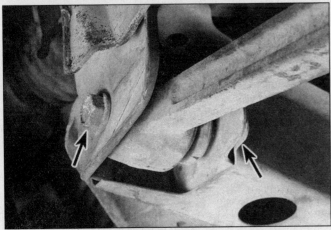

5.3 To disconnect the inner end of the control arm from the vehicle, remove this nut and bolt (arrows)

5.4a To disconnect the control arm from the steering knuckle, remove the ballstud pinch bolt and nut (arrows)

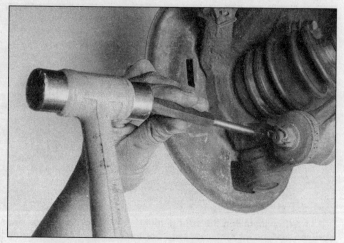

5.4b Knock out the ballstud pinch bolt with a punch

4 Remove the balljoint pinch bolt and nut from the steering knuckle (see illustrations).
5 Spread the joint slightly with a screwdriver or prybar, then pry the control arm down to separate it from the steering knuckle.
6 Disengage the control arm from the stabilizer bar and remove it from the vehicle.
7 Check the control arm for distortion and the pivot bushing for wear, damage and deterioration. If the control arm is damaged or bent, replace it. If the inner pivot bushing is worn, take the control arm to a dealer service department or automotive machine shop to have the old bushing removed and a new one installed. If the balljoint is worn or damaged, the control arm must be replaced.
8 Installation is the reverse of removal. Be sure to tighten all fasteners to the torque listed in this Chapter's Specifications.

6 Balljoints - check and replacement

The balljoints on this vehicle cannot be serviced separately. If a balljoint is worn or damaged, replace the control arm (see Section 5). Refer to the *Steering and suspension check* in Chapter 1 for the checking procedure.

7 Steering knuckle and hub - removal and installation

Removal

Warning: *Dust created by the brake system may contain asbestos, which is harmful to your health. Never blow it out with compressed air and do not inhale any of it. Do not, under any circumstances, use petroleum-based solvents to clean brake parts. Use brake system cleaner only.*

1 Loosen the driveaxle nut (see Chapter 8). Loosen the wheel lug nuts, raise the vehicle and support it securely on jackstands. Remove the wheel.
2 Remove the brake caliper and, on Aspire models, the brake disc (see Chapter 9), and disconnect the brake hose from the strut. On Festiva models, the brake disc cannot be removed from the hub until the hub itself has been removed (the heads of the hub-to-disc bolts face toward the steering knuckle).
3 If the vehicle is equipped with ABS, remove the wheel speed sensor.
4 Loosen, but don't remove the strut-to-steering knuckle nuts and bolts (see Section 2).
5 Separate the tie-rod end from the steering knuckle arm (see Section 12).
6 Remove the pinch bolt and nut holding the steering knuckle to the lower balljoint (see illustrations 5.4a and 5.4b). Pry the control arm balljoint from the steering knuckle.
7 Push the driveaxle from the hub (see Chapter 8). Support the end of the driveaxle with a piece of wire.
8 Remove the strut-to-knuckle bolts.
9 Carefully separate the steering knuckle from the strut and lift out the steering knuckle.
10 If you're replacing the steering knuckle, remove the hub and bearing assembly (see Section 8).
11 If you need to remove the disc on a Festiva model, have the hub and bearing assembly removed from the steering knuckle, then unbolt the hub from the disc.

Installation

12 If the disc was removed from the hub and bearing assembly on a Festiva model, install the disc and tighten the hub-to-disc bolts to the torque listed in the Chapter 9 Specifications.
13 If the hub and bearing assembly was removed, have it installed in the steering knuckle by a dealer service department or an automotive machine shop.
14 Guide the steering knuckle and hub assembly into position, inserting the driveaxle into the hub.
15 Push the steering knuckle into the strut flange and install the bolts and nuts, but do not tighten them yet.
16 Insert the balljoint into the steering knuckle and install the pinch bolt and nut. Tighten the ballstud pinch bolt and nut to the torque listed in this Chapter's Specifications.
17 Attach the tie-rod to the steering knuckle arm (see Section 12). Tighten the strut-to-steering knuckle bolts/nuts and the tie-rod end-to-steering knuckle nut to the torque listed in this Chapter's Specifications.
18 On Aspire models, install the brake disc on the hub and install the caliper as outlined in Chapter 9.
19 Install the driveaxle/hub nut and tighten it to the torque listed in the Chapter 8 Specifications.
20 Install the wheel and lug nuts.
21 Lower the vehicle and tighten the wheel lug nuts to the torque listed in the Chapter 1 Specifications.

8 Hub and bearing assembly (front) - removal and installation

Due to the special tools and expertise required to press the hub and bearing from the steering knuckle, this job should be left to a professional mechanic. However, the steering knuckle and hub may be removed and the assembly taken to a dealer service department or other repair shop. See Section 7 for the steering knuckle and hub removal procedure.

9 Shock absorber/coil spring (rear) - removal, inspection and installation

Removal

Refer to illustrations 9.3 and 9.5

1. Loosen the rear wheel lug nuts, raise the rear of the vehicle and support it securely on jackstands. Remove the wheels.
2. Support the axle beam near the lower shock mount with a floor jack.
3. Remove the shock-to-axle beam bolt **(see illustration)**.
4. To access the upper mounting nut, remove the luggage compartment side cover (see Chapter 11).
5. Remove the upper mounting nut **(see illustration)** while an assistant supports the shock so it doesn't fall. Guide the shock out of the fenderwell.

Inspection

6. Follow the inspection procedures described in Section 2. If you determine that the shock assembly must be disassembled for replacement of the shock or coil spring, refer to Section 3.

Installation

7. Maneuver the shock and coil spring assembly up into the fenderwell and insert the mounting studs through the holes in the body. Install the nut, but don't tighten it yet.
8. Position the lower end of the shock between the mounting bracket and the axle beam, install the bolt and tighten it to the torque listed in this Chapter's Specifications.
9. Install the wheel and lug nuts, lower the vehicle and tighten the lug nuts to the torque listed in the Chapter 1 Specifications.
10. Tighten the shock upper mounting nut to the torque listed in this Chapter's Specifications.

10 Axle beam - removal, inspection and installation

Warning: *Dust created by the brake system may contain asbestos, which is harmful to your health. Never blow it out with compressed air and don't inhale any of it. Do not, under any circumstances, use petroleum-based solvents to clean brake parts. Use brake cleaner or denatured alcohol only.*

Removal

Refer to illustration 10.7

1. Loosen the wheel lug nuts, raise the rear of the vehicle and place it securely on jackstands. Block the front wheels and remove the rear wheels.

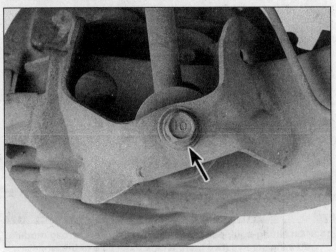

9.3 To detach the lower end of a rear shock from the axle beam, remove this bolt (arrow)

2. Remove the rear brake drums and the brake assemblies (see Chapter 9).
3. Disconnect the brake lines from the wheel cylinders and disconnect the brake hoses from the brake lines (see Chapter 9). Plug the brake lines to prevent moisture and contamination from entering the brake system.
4. Detach the brake backing plates from the axle beam and suspend them from the coil springs with pieces of wire. It isn't necessary to remove the parking brake cable from the backing plate.
5. Support the axle beam with a floor jack.
6. Disconnect the lower ends of the shock absorbers from the axle beam (see Section 9).
7. Remove the pivot bolts from the forward ends of the axle beam trailing arms **(see illustration)**.
8. Remove the axle beam assembly.

Inspection

9. Inspect the trailing arm bushings for cracks, deformation and wear. If they're damaged or worn out, take the axle beam assembly to a dealer service department or an automotive machine shop to have the old ones pressed out and new ones pressed in.

Installation

10. Installation is the reverse of removal. Be sure to tighten all fasteners to the torque listed in this Chapter's Specifications.
11. Lower the vehicle and tighten the lug nuts to the torque listed in

9.5 To detach the upper end of a rear shock from the body, remove this nut (arrow)

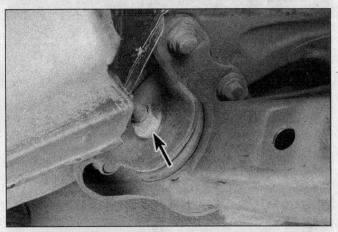

10.7 To detach the axle beam from the vehicle, remove the pivot nut (arrow) and bolt from the bracket for each trailing arm

10-8 Chapter 10 Suspension and steering systems

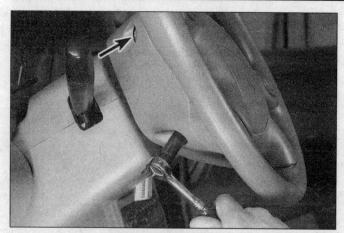

11.3 On airbag-equipped models, remove the four airbag module retaining screws (two more on the other side of the wheel, not visible in this photo)

11.5 To detach the steering wheel from the steering shaft, remove this nut

11.6 Be sure to mark the relationship of the steering wheel to the steering shaft to ensure that the steering is properly realigned during reassembly

the Chapter 1 Specifications.
12 Bleed the brakes (see Chapter 9).

11 Steering wheel - removal and installation

Refer to illustrations 11.3, 11.5, 11.6, 11.7 and 11.9

Warning: *Some models covered by this manual are equipped with Supplemental Restraint Systems (SRS), more commonly known as airbags. Always disconnect the negative battery cable, then the positive battery cable, then wait two minutes before working in the vicinity of the impact sensors, steering column or instrument panel to avoid the possibility of accidental deployment of the airbag, which could cause personal injury (see Chapter 12). Do not use electrical test equipment on any of the airbag system wiring or tamper with them in any way. Finally, the steering shaft must not be rotated while the steering wheel is removed; to do so could damage the airbag sliding contact assembly.*

1   Park the vehicle with the wheels pointing straight ahead. Disconnect the cable from the negative terminal of the battery. On airbag-equipped models, also disconnect the positive terminal, then wait two minutes before proceeding.
2   On models without an airbag, the horn pad is attached to the steering wheel with various combinations of clips and/or screws, depending on the model.
3   On models with an airbag, remove the four horn pad/airbag module retaining screws **(see illustration)**.
4   Lift off the horn pad and disconnect any wires for the horn, cruise control, etc.
5   Remove the steering wheel retaining nut **(see illustration)**.
6   Unless it's already marked, mark the relationship of the steering

11.7 Use a steering wheel puller, if necessary, to remove the steering wheel

11.9 Before installing the steering wheel on airbag models, make sure that the airbag sliding contact is properly aligned as shown

Chapter 10 Suspension and steering systems

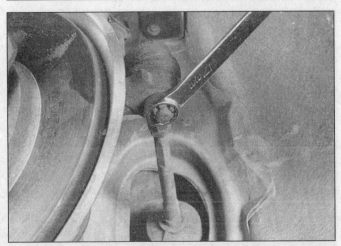

12.2 Remove the cotter pin and loosen - but don't remove - the castle nut on the tie-rod end ballstud; the nut needs to remain on the stud because sometimes the tie-rod end jumps out of the knuckle with considerable force when it pops loose

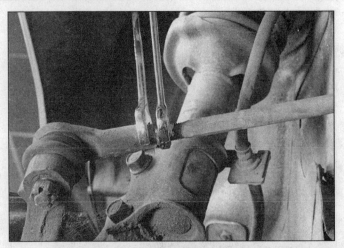

12.3a Using a back-up wrench on the tie-rod end, loosen the jam nut . . .

wheel to the steering shaft **(see illustration)** to ensure correct steering wheel alignment.

7 Use a puller, if necessary, to disconnect the steering wheel from the shaft **(see illustration)**.

8 On airbag models, tape the airbag sliding contact in place with a couple pieces of tape to prevent it from being accidentally rotated.

9 Make sure that the airbag sliding contact is correctly aligned **(see illustration)**. Note: *If the steering shaft has turned or sliding contact is not in alignment, turn the sliding contact clockwise until it locks, then turn it counterclockwise 2-3/4 turns and align the arrows on the sliding contact hub and housing.*

10 Installation is the reverse of removal. Be sure to tighten the steering wheel nut to the torque listed in this Chapter's Specifications.

12 Tie-rod ends - removal and installation

Removal

Refer to illustrations 12.2, 12.3a, 12.3b and 12.4

1 Loosen the wheel lug nuts. Raise the front of the vehicle, support it securely on jackstands, block the rear wheels and set the parking brake. Remove the front wheel.

2 Remove the cotter pin and *loosen* - don't remove - the nut on the tie-rod end stud **(see illustration)**.

3 Hold the tie-rod end with a back-up wrench and loosen the jam nut enough to mark the position of the tie-rod end in relation to the threads **(see illustrations)**.

4 Disconnect the tie rod from the steering knuckle arm with a puller **(see illustration)**.

5 Unscrew the tie-rod end from the tie-rod.

Installation

6 Thread the tie-rod end to the marked position on the tie-rod and insert the tie-rod stud into the steering knuckle arm. Tighten the jam nut securely.

7 Install the castle nut on the stud and tighten it to the torque listed in this Chapter's Specifications. Install a new cotter pin. If the hole for the cotter pin does not line up with one of the slots in the nut, turn the nut an additional amount until it slides through easily.

8 Install the wheel and lug nuts. Lower the vehicle and tighten the lug nuts to the torque listed in the Chapter 1 Specifications.

9 Have the alignment checked by a dealer service department or an alignment shop.

13 Steering gear boots - replacement

Refer to illustration 13.3

1 Loosen the lug nuts, raise the vehicle and support it securely on jackstands. Remove the wheel.

2 Remove the tie-rod end and jam nut (see Section 12).

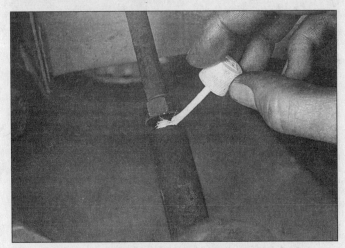

12.3b . . . then mark the position of the tie-rod end in relation to the threads

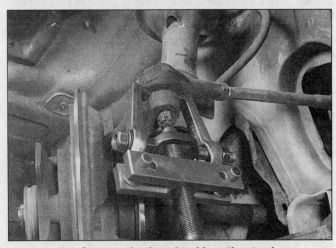

12.4 Separate the tie-rod end from the steering knuckle arm with a puller

Chapter 10 Suspension and steering systems

13.3 To remove an old steering gear boot, remove the clamps or cut off the wire-type retaining ring (arrows) and slide off the boot; that's one of the four steering gear bracket bolts (arrow) on the right

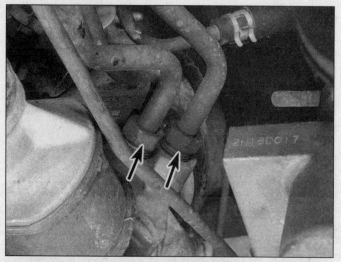

14.2 Disconnect the power steering line fittings (arrows)

3 Remove the steering gear boot clamps **(see illustration)**. Slide the boot off.
4 Before installing the new boot, wrap the threads and serrations on the end of the steering rod with a layer of tape so the small end of the new boot is not damaged during installation.
5 Place a new inner clamp over the steering gear.
6 Slide the new boot into position on the steering gear until it seats in the groove in the steering rod and install clamps. Tighten the clamps securely.
7 Remove the tape and install the tie-rod end jam nut and tie-rod end (see Section 12).
8 Install the wheel and lug nuts. Lower the vehicle and tighten the lug nuts to the torque listed in the Chapter 1 Specifications.
9 Have the alignment checked by a dealer service department or an alignment shop.

14 Steering gear - removal and installation

Removal

Refer to illustrations 14.2 and 14.7
Warning: *Some models are equipped with airbags. Make sure the steering shaft is not turned while the steering gear is removed or you could damage the airbag system. To prevent the shaft from turning, turn the ignition key to the lock position before beginning work or run the seat belt through the steering wheel and clip the seat belt into place. Make sure the ignition switch is OFF.*
1 Disconnect the cable from the negative terminal of the battery. If the vehicle is equipped with an airbag, also detach the positive cable and wait at least two minutes before proceeding. Loosen the front wheel lug nuts, raise the front of the vehicle and support it securely on jackstands. Apply the parking brake and remove the wheels. Remove the engine undercover splash shield.
2 If equipped with power steering, place a drain pan under the steering gear. Detach the power steering pressure and return lines **(see illustration)** and cap or cover the ends to prevent excessive fluid loss and contamination.
3 Remove the plastic protector that covers the intermediate steering shaft and the universal joint which connects the intermediate shaft to the steering gear input shaft.
4 Mark the relationship of the universal joint to the steering gear input shaft. Remove the U-joint pinch bolt.
5 Separate the tie-rod ends from the steering knuckle arms.

Remove the right tie-rod end (see Section 12).
6 **Note:** *Removing the catalytic converter will ease removal of the steering gear* (see Chapter 4).
7 Support the steering gear and remove the four steering gear bracket mounting fasteners **(see illustration)**.
8 Separate the intermediate steering shaft from the steering gear input shaft and move the steering gear to the left until the right tie-rod is clear of the fenderwell, then lower and remove the steering gear towards the right side of the vehicle. **Warning:** *Do NOT turn the steering wheel while the steering gear is removed. This is especially important on airbag models. If the steering wheel is inadvertently turned, remove the steering wheel and recenter the airbag sliding contact (see Section 11).*

Installation

9 Raise the steering gear into position, right end first, then insert the right tie-rod end through the right fender well opening far enough to allow insertion of the left tie-rod end through the left fender well opening. Don't snag the steering gear boots on any sharp edges. Connect the steering gear input shaft and the steering intermediate shaft U-joint. Make sure the marks are aligned.
10 Install the steering gear mounting brackets and tighten the bracket fasteners to the torque listed in this Chapter's Specifications.
11 Connect the tie-rod ends to the steering knuckle arms (see Section 12).

14.7 Remove the fasteners from the steering gear brackets (arrows)

Chapter 10 Suspension and steering systems

12 Install the universal joint pinch bolt and tighten it to the torque listed in this Chapter's Specifications.
13 If the vehicle is equipped with power steering, connect the power steering pressure and return lines to the steering gear.
14 Remove the jackstands and lower the vehicle.
15 If the vehicle is equipped with power steering, fill the power steering pump reservoir with the recommended fluid (see Chapter 1) and bleed the steering system (see Section 16).
16 Have the alignment checked by a dealer service department or an alignment shop.

15 Power steering pump - removal and installation

Removal

1 Disconnect the cable from the negative battery terminal.
2 Remove the air cleaner housing and intake duct (see Chapter 4).
3 Unplug the electrical connector for the power steering pressure switch.
4 Using a large syringe or suction gun, suck as much fluid out of the power steering fluid reservoir as possible. Place a drain pan under the vehicle to catch any fluid that spills out when the hoses are disconnected. Cap or cover the hoses to prevent entry of dirt or other contaminants.
5 Disconnect the reservoir-to-pump hose and plug it. Remove the power steering high-pressure hose bracket bolt and disconnect the high-pressure hose from the power steering pump. Plug the high-pressure hose.
6 Loosen and remove the power steering pump adjuster bolt locknut and the adjuster bolt and remove the drivebelt (see Chapter 1).
7 Remove the power steering pump pivot bolt and pump-to-mounting bracket bolts/nuts, then remove the pump from the vehicle.
8 While the pump is removed, inspect the power steering lines, hoses and connections for tears, cracks and any other deterioration and replace as necessary.
9 If access to engine components is required, remove the pump mounting bracket mounting bolts and remove the mounting bracket.

Installation

10 Installation is the reverse of removal. Be sure to tighten all pressure and return hose fittings securely.
11 Top off the fluid level in the reservoir (see Chapter 1) and bleed the power steering system (see Section 16).

16 Power steering system - bleeding

1 Following any operation in which the power steering fluid lines have been disconnected, the power steering system must be bled to remove all air and obtain proper steering performance.
2 Before starting the engine and with the front wheels in the straight ahead position, check the power steering fluid level and, if low, add fluid until it reaches the L (Low) mark on the dipstick.
3 Start the engine and allow it to run at fast idle. Recheck the fluid level and add more new, fresh power steering fluid if necessary to reach the L mark on the dipstick.
4 Bleed the system by turning the wheels from side to side, without hitting the stops. This will work the air out of the system. Continuously check the reservoir and keep the reservoir full of fluid as this is done.
5 When the air is worked out of the system, return the wheels to the straight ahead position and keep the vehicle running for several more minutes before shutting it off, or road test as follows before shutting the engine off.
6 Road test the vehicle to be sure the steering system is functioning normally and is noise-free.
7 Recheck and top off the power steering fluid level to the F (Full) mark on the dipstick while the engine is at normal operating temperature. Add fluid if necessary (see Chapter 1).

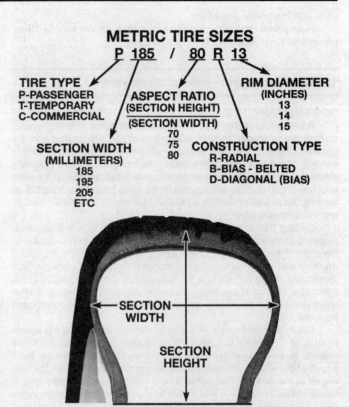

17.1 Metric tire size code

17 Wheels and tires - general information

Refer to illustration 17.1

All vehicles covered by this manual are equipped with metric-sized fiberglass or steel belted radial tires **(see illustration)**. Use of other sizes or types of tires may affect the ride and handling of the vehicle. Don't mix different types of tires, such as radials and bias belted, on the same vehicle as handling may be seriously affected. It is recommended that tires be replaced in pairs on the same axle, but if only one tire is being replaced, be sure it's the same size, type, structure and tread design as the other, with the same or better tread wear rating, traction rating, and temperature rating.

Because tire pressure has a substantial effect on handling and wear, the pressure on all tires should be checked at least once a month or before any extended trips (see Chapter 1). Make sure that tires are not worn below the tread depth wear indicators molded into the tread or depth measured is at least the recommended depth in your owner's manual or as recommended by the tire manufacturer.

These models are factory-equipped with either steel or aluminum wheels. If alkaline compounds (road salt or saltwater) get on aluminum wheels, flush both the outside and inside the wheels with water soon. Wheels must be replaced if they are bent, dented, leak air, have elongated bolt holes, are heavily rusted or corroded, have wobble that is noticeable visually (the radial runout is excessive) or if the lug nuts will not stay tight. Wheel repairs that use welding or peening are not recommended. Never use a temporary spare for more than the prescribed driving distance and speed, and do not use the temporary spare wheel with a standard tire.

When installing/demounting tires, make sure the tire shop uses a "non-contact" tire mounting machine if you have aluminum wheels. Tire and wheel balance is important in the overall handling, braking, and performance of the vehicle. Unbalanced wheels can adversely affect handling and ride characteristics as well as tire life. Whenever a tire is installed on a wheel, the tire and wheel should be balanced by a shop with the proper equipment.

Rotate tires front to back, and back to front. Do not include "Temporary Use Only" spare tire in the tire rotation.

18 Wheel alignment - general information

Refer to illustration 18.1

A wheel alignment refers to the adjustments made to the wheels so they are in proper angular relationship to the suspension and the ground. Wheels that are out of proper alignment not only affect vehicle control, but also increase tire wear. The front end angles normally measured are camber, caster and toe-in **(see illustration)**. On the front end, toe-in and camber are adjustable. No adjustments are possible on the rear. All angles should be measured, however, to check for bent or worn suspension parts.

Getting the proper wheel alignment is a very exacting process, requiring complicated alignment machines to perform the job properly. Because of this, you should have a shop with a four-wheel alignment machine and a technician with the proper training to properly perform these tasks. However, we give the following information describing the basic idea of what is involved with wheel alignment so you can better understand the process and deal intelligently with the shop that does the work.

Toe-in is the turning in of the wheels. The purpose of a toe specification is to ensure stability with controlled parallel rolling of the wheels. In a vehicle with zero toe-in, the distance between the front edges of the wheels will be the same as the distance between the rear edges of the wheels. The actual amount of toe-in is normally only a fraction of an inch. Toe-in is adjusted by the tie-rod end position on the tie-rod. Incorrect toe-in will cause the tires to wear, by making them scrub excessively on the road surface, and will cause the vehicle to be less stable, especially during straight line driving.

Camber is the tilt of the wheels from vertical when viewed from the end of the vehicle. When the wheels tilt out at the top, camber is positive (+). When the wheels tilt in at the top, camber is negative (-). The amount of tilt is measured in degrees from vertical; this measurement is the camber angle. This angle affects the amount of tire tread which contacts the road and compensates for changes in the suspension geometry when the vehicle is cornering or traveling over undulating surfaces. It is adjusted on the front end by turning the strut upper mount 180-degrees.

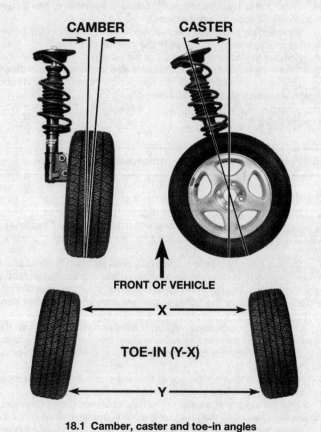

18.1 Camber, caster and toe-in angles

Caster is the tilting of the front steering axis from the vertical. A tilt toward the rear is positive caster and a tilt toward the front is negative caster. Caster is for directional stability by causing the steering to tend to return to center.

Chapter 11 Body

Contents

	Section		Section
Body - maintenance	2	Hood release latch and cable - removal and installation	10
Body repair - major damage	6	Instrument panel - removal and installation	26
Body repair - minor damage	5	Liftgate - removal, installation and adjustment	14
Bumpers - removal and installation	12	Liftgate support strut - replacement	15
Center console - removal and installation	24	Outside mirror (manual) - removal and installation	22
Dashboard trim panels - removal and installation	25	Radiator grille - removal and installation	11
Door - removal, installation and adjustment	17	Rear quarter window - replacement (1993 and earlier)	21
Door latch, lock cylinder and handle - removal and installation	18	Seat belts - check	28
Door trim panel - removal and installation	16	Seats - removal and installation	27
Door window glass - removal and installation	19	Steering column covers - removal and installation	23
Front fender - removal and installation	13	Upholstery and carpets - maintenance	4
General information	1	Vinyl trim - maintenance	3
Hinges and locks - maintenance	7	Window regulator - removal and installation	20
Hood - removal, installation and adjustment	9	Windshield and fixed glass - replacement	8

1 General information

The models covered by this manual feature a "unibody" construction, using a floor pan with front and rear frame side rails which support the body components, front and rear suspension systems and other mechanical components. Certain components are particularly vulnerable to accident damage and can be unbolted and repaired or replaced. Among these parts are the body moldings, bumpers, hood and trunk lids and all glass.

Only general body maintenance practices and body panel repair procedures within the scope of the do-it-yourselfer are included in this Chapter.

2 Body - maintenance

1 The condition of your vehicle's body is very important, because the resale value depends a great deal on it. It's much more difficult to repair a neglected or damaged body than it is to repair mechanical components. The hidden areas of the body, such as the wheel wells, the frame and the engine compartment, are equally important, although they don't require as frequent attention as the rest of the body.

2 Once a year, or every 12,000 miles, it is good to have the underside of the body steam cleaned. All traces of dirt and oil will be removed and the area can then be inspected carefully for rust, damaged brake lines, frayed electrical wires, damaged cables and other problems. The front suspension components should be greased after completion of this job.

3 At the same time, clean the engine and the engine compartment with a steam cleaner or water soluble degreaser.

4 The wheel wells should be given close attention, since undercoating can peel away and stones and dirt thrown up by the tires can cause the paint to chip and flake, allowing rust to set in. If rust is found, clean down to the bare metal and apply an anti-rust paint.

5 The body should be washed about once a week. Wet the vehicle thoroughly to soften the dirt, then wash it down with a soft sponge and plenty of clean soapy water. If the surplus dirt is not washed off very carefully, it can wear down the paint.

6 Spots of tar or asphalt thrown up from the road should be removed with a cloth soaked in solvent.

7 Once every six months, wax the body and chrome trim. If a chrome cleaner is used to remove rust from any of the vehicle's plated parts, remember that the cleaner also removes part of the chrome, so use it sparingly.

3 Vinyl trim - maintenance

Don't clean vinyl trim with detergents, caustic soap or petroleum-based cleaners. Plain soap and water works just fine, with a soft brush to clean dirt that may be ingrained. Wash the vinyl as frequently as the rest of the vehicle.

After cleaning, application of a high quality rubber and vinyl protectant will help prevent oxidation and cracks. The protectant can also be applied to weatherstripping, vacuum lines and rubber hoses, which often fail as a result of chemical degradation, and to the tires.

4 Upholstery and carpets - maintenance

1 Every three months remove the carpets or mats and clean the interior of the vehicle (more frequently if necessary). Vacuum the upholstery and carpets to remove loose dirt and dust.

2 Leather upholstery requires special care. Stains should be removed with warm water and a very mild soap solution. Use a clean, damp cloth to remove the soap, then wipe again with a dry cloth. Never use alcohol, gasoline, nail polish remover or thinner to clean leather upholstery.

3 After cleaning, regularly treat leather upholstery with a leather wax. Never use car wax on leather upholstery.

4 In areas where the interior of the vehicle is subject to bright sunlight, cover leather seats with a sheet if the vehicle is to be left out for any length of time.

These photos illustrate a method of repairing simple dents. They are intended to supplement *Body repair - minor damage* in this Chapter and should not be used as the sole instructions for body repair on these vehicles.

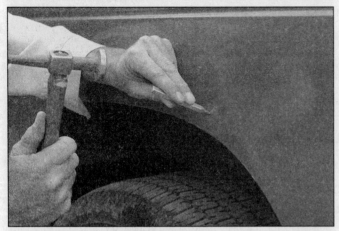

1   If you can't access the backside of the body panel to hammer out the dent, pull it out with a slide-hammer-type dent puller. In the deepest portion of the dent or along the crease line, drill or punch hole(s) at least one inch apart . . .

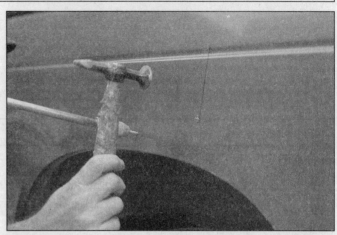

2   . . . then screw the slide-hammer into the hole and operate it. Tap with a hammer near the edge of the dent to help 'pop' the metal back to its original shape. When you're finished, the dent area should be close to its original contour and about 1/8-inch below the surface of the surrounding metal

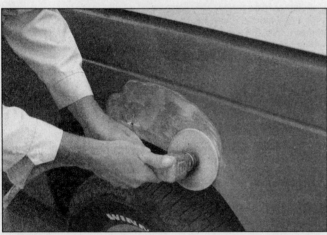

3   Using coarse-grit sandpaper, remove the paint down to the bare metal. Hand sanding works fine, but the disc sander shown here makes the job faster. Use finer (about 320-grit) sandpaper to feather-edge the paint at least one inch around the dent area

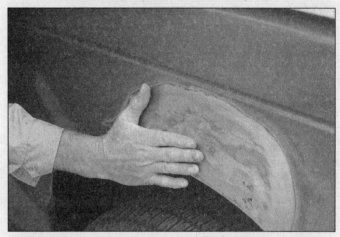

4   When the paint is removed, touch will probably be more helpful than sight for telling if the metal is straight. Hammer down the high spots or raise the low spots as necessary. Clean the repair area with wax/silicone remover

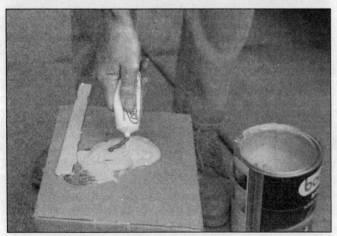

5   Following label instructions, mix up a batch of plastic filler and hardener. The ratio of filler to hardener is critical, and, if you mix it incorrectly, it will either not cure properly or cure too quickly (you won't have time to file and sand it into shape)

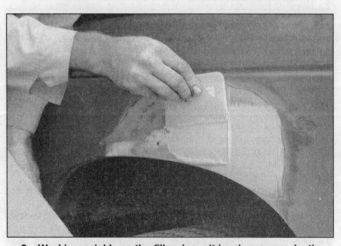

6   Working quickly so the filler doesn't harden, use a plastic applicator to press the body filler firmly into the metal, assuring it bonds completely. Work the filler until it matches the original contour and is slightly above the surrounding metal

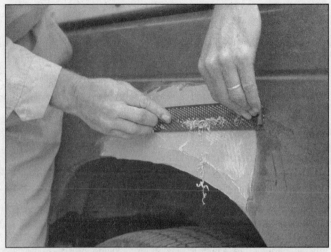
7 Let the filler harden until you can just dent it with your fingernail. Use a body file or Surform tool (shown here) to rough-shape the filler

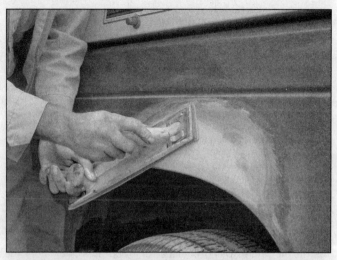
8 Use coarse-grit sandpaper and a sanding board or block to work the filler down until it's smooth and even. Work down to finer grits of sandpaper - always using a board or block - ending up with 360 or 400 grit

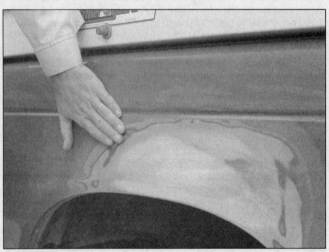
9 You shouldn't be able to feel any ridge at the transition from the filler to the bare metal or from the bare metal to the old paint. As soon as the repair is flat and uniform, remove the dust and mask off the adjacent panels or trim pieces

10 Apply several layers of primer to the area. Don't spray the primer on too heavy, so it sags or runs, and make sure each coat is dry before you spray on the next one. A professional-type spray gun is being used here, but aerosol spray primer is available inexpensively from auto parts stores

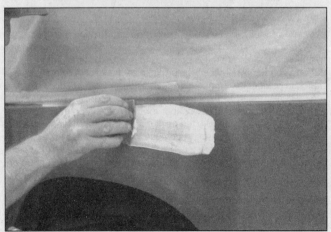
11 The primer will help reveal imperfections or scratches. Fill these with glazing compound. Follow the label instructions and sand it with 360 or 400-grit sandpaper until it's smooth. Repeat the glazing, sanding and respraying until the primer reveals a perfectly smooth surface

12 Finish sand the primer with very fine sandpaper (400 or 600-grit) to remove the primer overspray. Clean the area with water and allow it to dry. Use a tack rag to remove any dust, then apply the finish coat. Don't attempt to rub out or wax the repair area until the paint has dried completely (at least two weeks)

5 Body repair - minor damage

See photo sequence

Repair of minor scratches

1 If the scratch is superficial and does not penetrate to the metal of the body, repair is very simple. Lightly rub the scratched area with a fine rubbing compound to remove loose paint and built-up wax. Rinse the area with clean water.

2 Apply touch-up paint to the scratch, using a small brush. Continue to apply thin layers of paint until the surface of the paint in the scratch is level with the surrounding paint. Allow the new paint at least two weeks to harden, then blend it into the surrounding paint by rubbing with a very fine rubbing compound. Finally, apply a coat of wax to the scratch area.

3 If the scratch has penetrated the paint and exposed the metal of the body, causing the metal to rust, a different repair technique is required. Remove all loose rust from the bottom of the scratch with a pocket knife, then apply rust inhibiting paint to prevent the formation of rust in the future. Using a rubber or nylon applicator, coat the scratched area with glaze-type filler. If required, the filler can be mixed with thinner to provide a very thin paste, which is ideal for filling narrow scratches. Before the glaze filler in the scratch hardens, wrap a piece of smooth cotton cloth around the tip of a finger. Dip the cloth in thinner and then quickly wipe it along the surface of the scratch. This will ensure that the surface of the filler is slightly hollow. The scratch can now be painted over as described earlier in this Section.

Repair of dents

4 When repairing dents, the first job is to pull the dent out until the affected area is as close as possible to its original shape. There is no point in trying to restore the original shape completely as the metal in the damaged area will have stretched on impact and cannot be restored to its original contours. It is better to bring the level of the dent up to a point which is about 1/8-inch below the level of the surrounding metal. In cases where the dent is very shallow, it is not worth trying to pull it out at all.

5 If the back side of the dent is accessible, it can be hammered out gently from behind using a soft-face hammer. While doing this, hold a block of wood firmly against the opposite side of the metal to absorb the hammer blows and prevent the metal from being stretched.

6 If the dent is in a section of the body which has double layers, or some other factor makes it inaccessible from behind, a different technique is required. Drill several small holes through the metal inside the damaged area, particularly in the deeper sections. Screw long, self-tapping screws into the holes just enough for them to get a good grip in the metal. Now the dent can be pulled out by pulling on the protruding heads of the screws with locking pliers.

7 The next stage of repair is the removal of paint from the damaged area and from an inch or so of the surrounding metal. This is done with a wire brush or sanding disk in a drill motor, although it can be done just as effectively by hand with sandpaper. To complete the preparation for filling, score the surface of the bare metal with a screwdriver or the tang of a file, or drill small holes in the affected area. This will provide a good grip for the filler material. To complete the repair, see the subsection on filling and painting later in this Section.

Repair of rust holes or gashes

8 Remove all paint from the affected area and from an inch or so of the surrounding metal using a sanding disk or wire brush mounted in a drill motor. If these are not available, a few sheets of sandpaper will do the job just as effectively.

9 With the paint removed, you will be able to determine the severity of the corrosion and decide whether to replace the whole panel, if possible, or repair the affected area. New body panels are not as expensive as most people think and it is often quicker to install a new panel than to repair large areas of rust.

10 Remove all trim pieces from the affected area except those which will act as a guide to the original shape of the damaged body, such as headlight shells, etc. Using metal snips or a hacksaw blade, remove all loose metal and any other metal that is badly affected by rust. Hammer the edges of the hole in to create a slight depression for the filler material.

11 Wire brush the affected area to remove the powdery rust from the surface of the metal. If the back of the rusted area is accessible, treat it with rust inhibiting paint.

12 Before filling is done, block the hole in some way. This can be done with sheet metal riveted or screwed into place, or by stuffing the hole with wire mesh.

13 Once the hole is blocked off, the affected area can be filled and painted. See the following subsection on filling and painting.

Filling and painting

14 Many types of body fillers are available, but generally speaking, body repair kits which contain filler paste and a tube of resin hardener are best for this type of repair work. A wide, flexible plastic or nylon applicator will be necessary for imparting a smooth and contoured finish to the surface of the filler material. Mix up a small amount of filler on a clean piece of wood or cardboard (use the hardener sparingly). Follow the manufacturer's instructions on the package, otherwise the filler will set incorrectly.

15 Using the applicator, apply the filler paste to the prepared area. Draw the applicator across the surface of the filler to achieve the desired contour and to level the filler surface. As soon as a contour that approximates the original one is achieved, stop working the paste. If you continue, the paste will begin to stick to the applicator. Continue to add thin layers of paste at 20-minute intervals until the level of the filler is just above the surrounding metal.

16 Once the filler has hardened, the excess can be removed with a body file. From then on, progressively finer grades of sandpaper should be used, starting with a 180-grit paper and finishing with 600-grit wet-or-dry paper. Always wrap the sandpaper around a flat rubber or wooden block, otherwise the surface of the filler will not be completely flat. During the sanding of the filler surface, the wet-or-dry paper should be periodically rinsed in water. This will ensure that a very smooth finish is produced in the final stage.

17 At this point, the repair area should be surrounded by a ring of bare metal, which in turn should be encircled by the finely feathered edge of good paint. Rinse the repair area with clean water until all of the dust produced by the sanding operation is gone.

18 Spray the entire area with a light coat of primer. This will reveal any imperfections in the surface of the filler. Repair the imperfections with fresh filler paste or glaze filler and once more smooth the surface with sandpaper. Repeat this spray-and-repair procedure until you are satisfied that the surface of the filler and the feathered edge of the paint are perfect. Rinse the area with clean water and allow it to dry completely.

19 The repair area is now ready for painting. Spray painting must be carried out in a warm, dry, windless and dust-free atmosphere. These conditions can be created if you have access to a large indoor work area, but if you are forced to work in the open, you will have to pick the day very carefully. If you are working indoors, dousing the floor in the work area with water will help settle the dust which would otherwise be in the air. If the repair area is confined to one body panel, mask off the surrounding panels. This will help minimize the effects of a slight mismatch in paint color. Trim pieces such as chrome strips, door handles, etc., will also need to be masked off or removed. Use masking tape and several thickness of newspaper for the masking operations.

20 Before spraying, shake the paint can thoroughly, then spray a test area until the spray painting technique is mastered. Cover the repair area with a thick coat of primer. The thickness should be built up using several thin layers of primer rather than one thick one. Using 600-grit wet-or-dry sandpaper, rub down the surface of the primer until it is very smooth. While doing this, the work area should be thoroughly rinsed with water and the wet-or-dry sandpaper periodically rinsed as well. Allow the primer to dry before spraying additional coats.

21 Spray on the top coat, again building up the thickness by using several thin layers of paint. Begin spraying in the center of the repair area and then, using a circular motion, work out until the whole repair area and about two inches of the surrounding original paint is covered. Remove all masking material 10 to 15 minutes after spraying on the

Chapter 11 Body

9.4a On 1993 and earlier models, remove the hinge to hood retaining bolts (arrows) and lift off the hood with the help of an assistant

9.4b On 1994 and later models, remove the hinge to hood retaining nuts (arrows) and lift off the hood with the help of an assistant

final coat of paint. Allow the new paint at least two weeks to harden, then use a very fine rubbing compound to blend the edges of the new paint into the existing paint. Finally, apply a coat of wax.

6 Body repair - major damage

1   Major damage must be repaired by an auto body shop specifically equipped to perform unibody repairs. These shops have the specialized equipment required to do the job properly.
2   If the damage is extensive, the body must be checked for proper alignment or the vehicle's handling characteristics may be adversely affected and other components may wear at an accelerated rate.
3   Due to the fact that all of the major body components (hood, fenders, etc.) are separate and replaceable units, any seriously damaged components should be replaced rather than repaired. Sometimes the components can be found in a wrecking yard that specializes in used vehicle components, often at considerable savings over the cost of new parts.

7 Hinges and locks - maintenance

Once every 5000 miles, or every four months, the hinges and latch assemblies on the doors, hood and trunk should be given a few drops of light oil or lock lubricant. The door latch strikers should also be lubricated with a thin coat of grease to reduce wear and ensure free movement. Lubricate the door and trunk locks with spray-on graphite lubricant.

8 Windshield and fixed glass - replacement

Replacement of the windshield and fixed glass requires the use of special fast-setting adhesive/caulk materials and some specialized tools. It is recommended that these operations be left to a dealer or a shop specializing in glass work.

9 Hood - removal, installation and adjustment

Note: *The hood is heavy and somewhat awkward to remove and install - at least two people should perform this procedure.*

Removal and installation

Refer to illustrations 9.4a and 9.4b
1   Use blankets or pads to cover the cowl area of the body and fenders. This will protect the body and paint as the hood is lifted off.
2   Make marks or scribe a line around the hood hinge to ensure proper alignment during installation.
3   Disconnect any cables or wires that will interfere with removal.
4   Have an assistant support the hood. Remove the hinge-to-hood nuts or bolts **(see illustrations)**.
5   Lift off the hood.
6   Installation is the reverse of removal.

Adjustment

Refer to illustrations 9.10 and 9.11
7   Fore-and-aft and side-to-side adjustment of the hood is done by moving the hinge plate slot after loosening the bolts or nuts.
8   Scribe a line around the entire hinge plate so you can determine the amount of movement **(see illustration 9.4b)**.
9   Loosen the bolts or nuts and move the hood into correct alignment. Move it only a little at a time. Tighten the hinge bolts and carefully lower the hood to check the position.
10 If necessary after installation, the entire hood latch assembly can be adjusted up-and-down as well as from side-to-side on the radiator support so the hood closes securely and flush with the fenders. To make the adjustment, scribe a line or mark around the hood latch mounting bolts to provide a reference point, then loosen them and reposition the latch assembly, as necessary **(see illustration)**. Following adjustment, retighten the mounting bolts.

9.10 Scribe a line around the latch to use as a reference point. To adjust the hood latch, loosen the retaining bolts (arrows), move the latch and retighten bolts, then close the hood to check the fit

9.11 Adjust the hood closing height by turning the hood bumpers (arrow) in or out

10.2 Pry out the cable retainer from the backside of the hood latch assembly, then disengage the cable

11   Finally, adjust the hood bumpers on the radiator support so the hood, when closed, is flush with the fenders **(see illustration)**.
12   The hood latch assembly, as well as the hinges, should be periodically lubricated with white, lithium-base grease to prevent binding and wear.

10   Hood release latch and cable - removal and installation

Warning: *Some models covered by this manual are equipped with Supplemental Restraint systems (SRS), more commonly known as airbags. Always disconnect the negative battery cable, then the positive battery cable and wait two minutes before working in the vicinity of the impact sensors, steering column or instrument panel to avoid the possibility of accidental deployment of the airbag, which could cause personal injury (see Chapter 12). Do not use electrical test equipment on any of the airbag system wiring or tamper with them in any way.*

Latch

Refer to illustration 10.2
1   Scribe a line around the latch to aid alignment when installing, then detach the latch retaining bolts to the radiator support **(see illustration 9.10)** and remove the latch. **Note:** *If your vehicle is equipped with air bags it will be necessary to remove the crash sensor before removing the hood latch assembly.*
2   Disconnect the hood release cable by disengaging the cable from the latch assembly **(see illustration)**.
3   Installation is the reverse of the removal procedure. **Note:** *Adjust the latch so the hood engages securely when closed and the hood bumpers are slightly compressed.*

Cable

4   Disconnect the hood release cable from the latch assembly as described above.
5   Attach a piece of stiff wire to the end of the cable, then follow the cable back to the firewall and detach all the cable retaining clips.
6   Working in the passenger compartment, detach the screws securing the hood release lever.
7   Pull the cable and grommet rearward into the passenger compartment until you can see the wire. Ensure that the new cable has a grommet attached, then remove the old cable from the wire and replace it with the new cable.
8   Working from the engine compartment, pull the wire back through the firewall.
9   Installation is the reverse of the removal **Note:** *Push on the grommet with your fingers from the passenger compartment to seat the grommet in the firewall correctly.*

11   Radiator grille - removal and installation

Refer to illustration 11.2
Warning: *Some models covered by this manual are equipped with Supplemental Restraint systems (SRS), more commonly known as airbags. Always disconnect the negative battery cable, then the positive battery cable and wait two minutes before working in the vicinity of the impact sensors, steering column or instrument panel to avoid the possibility of accidental deployment of the airbag, which could cause personal injury (see Chapter 12). Do not use electrical test equipment on any of the airbag system wiring or tamper with them in any way.*
Note: *The radiator grille on 1994 and later SE models is part of the bumper assembly. See Section 12 for the removal procedure on these models.*
1   On 1993 and earlier models, remove the retaining screw located below the grille emblem. On 1994 and later models (except SE models), remove the retaining screw located above the grille emblem.
2   On all models, use a small screwdriver if necessary to disengage the grille retaining clips **(see illustration)**, then remove the radiator grille from the vehicle. Check the condition of the clips and replace as necessary.
3   Remove any of the grille retaining clips that remain fastened to the radiator support and reinstall the clips back into the grille.
4   To install, position the grille in place and press on the grille until the clips snap into place on the radiator support.

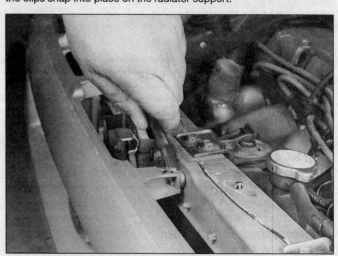

11.2 After removing the grille retaining screw, use a small screwdriver to disengage the grille retaining clips

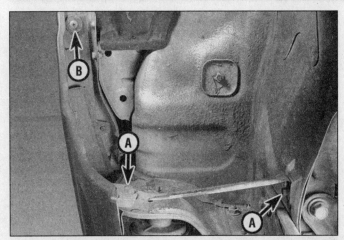

12.6 Detach the lower support bracket bolts (A) and the bumper cover to fender retaining screws (B) on each side of the vehicle

12.8 Remove the bumper retaining bolts (arrows) (1993 and earlier model shown)

12 Bumpers - removal and installation

Warning: *Some models covered by this manual are equipped with Supplemental Restraint systems (SRS), more commonly known as airbags. Always disconnect the negative battery cable, then the positive battery cable and wait two minutes before working in the vicinity of the impact sensors, steering column or instrument panel to avoid the possibility of accidental deployment of the airbag, which could cause personal injury (see Chapter 12). Do not use electrical test equipment on any of the airbag system wiring or tamper with them in any way.*

Note: *The bumper assembly is heavy and somewhat awkward to remove and install - at least two people should perform this procedure.*

Front bumper

Refer to illustrations 12.6 and 12.8

1 Apply the parking brake, raise the vehicle and support it securely on jackstands. Remove the under cover splash shield. Open the hood.
2 Disconnect the cable from the negative battery terminal and disconnect any wiring that would interfere with bumper removal.
3 Remove the radiator grille (see Section 11).
4 On 1994 and later models, remove the headlights and the front turn signal/running light assemblies (see Chapter 12).
5 On 1994 and later SE models, remove the bumper cover to fender retaining screws located in the front turn signal/running light opening.
6 Working under the vehicle, remove the bumper cover lower support brackets and the bumper cover to fender retaining screws **(see illustration)**.

13.3 Remove the inner splash shield retaining screws (arrows) (some retaining screws are not visible in this photo) and detach the inner splash shield from the vehicle

7 Support the bumper with a jack or jackstands. Alternatively, have an assistant support the bumper as the bolts are removed.
8 Remove the bumper retaining bolts and pull the bumper assembly out and away from the vehicle **(see illustration)**.
9 If replacing the bumper cover, simply remove the screws securing the bumper cover to the bumper.
10 Installation is the reverse of removal.

Rear bumper

11 Apply the parking brake, raise the vehicle and support it securely on jackstands.
12 Working under the vehicle, detach the plastic clips, screws and retaining brackets securing the lower edges of the bumper cover.
13 On 1994 and later models, remove the two lower bumper retaining nuts from the back side of the bumper.
14 On 1993 and earlier models, support the bumper with a jack or jackstands. Alternatively, have an assistant support the bumper as the bolts are removed. Remove the bumper retaining nuts from the backside of the bumper and pull the bumper assembly out and away from the vehicle.
15 On 1994 and later models, remove the plastic clips securing the upper edge of the bumper cover. Then working in the trunk, pry out the plastic clips securing the drivers side, passenger side, and rear inside trunk finishing panels to allow access to the bumper retaining bolts.
16 Support the bumper with a jack or jackstands. Alternatively, have an assistant support the bumper as the bolts are removed.
17 Detach the retaining nuts securing the bumper to the tail light panel. Pull the bumper assembly out and away from the vehicle.
18 If replacing the bumper cover, simply remove the screws securing the bumper cover to the bumper.
19 Installation is the reverse of removal.

13 Front fender - removal and installation

Refer to illustrations 13.3 and 13.5a, 13.5b and 13.5c

Warning: *Some models covered by this manual are equipped with Supplemental Restraint systems (SRS), more commonly known as airbags. Always disconnect the negative battery cable, then the positive battery cable and wait two minutes before working in the vicinity of the impact sensors, steering column or instrument panel to avoid the possibility of accidental deployment of the airbag, which could cause personal injury (see Chapter 12). Do not use electrical test equipment on any of the airbag system wiring or tamper with them in any way.*

1 Raise the vehicle, support it securely on jackstands and remove the front wheel.
2 Remove the front bumper assembly (see Section 12).
3 Detach the inner fenderwell splash shield and the front flair molding (if equipped) from the fender **(see illustration)**.

Chapter 11 Body

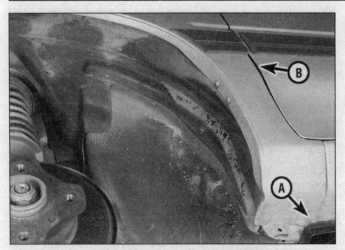

13.5a Remove the fender to rocker panel retaining bolt (A), then open the front door to access the fender to door pillar retaining bolt (B) (not visible in this photo)

13.5b Detach the fender retaining bolt (arrow) located behind front turn signal/running light assembly

4 On 1993 and earlier models, remove the front turn signal/running light assemblies (see Chapter 12).
5 Remove the remaining fender mounting bolts **(see illustrations)**.
6 Detach the fender. It's a good idea to have an assistant support the fender while it's being moved away from the vehicle to prevent damage to the surrounding body panels.
7 Installation is the reverse of removal.

14 Liftgate - removal, installation and adjustment

Note: *The liftgate is heavy and somewhat awkward to hold - at least two people should perform this procedure.*

Removal and installation

Refer to illustration 14.5
1 Open the liftgate and support it securely.
2 On 1993 and earlier models, remove the liftgate trim panel and disconnect all wiring harness connectors leading to the liftgate. Remove the cargo cover straps if equipped.
3 On 1994 and later models, remove the right side rear quarter trim panels from the trunk compartment and disconnect the all wiring harness connectors leading to the liftgate. Disconnect the rear window washer hose from the liftgate (if equipped).
4 While an assistant supports the liftgate, detach the support struts from the liftgate assembly (see Section 15).
5 Scribe a line around the liftgate hinges for a reference point to aid

13.5c Remove the remaining bolts (arrows) securing the top of the fender and detach the fender from the vehicle

the installation procedure. Then detach the hinge-to-liftgate bolts or nuts **(see illustration)** and remove the liftgate from the vehicle.
6 Installation is the reverse of removal.

Adjustment

Refer to illustration 14.8
7 Adjustments are made by loosening the hinge-to-liftgate bolts and moving the liftgate. Proper alignment is achieved when the edges of the liftgate are parallel with the rear quarter panel and the top of the tailgate.

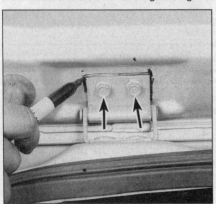

14.5 Before loosening the liftgate retaining bolts (arrows), draw a line around the hinge plate for a reinstallation reference

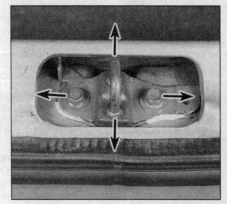

14.8 Adjust the liftgate lock striker by loosening the mounting bolts and gently tapping the striker in the desired direction (arrows)

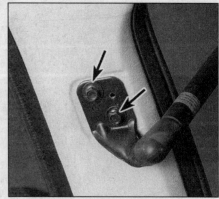

15.1 After supporting the liftgate, remove the bolts (arrows) at each end and detach the support strut

Chapter 11 Body

16.1 Remove the inside door handle retaining screw, then pull out the handle, detach the latch rod and remove the door handle - 1994 and later shown

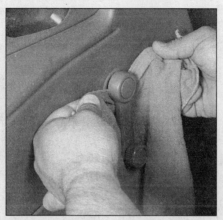

16.2 Work a cloth up behind the window regulator handle, then move it back-and-forth until the handle retaining clip releases itself from the regulator shaft

16.3a On 1993 and earlier models, remove retaining screws from the lower half of armrest/pull handle

8   Finally, adjust the latch striker assembly as necessary to provide positive engagement with the latch mechanism **(see illustration)**.

15 Liftgate support strut - replacement

Refer to illustration 15.1
Warning: *The support strut is filled with pressurized gas - do not disassemble this component. If it is faulty replace it with a new one.*
Note: *The rear liftgate is heavy and somewhat awkward to hold securely while replacing the struts - at least two people should perform this procedure.*
1   Remove the bolts at the ends and detach the strut from the liftgate and the chassis **(see illustration)**.
2   Installation is the reverse of the removal.

16 Door trim panel - removal and installation

Refer to illustrations 16.1, 16.2, 16.3a, 16.3b, 16.4a, 16.4b, 16.4c, 16.5 and 16.7
1   Remove the inside door handle retaining screw **(see illustration)**. Then disengage the inside door handle from the latch rod and remove it from the door assembly.
2   Remove the window crank by working a cloth back-and-forth behind the handle to dislodge the retainer **(see illustration)**. A special

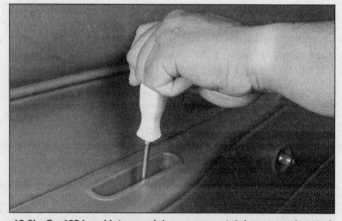

16.3b On 1994 and later models, remove retaining screw located in the center of the armrest/pull handle

tool is available for this purpose, but it is not essential. With the retainer removed, pull off the handle.
3   Detach the armrest pull handle retaining screw(s) **(see illustrations)**.
4   On front doors, remove the remaining door panel retaining screws securing the outer edges of the door trim panel **(see illustrations)**.

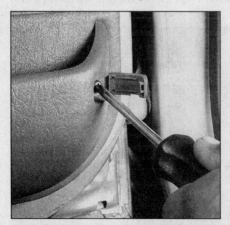

16.4a On 1993 and earlier models, remove the retaining screw from the lower front corner of the front door trim panel

16.4b On 1994 and later models, remove the retaining screws (arrows) securing the front edge of the front door trim panel

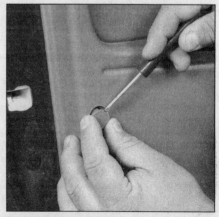

16.4c On 1994 and later two door models, use a small screwdriver to pry out the trim cap and remove the retaining screw securing the rear edge of the front door trim panel

11-10   Chapter 11 Body

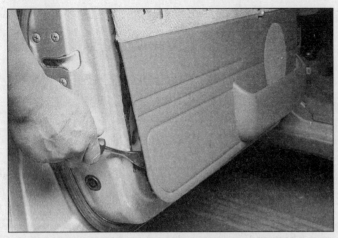

16.5 Use a trim panel removal tool to detach the trim panel retaining clips, then pull the door trim up and out to remove it - 1993 and earlier medal shown

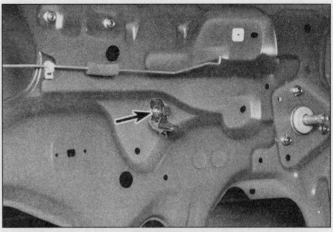

16.7 On 1994 and later models, remove the door trim panel support bracket (arrow) before peeling back the watershield

5   Insert a wide putty knife, a screwdriver or a special trim panel removal tool between the trim panel and the head of the retaining clip to disengage the door panel retaining clips **(see illustration)**. **Note:** *Door trim panel retaining clips are approximately six to ten inches apart. Pry at the clip location only. Prying in between clips will result in distorted or damaged door trim panels.*
6   Once all of the clips and screws are disengaged, detach the trim panel and remove the trim panel from the vehicle by gently pulling it up and out.
7   On 1993 and earlier models, carefully peel back the plastic watershield for access to the inner door. On 1994 and later vehicles, remove the door panel support bracket **(see illustration)**, then peel back the watershield for access to the inner door.
8   Prior to installation of the door panel, be sure to reinstall any clips in the panel which may have come out during the removal procedure and remain in the door itself.
9   Installation is the reverse of the removal procedure. **Note:** *When installing door trim panel retaining clips, make sure the clips are lined up with their mating holes first, then gently tap the clips in with the palm of your hand.*
10 Installation is the reverse of removal.

17 Door - removal, installation and adjustment

Note: *The door is heavy and somewhat awkward to remove and install - at least two people should perform this procedure.*

Removal and installation

Refer to illustrations 17.4 and 17.8

1   Raise the window completely in the door and then disconnect the negative cable from the battery.
2   Open the door all the way and support it on jacks or blocks covered with rags to prevent damaging the paint.
3   Remove the door trim panel and water deflector as described in Section 16.
4   Remove the door stop strut center pin **(see illustration)**.
5   Disconnect all electrical connections, ground wires and harness retaining clips from the door. **Note:** *It is a good idea to label all connections to aid the reassembly process.*
6   From the door side, detach the rubber conduit between the body and the door. Then pull the wiring harness through the conduit hole and remove the wiring from the door.
7   Mark around the door hinges with a pen or a scribe to facilitate realignment during reassembly.
8   With an assistant holding the door, remove the hinge-to-door bolts **(see illustrations)** and lift the door off.
9   Installation is the reverse of removal.

17.4 Gently tap the pin for the door stop strut in the direction shown

Adjustment

Refer to illustration 17.13

10 Having proper door-to-body alignment is a critical part of a well functioning door assembly. First check the door hinge pins for excessive play. Fully open the door and lift up and down on the door without lifting the body. If a door has 1/16-inch or more play, the hinges should be replaced.

17.8 Before loosening the door retaining bolts (arrows), draw a line around the hinge plate for a reinstallation reference (upper hinge bolts not visible in this photo)

Chapter 11 Body

17.13 Adjust the door lock striker by loosening the mounting screws and gently tapping the striker in the desired direction (arrows)

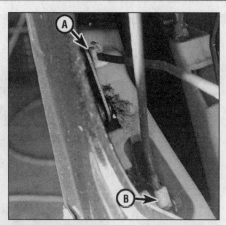

18.2 Detach the plastic clip(s) on the inside handle-to-latch rod (A) and the inside door lock-to-latch rod (B)

18.4 Remove the latch screws (arrows) from the end of the door and pull the latch assembly through the access hole

11 Door-to-body alignment adjustments are made by loosening the hinge-to-body bolts or hinge-to-door bolts and moving the door **(see illustration 17.8)**. Proper body alignment is achieved when the top of the doors are parallel with the roof section, the front door is flush with the fender, the rear door is flush with the rear quarter panel and the bottom of the doors are aligned with the lower rocker panel. If these goals can't be reached by adjusting the hinge-to-body or hinge-to-door bolts, body alignment shims may have to be purchased and inserted behind the hinges to achieve correct alignment.

12 To adjust the door closed position, scribe a line or mark around the striker plate to provide a reference point, then check that the door latch is contacting the center of the latch striker. If not adjust the up and down position first.

13 Finally adjust the latch striker sideways position, so that the door panel is flush with the center pillar or rear quarter panel and provides positive engagement with the latch mechanism **(see illustration)**.

18 Door latch, lock cylinder and handle - removal and installation

Door latch

Refer to illustrations 18.2 and 18.4

1 Raise the window then remove the door trim panel and watershield as described in (Section 16).

2 Working through the large access hole, disengage the inside handle-to-latch rod and the inside lock-to-latch rod **(see illustration)**. Disconnect any electrical connectors which would interfere with removal of the latch assembly.

3 All door locking rods are attached by plastic clips. The plastic clips can be removed by unsnapping the portion engaging the rod and then by pulling the rod out of its locating hole.

4 Remove the screws securing the latch to the door **(see illustration)**, then remove the latch assembly from the door.

5 Installation is the reverse of removal.

Lock cylinder and outside handle

Refer to illustrations 18.8 and 18.10

6 To remove the lock cylinder, raise the window and remove the door trim panel and watershield as described in Section 16.

7 Working through the large access hole, disengage the plastic clip that secures the lock cylinder-to-latch rod.

8 Using a screwdriver, slide the lock cylinder retaining clip out of engagement and remove the lock cylinder from the door **(see illustration)**.

9 To remove the outside handle, work through the access hole and disengage the plastic clip that secures the outside handle-to-latch rod.

10 Remove the outside handle retaining nuts **(see illustration)** and pull the handle from the door.

11 Installation is the reverse of removal.

18.8 To remove the lock cylinder, detach the plastic clip securing the lock rod (A) then pry off the lock cylinder retaining clip (B)

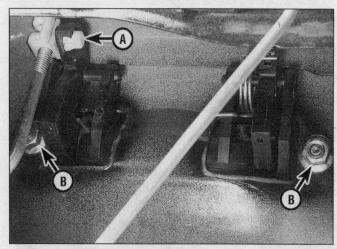

18.10 Detach the actuating rod clip (A) and the outside handle retaining nuts (B) - both items can be reached through the access hole in the door

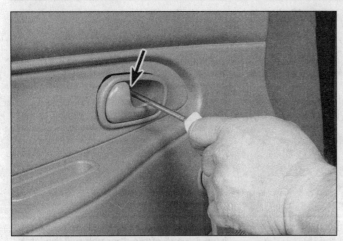

18.12 Remove the inside handle retaining screws (arrow), then rotate the handle outward and detach the actuating rods from the backside

19.3 On rear doors, carefully peel back the weatherstripping, detach the run channel retaining screw, then remove the glass run channel from the door to allow clearance for removal of the window glass

Inside handle

Refer to illustration 18.12

12 Remove the retaining screw **(see illustration)**.
13 Pull the handle away from the door trim panel, then disengage the inside door handle from the latch rod and remove the handle from the door.
14 Installation is the reverse of removal.

19 Door window glass - removal and installation

Refer to illustrations 19.3 and 19.4

1 Remove the door trim panel and the plastic watershield (see Section 16).
2 Lower the window glass all the way down into the door.
3 If you're replacing the rear door glass on a four door model, pry out the weather stripping from the top of the door frame. Detach the glass run channel retaining screw, then remove the glass run channel from the door **(see illustration)**.
4 Raise the window just enough to access the window retaining bolts through the hole in the door **(see illustration)**.
5 Place a rag over the glass to help prevent scratching the glass and remove the two glass mounting bolts.
6 Remove the glass by pulling it up and out.
7 Installation is the reverse of removal.

20 Window regulator - removal and installation

Refer to illustration 20.3

1 Remove the door trim panel and the plastic watershield (see Section 16).
2 Remove the window glass assembly (see Section 19).
3 Remove the window regulator assembly mounting bolts **(see illustration)**.
4 Pull the regulator assembly through the service hole in the door to remove it.
5 Installation is the reverse of removal.

21 Rear quarter window - replacement (1993 and earlier)

Refer to illustrations 21.1, 21.2a and 21.2b
Note: *The rear quarter window glass is fragile and somewhat awkward to remove and install - at least two people should perform this procedure.*

1 Open the window latch and remove the retaining screws securing the latch to the body **(see illustration)**.
2 With an assistant holding the window glass, remove the retaining screws securing the hinges to the body **(see illustrations)**.
3 Remove the rear quarter window glass from the vehicle.
4 Installation is the reverse of removal.

19.4 Raise the window just enough to access the glass retaining bolts (arrows) through the hole in the door

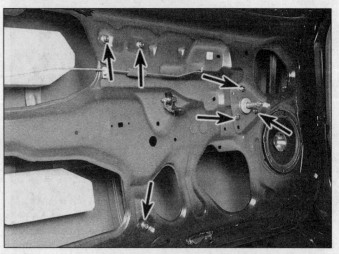

20.3 Detach the regulator mounting bolts (arrows) and remove it through the access hole in the door

Chapter 11 Body

21.1 With the rear quarter window latch in the open position, remove the retaining screws securing the latch to the body

21.2a Use a small screwdriver to pry off the hinge trim covers

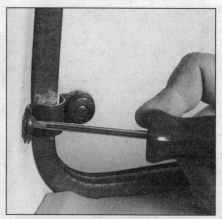

21.2b While an assistant holds the rear quarter window glass remove the hinge retaining screws

22.1a On 1994 and later models, pry off the mirror control handle trim cover . . .

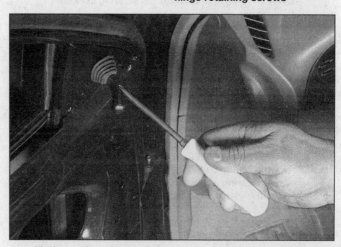

22.1b . . . then detach the control handle retaining screw

22 Outside mirror (manual) - removal and installation

Refer to illustrations 22.1a, 22.1b, 22.2 and 22.3

1 On 1994 and later models, pry off the mirror control handle trim cover. Then detach the control handle retaining screw and remove the control handle and the rubber boot assembly **(see illustrations)**.

2 On 1993 and earlier models, detach the mirror trim cover retaining screw and remove the mirror trim cover. On 1994 and later models, simply pry off the mirror trim cover **(see illustration)**.

3 Remove the mirror retaining bolts or screws and detach the mirror from the vehicle **(see illustration)**.

4 Installation is the reverse of removal.

22.2 Pry off the mirror trim cover - 1994 and later model shown

22.3 Remove the mirror retaining screws (arrows) to remove the mirror - 1994 and later model shown

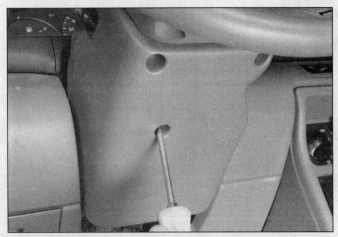

23.3 Detach the screws, then separate the steering column cover halves to remove them from the steering column

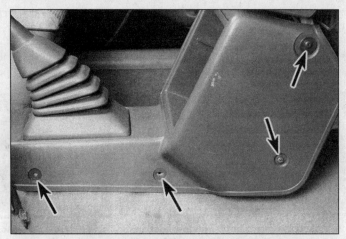

24.7a On 1993 and earlier models, detach the retaining screws (arrows) from each side of the front console

23 Steering column covers - removal and installation

Refer to illustration 23.3

Warning: *Some models covered by this manual are equipped with Supplemental Restraint systems (SRS), more commonly known as airbags. Always disconnect the negative battery cable, then the positive battery cable and wait two minutes before working in the vicinity of the impact sensors, steering column or instrument panel to avoid the possibility of accidental deployment of the airbag, which could cause personal injury (see Chapter 12). Do not use electrical test equipment on any of the airbag system wiring or tamper with them in any way.*

1   Remove the steering wheel (see Chapter 10).
2   On 1993 and earlier models, remove the instrument cluster bezel (see Section 25).
3   Remove the steering column cover screws. **(see illustration)**.
4   Separate the cover halves and detach them from the steering column.
5   Installation is the reverse of removal.

24 Center console - removal and installation

Warning: *Some models covered by this manual are equipped with Supplemental Restraint systems (SRS), more commonly known as airbags. Always disconnect the negative battery cable, then the positive battery cable and wait two minutes before working in the vicinity of the impact sensors, steering column or instrument panel to avoid the possibility of accidental deployment of the airbag, which could cause personal injury (see Chapter 12). Do not use electrical test equipment on any of the airbag system wiring or tamper with them in any way.*

1   If the vehicle is equipped with an airbag, disconnect the negative battery cable, then the positive battery cable and wait two minutes before proceeding any further.

Parking brake console

2   On 1993 and earlier models, position the front seats all the way forward and remove the safety belt buckle retaining bolts on each side of the rear console. Remove two screws at the rear of the console and one screw located at the front of the console.
3   On 1994 and later models, remove the rear console ash tray then detach the console retaining screw located below the ash tray.
4   On all models, apply the parking brake lever and remove the rear console access cover.
5   Gently lift the rear console up and over the parking brake lever and remove it from the vehicle.

Shift lever console

Refer to illustrations 24.7a, 24.7b and 24.7c

6   On vehicles equipped with manual transmissions, unscrew the shift lever knob.
7   Detach the retaining screws securing the front half of the console **(see illustrations)**.
8   Lift the console up and over the shift lever. Disconnect any electrical connections and remove the console from the vehicle.
9   Installation is the reverse of removal.

24.7b On 1994 and later models, remove two screws (arrows) at the rear of the front console . . .

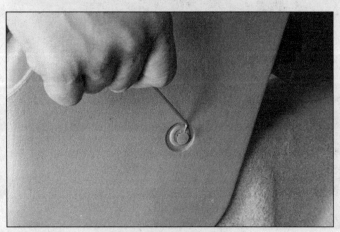

24.7c . . . then remove the plastic fasteners from each side of the console

Chapter 11 Body

25.3a On 1994 and later models, pry off the bezel trim cover . . .

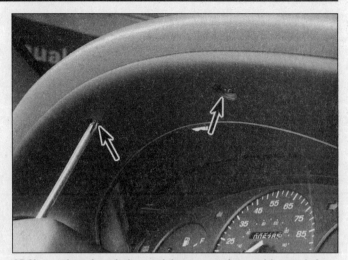

25.3b . . . then detach the retaining screws (arrows) located along the upper edge of the instrument cluster bezel - grasp the bezel securely and pull back sharply to detach the remaining clips securing the bezel

25 Dashboard trim panels - removal and installation

Warning: *Some models covered by this manual are equipped with Supplemental Restraint systems (SRS), more commonly known as airbags. Always disconnect the negative battery cable, then the positive battery cable and wait two minutes before working in the vicinity of the impact sensors, steering column or instrument panel to avoid the possibility of accidental deployment of the airbag, which could cause personal injury (see Chapter 12). Do not use electrical test equipment on any of the airbag system wiring or tamper with them in any way.*

Instrument cluster bezel

Refer to illustrations 25.3a and 25.3b

1 Disconnect the negative battery cable. If the vehicle is equipped with an airbag, also disconnect the positive battery cable and wait two minutes before proceeding any further.
2 On 1993 and earlier models, remove five bezel retaining screws. Pull the instrument cluster bezel outward to access the electrical connections on the backside. Disconnect all electrical connections from the back of the cluster bezel and remove the bezel from the vehicle.
3 On 1994 and later models, carefully pry upward on the instrument cluster bezel trim cover and remove it from the instrument panel **(see illustration)**. Remove two screws at the top of the instrument cluster bezel **(see illustration)**. Then grasp the bezel securely and pull back sharply to detach the clips securing the bezel to the instrument panel.
4 Installation is the reverse of the removal.

Radio trim bezel (1993 and earlier)

5 Remove the two retaining screws located along the lower edge of the bezel, then grasp the bezel securely and detach it from the instrument panel by pulling it straight back.
6 Installation is the reverse of the removal.

Glove box

Refer to illustration 25.7

7 To remove the glove box, simply open the glove box door and detach the retaining screws from the hinge **(see illustration)**.
8 Lower the glove box from the instrument panel.
9 Installation is the reverse of the removal.

26 Instrument panel - removal and installation

Refer to illustrations 26.8, 26.9, 26.11, 26.13a, 26.13b, 26.13c, 26.14a, 26.14b, 26.15a and 26.15b

Warning: *Some models covered by this manual are equipped with Supplemental Restraint systems (SRS), more commonly known as airbags. Always disconnect the negative battery cable, then the positive battery cable and wait two minutes before working in the vicinity of the impact sensors, steering column or instrument panel to avoid the possibility of accidental deployment of the airbag, which could cause personal injury (see Chapter 12). Do not use electrical test equipment on any of the airbag system wiring or tamper with them in any way.*
Note: *The instrument panel is heavy and somewhat awkward to remove and install - at least two people should perform this procedure.*

1 Disconnect the negative battery cable. If the vehicle is equipped with an airbag, also disconnect the positive battery cable and wait two minutes before proceeding any further. **Warning:** *Always use extreme care when working around airbag modules. Never strike, pry or bump airbag modules to avoid the possibility of accidental deployment.*
2 Remove the center floor console (see Section 24).
3 Remove all of the dashboard trim panels described in Section 25.
4 Remove the air conditioning and heater control panel (see Chapter 3).
5 On 1993 and earlier models remove the radio (see Chapter 12).
6 Remove the instrument cluster (see Chapter 12).
7 Remove the driver's side airbag (if equipped) and the steering wheel (see Chapter 10).
8 On 1993 and earlier models, remove the left and right side heater ducts from underneath the instrument panel, then remove the lower steering column cover, shield and the shield retaining bracket **(see illustration)**.

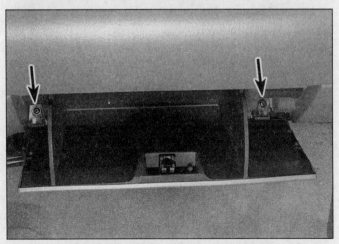

25.7 Glove box retaining screws (arrows)

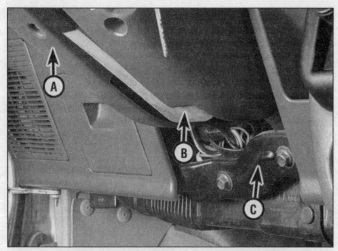

26.8 On 1993 and earlier models, remove the lower steering column cover (A), steering column shield (B) and the shield retaining bracket (C)

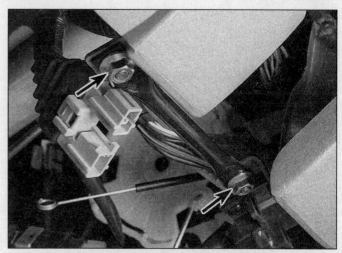

26.9 On 1994 and later models, remove the instrument panel support brace retaining nuts (arrows)

26.11 Remove the interior fuse panel retaining screws (arrows)

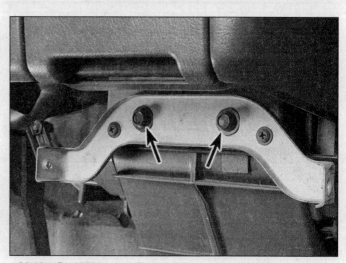

26.13a On 1993 and earlier models, remove the lower retaining bolts (arrows) located in the center of the instrument panel . . .

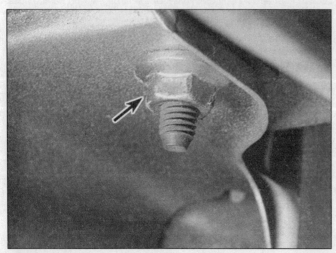

26.13b . . . and the nuts (arrow) at each of the lower corners of the instrument panel

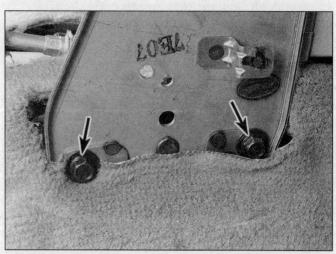

26.13c On 1994 and later models, remove the two bolts (arrows) from each side of the instrument panel lower brace

Chapter 11 Body

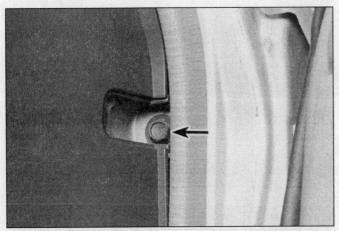

26.14a On 1993 and earlier models, pry out the retaining bolt trim covers and remove the bolt (arrow) securing each side of the instrument panel

26.14b On 1994 and later models, pry out the trim covers and remove the bolts (arrows) securing each side of the instrument panel

26.15a On 1993 and earlier models, pry out the trim covers and remove the three bolts (arrow) retaining the upper half of the instrument panel

26.15b On 1994 and later models, pry out the trim cover and remove one bolt (arrow) retaining the upper half of the instrument panel

9 On 1994 and later models, remove the instrument panel support brace from below the steering column **(see illustration)**.
10 Disconnect the electrical connectors leading to the steering column, then detach the steering column retaining bolts and lower the column to the floor.
11 Detach the screws securing the fuse box and the hood release handle **(see illustration)**.
12 Working through the glove box opening, detach the four retaining bolts securing the passenger side air bag module (if equipped). Lift the air bag module out and away from the instrument panel, then disconnect the electrical connectors and remove the air bag module.
13 Remove the bolts securing the lower half of the instrument panel **(see illustrations)**.
14 Pry out the trim covers and remove the screws securing the left and right sides of the instrument panel **(see illustrations)**.
15 Pry out the trim covers and remove the screws securing the upper edge of the instrument panel **(see illustrations)**.
16 Pull the instrument panel towards the rear of the vehicle and detach any electrical connectors interfering with removal.
17 With the help of an assistant remove the instrument panel from the vehicle.
18 Installation is the reverse of removal.

27 Seats - removal and installation

Front seat

Refer to illustration 27.2

1 Position the seat all the way forward or all the way to the rear to access the front seat retaining bolts.
2 Detach any bolt trim covers and remove the retaining bolts **(see illustration)**.
3 Tilt the seat upward to access the underneath, then disconnect any electrical connectors and lift the seat from the vehicle.
4 Installation is the reverse of removal.

27.2 The front seat retaining bolts (arrows) can be accessed by moving the seat forward or backward

27.5 Remove the bolts (arrow) securing the lower edge of the seat cushion

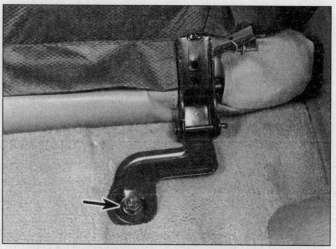

27.6 Fold the seat back downward to access the seat back hinge retaining bolts (arrow)

Rear seat

Refer to illustrations 27.5 and 27.6

5   Remove the bolts securing the lower front edge of the cushion **(see illustration)**.
6   Fold the rear seat back down and remove the retaining bolts at the lower edge of the seat back **(see illustration)**.
7   Lift up on the lower edge of the seat back and remove the seat from the vehicle.
8   Installation is the reverse of removal.

28 Seat belts - check

1   Check the seat belts, buckles, latch plates and guide loops for any obvious damage or signs of wear.
2   Make sure the seat belt reminder light comes on when the key is turned on.
3   The seat belts are designed to lock up during a sudden stop or impact, yet allow free movement during normal driving. The retractors should hold the belt against your chest while driving and rewind the belt when the buckle is unlatched.
4   If any of the above checks reveal problems with the seat belt system, replace parts as necessary.

Chapter 12
Chassis electrical system

Contents

Section		Section
Airbag system - general information	24	Headlights - adjustment	13
Antenna - removal and installation	17	Horn - check and replacement	20
Bulb replacement	14	Ignition key lock cylinder - removal and installation	10
Combination switch - check	8	Ignition switch - check and replacement	9
Combination switch - removal and installation	7	Instrument cluster - removal and installation	18
Daytime Running Lights (DRL) - general information	15	Radio and speakers - removal and installation	16
Electric side view mirrors - description and check	23	Rear window defogger - check and repair	22
Electrical troubleshooting - general information	2	Rear window defogger switch - check and replacement	21
Fuses - general information	3	Relays - general information and testing	5
Fusible links - general information	4	Turn signal/hazard flashers - check and replacement	6
General information	1	Wiper motor - check and replacement	19
Headlight bulb - replacement	11	Wiring diagrams	25
Headlight housing - removal and installation	12		

1 General information

The electrical system is a 12-volt, negative ground type. Power for the lights and all electrical accessories is supplied by a lead/acid-type battery which is charged by the alternator.

This Chapter covers repair and service procedures for the various electrical components not associated with the engine. Information on the battery, alternator, distributor and starter motor can be found in Chapter 5.

It should be noted that when portions of the electrical system are serviced, the cable should be disconnected from the negative battery terminal to prevent electrical shorts and/or fires.

2 Electrical troubleshooting - general information

A typical electrical circuit consists of an electrical component, any switches, relays, motors, fuses, fusible links or circuit breakers related to that component and the wiring and electrical connectors that link the component to both the battery and the chassis. To help you pinpoint an electrical circuit problem, wiring diagrams are included at the end of this book.

Before tackling any troublesome electrical circuit, first study the appropriate wiring diagrams to get a complete understanding of what makes up that individual circuit. Trouble spots, for instance, can often be narrowed down by noting if other components related to the circuit are operating properly. If several components or circuits fail at one time, chances are the problem is in a fuse or ground connection, because several circuits are often routed through the same fuse and ground connections.

Electrical problems usually stem from simple causes, such as loose or corroded connections, a blown fuse, a melted fusible link or a bad relay. Visually inspect the condition of all fuses, wires and connections in a problem circuit before troubleshooting it.

If testing instruments are going to be utilized, use the diagrams to plan ahead of time where you will make the necessary connections in order to accurately pinpoint the trouble spot.

The basic tools needed for electrical troubleshooting include a circuit tester or voltmeter (a 12-volt bulb with a set of test leads can also be used), a continuity tester, which includes a bulb, battery and set of test leads, and a jumper wire, preferably with a circuit breaker incorporated, which can be used to bypass electrical components. Before attempting to locate a problem with test instruments, use the wiring diagram(s) to decide where to make the connections.

Voltage checks

Voltage checks should be performed if a circuit is not functioning properly. Connect one lead of a circuit tester to either the negative battery terminal or a known good ground. Connect the other lead to a electrical connector in the circuit being tested, preferably nearest to the battery or fuse. If the bulb of the tester lights, voltage is present, which means that the part of the circuit between the electrical connector and the battery is problem free. Continue checking the rest of the circuit in the same fashion. When you reach a point at which no voltage is present, the problem lies between that point and the last test point with voltage. Most of the time the problem can be traced to a loose connection. **Note:** *Keep in mind that some circuits receive voltage only when the ignition key is in the Accessory or Run position.*

3.1a The interior fuse box is located on the driver's side of the instrument panel, behind the fuse panel cover

3.1b The engine compartment fuse box on 1994 and later models is located next to the battery - it contains cartridge type fusible links similar to large fuses

Finding a short

One method of finding shorts in a circuit is to remove the fuse and connect a test light or voltmeter in its place to the fuse terminals. There should be no voltage present in the circuit. Move the wiring harness from side to side while watching the test light. If the bulb goes on, there is a short to ground somewhere in that area, probably where the insulation has rubbed through. The same test can be performed on each component in the circuit, even a switch.

Ground check

Perform a ground test to check whether a component is properly grounded. Disconnect the battery and connect one lead of a self-powered test light, known as a continuity tester, to a known good ground. Connect the other lead to the wire or ground connection being tested. If the bulb goes on, the ground is good. If the bulb does not go on, the ground is not good.

Continuity check

A continuity check is done to determine if there are any breaks in a circuit - if it is passing electricity properly. With the circuit off (no power in the circuit), a self-powered continuity tester can be used to check the circuit. Connect the test leads to both ends of the circuit (or to the "power" end and a good ground), and if the test light comes on the circuit is passing current properly. If the light doesn't come on, there is a break somewhere in the circuit. The same procedure can be used to test a switch, by connecting the continuity tester to the power in and power out sides of the switch. With the switch turned On, the test light should come on.

Finding an open circuit

When diagnosing for possible open circuits, it is often difficult to locate them by sight because oxidation or terminal misalignment are hidden by the electrical connectors. Merely wiggling an electrical connector on a sensor or in the wiring harness may correct the open circuit condition. Remember this when an open circuit is indicated when troubleshooting a circuit. Intermittent problems may also be caused by oxidized or loose connections.

Electrical troubleshooting is simple if you keep in mind that all electrical circuits are basically electrical current running from the battery, through the wires, switches, relays, fuses and fusible links to each electrical component (light bulb, motor, etc.) and to ground, from which it is passed back to the battery. Any electrical problem is an interruption in the flow of electricity to the electrical component and back to the battery.

3 Fuses - general information

Refer to illustrations 3.1a, 3.1b and 3.2

The electrical circuits of the vehicle are protected by a combination of fuses, circuit breakers and fusible links. 1993 and earlier models are equipped with an interior fuse block which is located under the instrument panel on the left side of the dashboard. While 1994 and later models are equipped with two fuse blocks, one for standard fuses located under the instrument panel on the left side of the dashboard and one for high current fuses in the engine compartment, adjacent to the battery **(see illustrations)**. See Section 4 for the replacement procedure on high current fuses.

Miniaturized fuses are employed in the passenger compartment fuse block. These compact fuses, with blade terminal design, allow fingertip removal and replacement. If an electrical component fails, always check the fuse first. The best way to check the fuses is with a test light. Check for power at the exposed terminal tips of each fuse. If power is present at one side of the fuse but not the other, the fuse is blown. A blown fuse can also be identified by visually inspecting it **(see illustration)**.

Be sure to replace blown fuses with the correct type. Fuses of different ratings are physically interchangeable, but only fuses of the proper rating should be used. Replacing a fuse with one of a higher or lower value than specified is not recommended. Each electrical circuit needs a specific amount of protection. The amperage value of each fuse is molded into the fuse body.

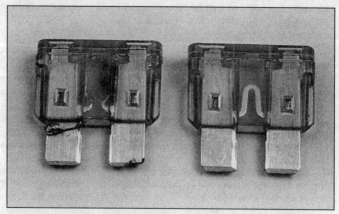

3.2 The fuses can easily be checked visually to see if they are blown

Chapter 12 Chassis electrical system

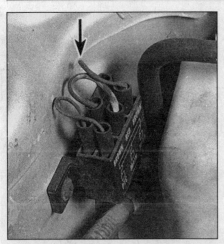

4.2 On 1993 and earlier models, the fusible link panel is mounted on the left shock/strut tower in the engine compartment

5.4 Most relays are marked on the outside to easily identify the control and power circuits

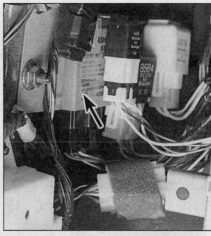

6.1 On 1994 and later models, the flasher module (arrow) is mounted on the driver's side door pillar under the instrument panel

If the replacement fuse immediately fails, don't replace it again until the cause of the problem is isolated and corrected. In most cases, the cause will be a short circuit in the wiring caused by a broken or deteriorated wire.

4 Fusible links - general information

Refer to illustration 4.2

Some circuits are protected by fusible links. The links are used in circuits with high current draws which are not ordinarily fused, such as the fuel injection, headlights, cooling fan, air conditioner, etc.

In addition to the conventional type of fusible link (described below) which is used on 1993 and earlier models **(see illustration)**, there is also a cartridge type fusible link used on 1994 and later models. Cartridge type fusible links are similar to a large fuses **(see illustration 3.1b)**, and, after disconnecting the negative battery cable, are simply unplugged and replaced by a unit of the same amperage. Some fusible links are held in place by a screw which must be loosened before removing the link.

Conventional type fusible links cannot be repaired, a new link of the same size wire should be installed in its place. The procedure is as follows:

a) Disconnect the cable from the negative battery terminal.
b) Disconnect the fusible link from the wiring harness.
c) Insert a new fusible link of the same amperage onto the connectors of the fuse panel.
d) Connect the battery ground cable. Test the circuit for proper operation.

5 Relays - general information and testing

General information

1 Several electrical accessories in the vehicle, such as the fuel injection system, horn, starter, and the cooling fan use relays to transmit the electrical signal to the component. Relays use a low-current circuit (the control circuit) to open and close a high-current circuit (the power circuit). If the relay is defective, that component will not operate properly. The various relays are mounted in engine compartment and several locations throughout the vehicle. If a faulty relay is suspected, it can be removed and tested using the procedure below or by a dealer service department or a repair shop. Defective relays must be replaced as a unit.

Testing

Refer to illustration 5.4

2 It's best to refer to the wiring diagram for the circuit to determine the proper hook-ups for the relay you're testing. However, if you're not able to determine the correct hook-up from the wiring diagrams, you may be able to determine the test hook-ups from the information that follows.

3 On most relays, two of the terminals are the relay's control circuit (they connect to the relay coil which, when energized, closes the large contacts to complete the circuit). The other terminals are the power circuit (they are connected together within the relay when the control-circuit coil is energized).

4 Most relays are marked as an aid to help you determine which terminals are the control circuit and which are the power circuit **(see illustration)**.

5 Connect a fused jumper wire between one of the two control circuit terminals and the positive battery terminal. Connect another jumper wire between the other control circuit terminal and ground. When the connections are made, the relay should click. On some relays, polarity may be critical, so, if the relay doesn't click, try swapping the jumper wires on the control circuit terminals.

6 With the jumper wires connected, check for continuity between the power circuit terminals as indicated by the markings on the relay.

7 If the relay fails any of the above tests, replace it.

6 Turn signal/hazard flashers - check and replacement

Refer to illustration 6.1

Warning: *Some models covered by this manual are equipped with Supplemental Restraint systems (SRS), more commonly known as airbags. Always disconnect the negative battery cable, then the positive battery cable and wait two minutes before working in the vicinity of the impact sensors, steering column or instrument panel to avoid the possibility of accidental deployment of the airbag, which could cause personal injury (see Section 24). Do not use electrical test equipment on any of the airbag system wiring or tamper with them in any way.*

1 The turn signal and hazard flashers are controlled from a single electronic flasher unit which is located under the instrument panel **(see illustration)**. **Note:** *On 1993 and earlier models, the flasher module is mounted to the inner firewall slightly above the interior fuse panel.*

2 When the flasher unit is functioning properly, an audible click can be heard during its operation. If the turn signals fail on one side or the other and the flasher unit does not make its characteristic clicking sound, a faulty turn signal bulb is indicated.

Chapter 12 Chassis electrical system

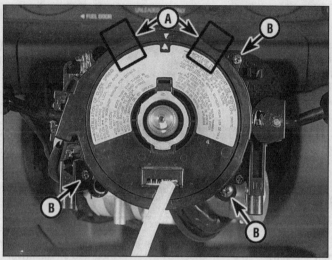

7.5 Apply two pieces of tape (A) across the airbag sliding contact inner and outer hub to prevent accidental rotation, then detach the retaining screws (B), and the electrical connectors to remove the airbag sliding contact assembly - notice the airbag sliding contact alignment marks

7.6 Remove the combination switch mounting screws - 1994 and later model shown

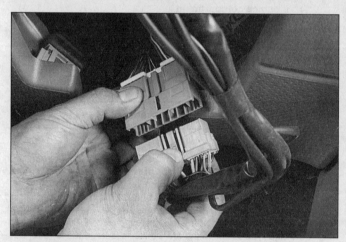

7.7 Slip the combination switch off the steering column and unplug electrical connector(s) at the base of the steering column

3   If both turn signals fail to blink, the problem may be due to a blown fuse, a faulty flasher unit, a broken switch or a loose or open connection. If a quick check of the fuse box indicates that the turn signal fuse has blown, check the wiring for a short before installing a new fuse. If one side only fails to flash properly, check for burnt out bulbs.
4   To replace the flasher, simply disconnect the electrical connectors, then remove the flasher from its mounting bracket.
5   Make sure that the replacement unit is identical to the original. Compare the old one to the new one before installing it.
6   Installation is the reverse of removal.

7   Combination switch - removal and installation

Refer to illustrations 7.5, 7.6 and 7.7
Warning: *Some models covered by this manual are equipped with Supplemental Restraint systems (SRS), more commonly known as airbags. Always disconnect the negative battery cable, then the positive battery cable and wait two minutes before working in the vicinity of the impact sensors, steering column or instrument panel to avoid the possibility of accidental deployment of the airbag, which could cause personal injury (see Section 24). Do not use electrical test equipment on any of the airbag system wiring or tamper with them in any way.*
Caution: *Always use extreme care when working around airbag modules. Never strike, pry or bump airbag modules to avoid the possibility of accidental deployment.*

1   Position the front wheels pointing straight ahead.
2   Disconnect the negative battery cable. then the positive battery cable and wait two minutes before proceeding any further.
3   Remove the driver's side airbag (if equipped) and the steering wheel (see Chapter 10).
4   Remove the steering column upper and lower covers (see Chapter 11).
5   On airbag equipped models, apply several strips of masking tape across the airbag sliding contact inner and outer hub. This will prevent accidental rotation of the airbag sliding contact and aid in the reinstallation process. Detach the airbag sliding contact retaining screws and electrical connections, then remove the airbag sliding contact from the steering column **(see illustration)**. **Warning:** *Handle the airbag sliding contact assembly very carefully. Damage to the sliding contact could cause an airbag system failure resulting in serious personal injury.*
6   Remove the combination switch retaining screws **(see illustration)**.

7   Disconnect the electrical connectors and slide the combination switch off the column. **(see illustration)**.
8   Individual switches can now be removed from the switch body and replaced as necessary. Detach the retaining screws or pins and remove the defective switch.
9   Installation is the reverse of removal. **Caution:** *If accidental rotation of the airbag sliding contact has occurred, the airbag sliding contact will need to be aligned properly before installing the airbag sliding contact assembly onto the steering shaft. Center the airbag sliding contact as follows:*

a) *Position the front wheels pointing straight ahead.*
b) *Gently rotate the sliding contact inner hub clockwise to the end of its stop. Do not force it against the stop.*
c) *Rotate the clockspring 2-3/4 turns counterclockwise.*
d) *Align the mark on the inner hub with the mark on the housing* **(see illustration 7.5)**.

8   Combination switch - check

Refer to illustrations 8.4a through 8.4l
Warning: *Some models covered by this manual are equipped with Supplemental Restraint systems (SRS), more commonly known as airbags. Always disconnect the negative battery cable, then the positive battery cable and wait two minutes before working in the vicinity of*

Chapter 12 Chassis electrical system

the impact sensors, steering column or instrument panel to avoid the possibility of accidental deployment of the airbag, which could cause personal injury (see Section 24). *Do not use electrical test equipment on any of the airbag system wiring or tamper with them in any way.*
Caution: *Always use extreme care when working around airbag modules. Never strike, pry or bump airbag modules to avoid the possibility of accidental deployment.*

1 Disconnect the negative battery cable. then the positive battery cable and wait two minutes before proceeding any further.
2 On 1993 and earlier models it will be necessary to remove the multi-function switch (see Section 7). Disconnect the electrical connectors from the back of the switch and proceed to step 4.
3 On 1994 and later models, trace the wire from the combination switch to the main wiring harness connector and disconnect the connectors.
4 Using an ohmmeter or self-powered test light and the accompa-

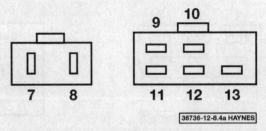

8.4a **1988 and 1989 headlight switch terminal identification guide**

nying diagrams, check for continuity between the indicated switch terminals with the switch in each of the indicated positions **(see illustrations)**. If the continuity isn't as specified, replace the switch.

Switch positions	Test terminals	Ohmmeter readings
OFF	8 and 12; 8 and 13; 7 and 11	open circuit
PARK	7 and 11	closed circuit
LO beam	8 and 12; 7 and 11	closed circuit
Hi beam Flash to pass	8 and 13; 7 and 11	closed circuit

8.4b **1988 and 1989 headlight switch continuity chart**

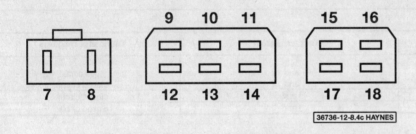

8.4c **1990 through 1993 headlight switch terminal identification guide**

Switch positions	Test terminals	Ohmmeter readings
OFF	8 and 13; 8 and 14; 12 and 16	open circuit
PARK	12 and 16	closed circuit
LO beam	8 and 13; 12 and 16	closed circuit
Hi beam Flash to pass	8 and 14; 12 and 16	closed circuit

8.4d **1990 through 1993 headlight switch continuity chart**

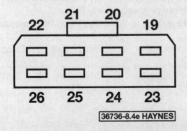

8.4e **1993 and earlier turn signal and hazard switch terminal identification guide**

Switch positions	Test terminals	Ohmmeter readings
Hazard OFF	20 and 21	closed circuit
ON	19 and 20; 22 and 26; 25 and 26	closed circuit
Left	22 and 26	closed circuit
Right	25 and 26	closed circuit

8.4f **1993 and earlier turn signal and hazard switch continuity chart**

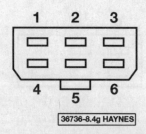

8.4g 1993 and earlier windshield wiper switch terminal identification guide

Switch positions	Test terminals	Ohmmeter readings
Wiper interval (if equipped)	3 and 2	closed circuit
Wiper LO	4 and 3	closed circuit
Wiper HI	4 and 1	closed circuit
Wash ON	4 and 6	closed circuit

8.4h 1993 and earlier windshield wiper switch continuity chart

8.4i 1994 and later combination switch terminal identification guide

Switch positions	Test terminals	Ohmmeter readings
OFF	6 and 13; 2 and 12; 14 and 17; 15 and 17	open circuit
PARK	2 and 12	closed circuit
LO beam	14 and 17; 2 and 12	closed circuit
Hi beam	15 and 17; 2 and 12	closed circuit
Flash to pass	15 and 16	closed circuit

8.4j 1994 and later headlight switch continuity chart

Switch positions	Test terminals	Ohmmeter readings
Hazard OFF	1 and 12	open circuit
ON	1 and 12	closed circuit
Left	3 and 5	closed circuit
Right	3 and 4	closed circuit

8.4k 1994 and later turn signal and hazard switch continuity chart

Switch positions	Test terminals	Ohmmeter readings
Wiper interval (if equipped)	11 and 18	open circuit
Wiper LO	7 and 18	closed circuit
Wiper HI	8 and 18	closed circuit
Wash ON	9 and 18	closed circuit

8.4l 1994 and later windshield wiper switch continuity chart

9 Ignition switch - check and replacement

Warning: *Some models covered by this manual are equipped with Supplemental Restraint systems (SRS), more commonly known as airbags. Always disconnect the negative battery cable, then the positive battery cable and wait two minutes before working in the vicinity of the impact sensors, steering column or instrument panel to avoid the possibility of accidental deployment of the airbag, which could cause personal injury (see Section 24). Do not use electrical test equipment on any of the airbag system wiring or tamper with them in any way.*
Caution: *Always use extreme care when working around airbag modules. Never strike, pry or bump airbag modules to avoid the possibility of accidental deployment.*

1 Disconnect the negative battery cable. then the positive battery cable and wait two minutes before proceeding any further.
2 Remove the driver's side airbag (if equipped) and the steering wheel (see Chapter 10).
3 Remove the steering column upper and lower covers (see Chapter 11).
4 On 1993 and earlier models, remove the lower steering column shield and the shield retaining bracket (see Chapter 11).

Check

Refer to illustrations 9.7a, 9.7b, 9.7c and 9.7d

5 On 1993 and earlier models, trace the wire from the ignition switch to the main wiring harness connector and disconnect the connectors.

Chapter 12 Chassis electrical system

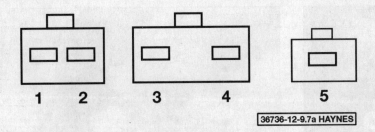

9.7a 1993 and earlier ignition switch terminal identification guide

Switch positions	Test terminals	Ohmmeter readings
ACC	2 and 4	closed circuit
OFF/LOCK	1 and 2; 2 and 3 2 and 4; 2 and 5	open circuit
RUN	1 and 2; 2 and 3 2 and 4	closed circuit
START	1 and 2; 2 and 5	closed circuit

9.7b 1993 and earlier ignition switch continuity chart

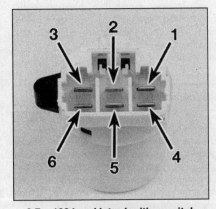

9.7c 1994 and later ignition switch terminal identification guide

Switch positions	Test terminals	Ohmmeter readings
ACC	2 and 6	closed circuit
OFF/LOCK	1 and 5; 2 and 4 2 and 6; 3 and 5	open circuit
RUN	1 and 5; 2 and 4 2 and 6	closed circuit
START	2 and 4; 3 and 5	closed circuit

9.7d 1994 and later ignition switch continuity chart

6 On 1994 and later models, the ignition switch must first be removed (see Step 8)
7 Using an ohmmeter or self-powered test light and the accompanying diagrams, check for continuity between the indicated switch terminals with the switch in each of the indicated positions (see illustrations). If the continuity isn't as specified, replace the switch.

Replacement

Refer to illustrations 9.8 and 9.9

8 Unplug the ignition switch wiring harness (see illustration).
9 Detach the switch retaining screw (see illustration) remove the switch assembly from the steering column. **Note:** *On 1993 and earlier models it may be necessary to lower the steering column to allow*

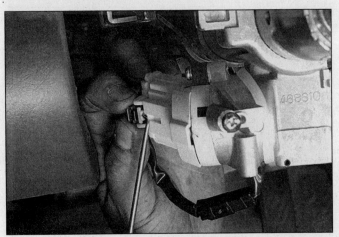

9.8 On 1994 and later models disconnect the electrical connector at the ignition switch

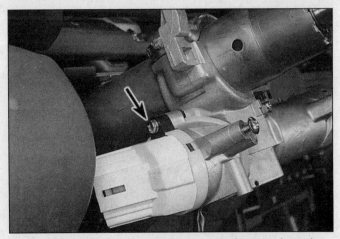

9.9 The ignition switch is retained by a single screw (arrow) - 1994 and later shown, 1993 and earlier similar

10.4 To remove the ignition/lock cylinder assembly, detach the shift lock cable (A) (automatic transaxles only), then drill out the two retaining bolts ((B))

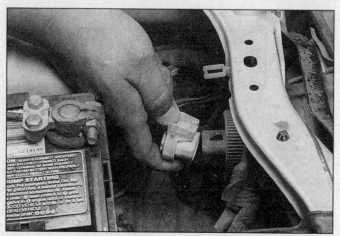

11.1 Squeeze the tabs and then pull the headlight bulb electrical connector from the headlight bulb

removal of the ignition switch wiring harness and connectors.
10 Installation is reverse of the removal.

10 Ignition key lock cylinder - removal and installation

Refer to illustration 10.4
Warning: *Some models covered by this manual are equipped with Supplemental Restraint systems (SRS), more commonly known as airbags. Always disconnect the negative battery cable, then the positive battery cable and wait two minutes before working in the vicinity of the impact sensors, steering column or instrument panel to avoid the possibility of accidental deployment of the airbag, which could cause personal injury (see Section 24). Do not use electrical test equipment on any of the airbag system wiring or tamper with them in any way.*
Caution: *Always use extreme care when working around airbag modules. Never strike, pry or bump airbag modules to avoid the possibility of accidental deployment.*

1 Disconnect the negative battery cable. then the positive battery cable and wait two minutes before proceeding any further.
2 Remove the combination switch (see Section 7).
3 Remove the ignition switch (see Section 9).
4 If equipped with an automatic transaxle, remove the shift lock cable from the lock cylinder housing **(see illustration)**.
5 Remove the shear-head bolts retaining the ignition switch/lock cylinder assembly and separate the bracket halves from the steering column **(see illustration 10.4)**. This can be accomplished by drilling out the screws or using a hacksaw to make slots in the them so they can be unscrewed with a screwdriver.
6 Place the new switch/lock cylinder assembly in position, install the new shear-head bolts and tighten them until the heads snap off. If equipped with an airbag, be sure to center the airbag sliding contact before installing the steering wheel (see Section 7).
7 The remainder of the installation is the reverse of removal.

11 Headlight bulb - replacement

Refer to illustrations 11.1, 11.2 and 11.3
Warning: *Halogen gas filled bulbs are under pressure and may shatter if the surface is scratched or the bulb is dropped. Wear eye protection and handle the bulbs carefully, grasping only the base whenever possible. Do not touch the surface of the bulb with your fingers because the oil from your skin could cause it to overheat and fail prematurely. If you do touch the bulb surface, clean it with rubbing alcohol.*
1 Reach behind the headlight assembly and unplug the electrical connector **(see illustration)**.
2 Rotate the headlight bulb retaining ring counterclockwise as viewed from the rear and remove it from the bulb holder assembly **(see illustration)**.
3 On 1993 and earlier models, grasp the bulb holder securely and pull it out of the housing **(see illustration)**.
4 On 1994 and later models, remove the rubber boot and detach the bulb holder retaining clip, then grasp the bulb holder securely and

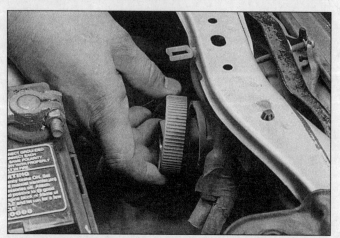

11.2 Rotate the bulb retaining ring counterclockwise about one-eighth of a turn and remove the retaining ring

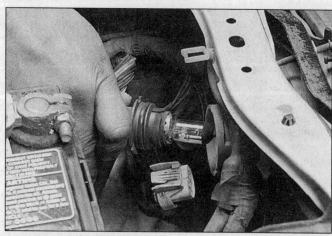

11.3 Pull the headlight bulb assembly straight out of the housing

Chapter 12 Chassis electrical system

12.5a 1993 and earlier headlight housing mounting details

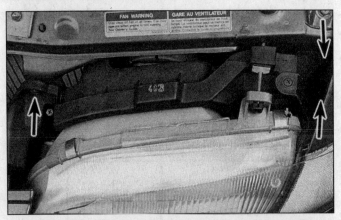

12.5b 1994 and later headlight housing mounting details

13.1a 1993 and earlier headlight adjustment screw locations

   A   Horizontal adjustment screw
   B   Vertical adjustment screw

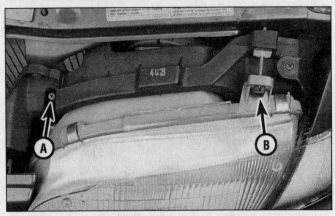

13.1b 1994 and later headlight adjustment screw locations

   A   Horizontal adjustment screw
   B   Vertical adjustment screw

pull it out of the housing.
5 Remove the bulb from the socket assembly by pulling it straight out.
6 Without touching the glass with your bare fingers, insert the new bulb into the socket assembly, then align the tabs on the bulb holder flange with the tabs on the headlight housing and insert the bulb assembly into the headlight housing.
7 The remainder of the installation is the reverse of removal.
8 Test headlight operation, then close the hood.

12 Headlight housing - removal and installation

Refer to illustrations 12.5a and 12.5b

Warning: *Some models covered by this manual are equipped with Supplemental Restraint systems (SRS), more commonly known as airbags. Always disconnect the negative battery cable, then the positive battery cable and wait two minutes before working in the vicinity of the impact sensors, steering column or instrument panel to avoid the possibility of accidental deployment of the airbag, which could cause personal injury (see Section 24). Do not use electrical test equipment on any of the airbag system wiring or tamper with them in any way.*

1 Disconnect the negative battery cable, then the positive battery cable and wait two minutes before proceeding any further.
2 Remove the radiator grille (see Chapter 11).
3 Remove the front combination light (turn signal/side marker light) retaining screw(s), remove the combination light bulb electrical connector, and pull the combination light assembly off the vehicle (see Section 14).
4 Remove the headlight bulb electrical connector and remove the headlight bulb (see Section 11). **Warning:** *Halogen gas-filled bulbs are under pressure and may shatter if the surface is scratched or the bulb is dropped. Wear eye protection and handle the bulbs carefully, grasping only the base whenever possible. Keep bulbs away from children. Allow the bulb to cool before removal. Do not touch the surface of the bulb with your fingers because the oil from your skin could cause it to overheat and fail prematurely. If you do touch the bulb surface, wipe the bulb clean with rubbing alcohol.*
5 Remove the headlight housing mounting bolts **(see illustrations)**.
6 Installation is the reverse of removal. Refer to Section 11 for headlight bulb installation.

13 Headlights - adjustment

Refer to illustrations 13.1a, 13.1b and 13.3

Note: *The headlights must be aimed correctly. If adjusted incorrectly they could blind the driver of an oncoming vehicle and cause a serious accident or seriously reduce your ability to see the road. The headlights should be checked for proper aim every 12 months and any time a new headlight is installed or front end body work is performed. It should be emphasized that the following procedure is only an interim step which will provide temporary adjustment until the headlights can be adjusted by a properly equipped shop.*

1 The headlights have two adjusting screws, one on the top controlling up-and-down (vertical) movement and one on the side controlling left-and-right (horizontal) movement **(see illustrations)**.
2 There are several methods of adjusting the headlights. The simplest method requires masking tape, a blank wall and a level floor.
3 Position masking tape vertically on the wall in reference to the

12-10   Chapter 12 Chassis electrical system

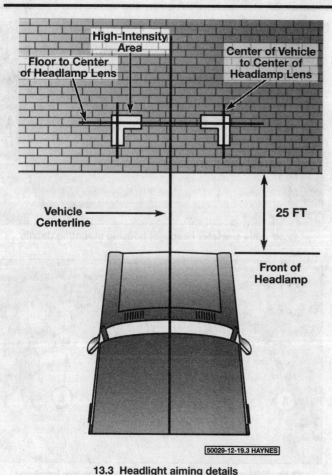

13.3 Headlight aiming details

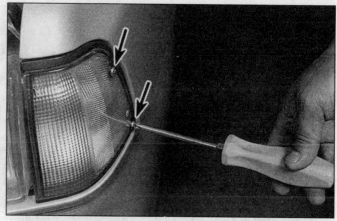

14.1a The parking/side marker light housing on 1993 and earlier models can be removed by detaching the retaining screws (arrows) located on the side of the lens

vehicle centerline and the centerlines of both headlights (see illustration).

4   Position a horizontal tape line in reference to the centerline of all the headlights. **Note:** *It may be easier to position the tape on the wall with the vehicle parked only a few inches away.*

5   Adjustment should be made with the vehicle parked 25 feet from the wall, sitting level, the gas tank half-full and no unusually heavy load in the vehicle.

6   Starting with the low beam adjustment, position the high intensity zone so it is two inches below the horizontal line and two inches to the right of the headlight vertical line. Adjustment is made by turning the top adjusting screw clockwise to raise the beam and counterclockwise to lower the beam. The adjusting screw on the side should be used in the same manner to move the beam left or right.

7   With the high beams on, the high intensity zone should be vertically centered with the exact center just below the horizontal line.
Note: *It may not be possible to position the headlight aim exactly for both high and low beams. If a compromise must be made, keep in mind that the low beams are the most used and have the greatest effect on safety.*

8   Have the headlights adjusted by a dealer service department or service station at the earliest opportunity.

14 Bulb replacement

Front turn signal lights (1994 and later) and parking/side marker lights (1993 and earlier)

Refer to illustrations 14.1a, 14.1b and 14.2

1   The front turn signal light housing (1994 and later) and the parking/side marker light housing (1993 and earlier) are mounted on the outboard side of the headlight assemblies. To replace the bulbs remove the housing retaining screw(s) **(see illustrations)**, then pull the

14.1b The front turn signal housing on 1994 and later models can be removed by detaching the retaining screw (arrow) located at the top of the housing

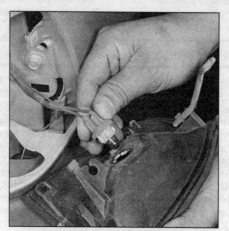

14.2 Rotate the bulb holder counterclockwise to release the bulb from the housing

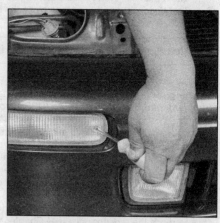

14.5 The front turn signal lights on 1993 and earlier models and the side marker lights on 1994 and later models are accessible after removing the lens retaining screws

Chapter 12 Chassis electrical system

14.8 The front parking light bulbs (arrow) on 1994 and later models are accessible from the rear of the headlight housing

lamp assembly outward.
2 Twist the bulb socket a quarter turn counterclockwise, then remove the bulb assembly from the housing **(see illustration)**.
3 The defective bulb can then be pulled straight out of the socket and replaced.
4 Installation is the reverse of removal.

Front turn signal lights (1993 and earlier) and side marker lights (1994 and later)

Refer to illustration 14.5

5 The front turn signal lights (1993 and earlier) and the side marker lights (1994 and later) are mounted in the front bumper. Detach the two screws securing the lens and remove the lens **(see illustration)**.
6 The defective bulb can then be twisted out of the socket and replaced.
7 Installation is the reverse of removal.

Front parking lights (1994 and later)

Refer to illustration 14.8

8 The front parking lights (1994 and later) is incorporated into the headlight housing assembly **(see illustration)**.
9 Twist the bulb socket a quarter turn counterclockwise, then remove the bulb assembly from the housing.

10 The defective bulb can then be pulled straight out of the socket and replaced.
11 Installation is the reverse of removal.

Rear brake, turn signal, tail, back-up lights, and side marker

Refer to illustrations 14.13a and 14.13b

12 Open the rear lift gate.
13 On 1993 and earlier models, detach the rear quarter trim panel access cover. Twist the defective bulb socket a quarter turn counterclockwise, then remove the bulb assembly from the housing **(see illustrations)**. The defective bulb can then be twisted out of the socket and replaced.
14 On 1994 and later models, remove the screws securing the inboard side of the tail light housing. Pull outward on the inboard side of the tail light assembly while lifting upward on the outboard side and detach the tail light housing from the vehicle. Twist the defective bulb socket a quarter turn counterclockwise, then remove the bulb assembly from the housing. The defective bulb can then be twisted out of the socket and replaced.
15 Installation is the reverse of removal.

License plate light

16 Detach the retaining screws which secure the lens.
17 The defective bulb can then be pulled straight out of the socket and replaced.
18 Installation of the lens is the reverse of removal.

High-mounted brake light

19 Open the rear lift gate.
20 On 1993 and earlier models, the brake light cover is retained by screws. Remove the cover and replace the bulb.
21 On 1994 and later models, remove the liftgate trim panel see Chapter 11 if necessary. Twist the bulb socket a quarter turn counterclockwise, then remove the bulb assembly from the housing. The defective bulb can then be pulled straight out of the socket and replaced.

Interior light

Refer to illustration 14.22

22 Pry the interior lens off the interior light housing **(see illustration)**.
23 Detach the bulb from the terminals. It may be necessary to pry the bulb out - if this is the case, pry only on the ends of the bulb (otherwise the glass may shatter).
24 Installation is the reverse of removal.

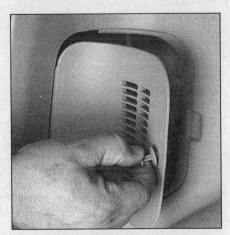

14.13a On 1993 and earlier models, remove the access cover from the rear quarter trim panel to access the tail light bulbs

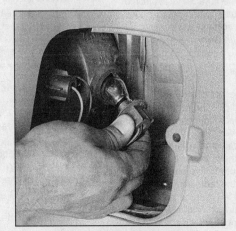

14.13b Rotate the bulb holder counterclockwise to release the bulb from the housing

14.22 Pry off the interior light lens to access the bulb

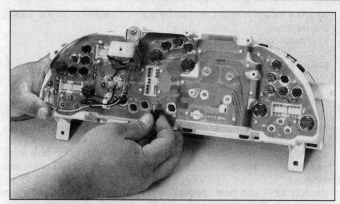

14.25 Remove the instrument cluster bulbs by rotating them 1/4-turn counterclockwise and pulling straight out

Instrument cluster illumination

Refer to illustration 14.25

25 To gain access to the instrument cluster illumination lights, the instrument cluster will have to be removed (see Section 18). The bulbs can then be removed and replaced from the rear of the cluster **(see illustration)**.

15 Daytime Running Lights (DRL) - general information

The Daytime Running Lights (DRL) system used on Canadian models illuminates the headlights whenever the engine is running. The only exception is with the engine running and the parking brake engaged. Once the parking brake is released, the lights will remain on as long as the ignition switch is on, even if the parking brake is later applied.

The DRL system supplies reduced power to the headlights so they won't be too bright for daytime use, while prolonging headlight life.

16 Radio and speakers - removal and installation

Warning: *Some models covered by this manual are equipped with Supplemental Restraint systems (SRS), more commonly known as airbags. Always disconnect the negative battery cable, then the positive battery cable and wait two minutes before working in the vicinity of the impact sensors, steering column or instrument panel to avoid the possibility of accidental deployment of the airbag, which could cause personal injury (see Section 24). Do not use electrical test equipment on any of the airbag system wiring or tamper with them in any way.*

Radio

Refer to illustrations 16.3a and 16.3b

1 On 1993 and earlier models, remove the radio trim bezel (see Chapter 11). Remove the retaining screws and pull the radio outward to access the backside, then disconnect the electrical connectors and the antenna lead and remove the unit from the vehicle.

2 On 1994 and later models, the radio receiver is retained in the instrument panel by special clips. Releasing these clips requires the use of two special removal tools (available at automotive parts stores). Insert the tools into the holes at the corners of the radio assembly until you feel the internal clips release.

3 With the clips released, push outward simultaneously on both tools and pull the assembly out of the instrument panel, disconnect the antenna and electrical connectors and remove the unit from the vehicle **(see illustrations)**.

4 On 1994 and later models, install the radio by plugging in the electrical connectors, then sliding the radio along the track and into the

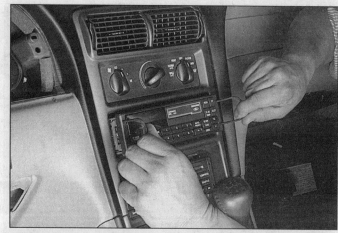

16.3a On 1994 and later models, insert the removal tools until they seat, then push outward simultaneously on both tools to release the clips and withdraw the radio from the dash

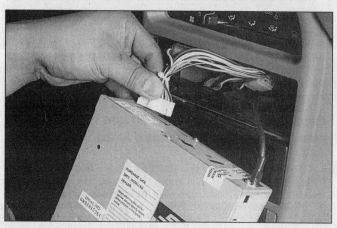

16.3b Disconnect the electrical connectors and the antenna lead, then remove the unit

instrument panel until the clips snap into place.

5 On 1993 and earlier models, installation is the reverse of removal.

Speakers

Instrument panel mounted

6 On 1993 and earlier models, the front speakers are mounted in the instrument panel. Using a small screwdriver pry off the speaker grille to access the speakers.

7 Remove the speaker retaining screws, withdraw the speaker, unplug the electrical connector and remove the speaker from the vehicle.

8 Installation is the reverse of removal

Door mounted

Refer to illustration 16.10

9 On 1994 and later models, the front speakers are mounted in the door. Remove the door trim panel to access the speakers (see Chapter 11).

10 Remove the mounting screws, withdraw the speaker, unplug the electrical connector and remove the speaker from the vehicle **(see illustration)**.

11 Installation is the reverse of removal.

Rear quarter panel mounted

12 On 1993 and earlier models, simply remove the speaker retaining screws from the rear quarter trim panels, withdraw the speaker, unplug the electrical connector and remove the speaker from the vehicle.

Chapter 12 Chassis electrical system

16.10 Remove the retaining screws (arrows), pull the speaker out and disconnect the electrical connector

13 On 1994 and later models, remove the plastic clips and screws securing the speaker trim panel to the rear quarter trim panel, withdraw the speaker trim panel, unplug the electrical connector and remove the speaker retaining screws/nuts.
14 Installation is the reverse of removal

17 Antenna - removal and installation

Refer to illustration 17.3
Warning: *Some models covered by this manual are equipped with Supplemental Restraint systems (SRS), more commonly known as airbags. Always disconnect the negative battery cable, then the positive battery cable and wait two minutes before working in the vicinity of the impact sensors, steering column or instrument panel to avoid the possibility of accidental deployment of the airbag, which could cause personal injury (see Section 24). Do not use electrical test equipment on any of the airbag system wiring or tamper with them in any way.*

1 Detach the radio and disconnect the antenna lead from the backside of the radio (see Section 16). Attach a piece of stiff wire to the end of the antenna lead as an aid for reinstallation.
2 Remove the instrument cluster (see Section 18) then remove any retaining clips under the instrument panel securing the antenna lead.
3 Working on the outside of the vehicle, remove the antenna retaining screws **(see illustration)**.
4 Working in the door opening, grasp the antenna base with two hands and slowly maneuver it upward until the antenna insulating tube/drain tube clears the outer sheetmetal.

17.3 Remove the retaining screws (arrows) securing the antenna base to the door pillar

5 Pull the remaining antenna lead up and out until the installation wire protrudes past the outer sheet metal. Then remove the old antenna lead from the wire and replace it with the new antenna lead.
6 Working from passenger compartment pull the wire back through the door pillar until the antenna base assembly seats on the outer sheet metal. **Note:** *Make sure the insulating tube/drain tube lines up with the lower drain hole.*
7 The remainder of the installation is the reverse of removal.

18 Instrument cluster - removal and installation

Refer to illustrations 18.3, 18.4a and 18.4b
Warning: *Some models covered by this manual are equipped with Supplemental Restraint systems (SRS), more commonly known as airbags. Always disconnect the negative battery cable, then the positive battery cable and wait two minutes before working in the vicinity of the impact sensors, steering column or instrument panel to avoid the possibility of accidental deployment of the airbag, which could cause personal injury (see Section 24). Do not use electrical test equipment on any of the airbag system wiring or tamper with them in any way.*

1 Disconnect the negative battery cable, then the positive battery cable and wait two minutes before proceeding any further.
2 Tilt the steering wheel to its lowest position and remove the instrument cluster bezel (see Chapter 11).
3 Remove the instrument cluster retaining screws **(see illustration)**.
4 Pull the instrument cluster out and unplug the electrical connectors and the speedometer cable from the backside, then remove the

18.3 The instrument cluster is held in place by screws (arrows) at both sides of the housing (1994 and later shown, 1993 and earlier similar)

18.4a Disconnect the electrical connectors . . .

12-14   Chapter 12 Chassis electrical system

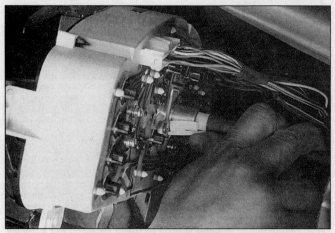

18.4b ... and the speedometer cable

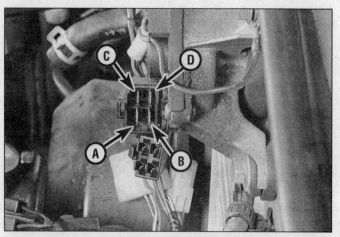

19.2 To check the wiper motor, apply battery voltage to terminal B and ground to terminal A to check the low speed operation, then move the ground lead to terminal C to check the high speed operation

cluster from the instrument panel **(see illustrations)**.
5   Installation is the reverse of removal.

19   Wiper motor - check and replacement

Wiper motor circuit check

Refer to illustration 19.2

Note: *Refer to the wiring diagrams for wire colors and locations in the following checks. Keep in mind that power wires are generally larger in diameter and brighter colors, where ground wires are usually smaller in diameter and darker colors. When checking for voltage, probe a grounded 12-volt test light to each terminal at a connector until it lights; this verifies voltage (power) at the terminal.*

1   If the wipers work slowly, make sure the battery is in good condition and has a strong charge (see Chapter 1). If the battery is in good condition, remove the wiper motor (see below) and operate the wiper arms by hand. Check for binding linkage and pivots. Lubricate or repair the linkage or pivots as necessary. Reinstall the wiper motor. If the wipers still operate slowly, check for loose or corroded connections, especially the ground connection. If all connections look OK, replace the motor.

2   If the wipers fail to operate when activated, check the fuse. If the fuse is OK, connect a jumper wire between the wiper motor and ground, then retest. If the motor works now, repair the ground connection. If the motor still doesn't work, turn the wiper switch to the HI position and check for voltage at the motor. If there's voltage at the motor, remove the motor and check it off the vehicle with fused jumper wires from the battery **(see illustration)**. If the motor now works, check for binding linkage (see Step 1 above). If the motor still doesn't work, replace it. If there's no voltage at the motor check the switch.

3   If the interval (delay) function is inoperative, check the continuity of all the wiring between the switch and wiper motor. If the wiring is OK, check the resistance of the delay control knob of the multi-function switch (see Section 8). If the delay control knob is within the specified resistance, replace the wiper motor.

4   If the wipers stop at the position they're in when the switch is turned off (fail to park), check for voltage at the park feed wire of the wiper motor connector when the wiper switch is OFF but the ignition is ON. If no voltage is present, check for an open circuit between the wiper motor and the fuse panel.

5   If the wipers won't shut off unless the ignition is OFF, disconnect the wiring from the wiper control switch. If the wipers stop, replace the switch. If the wipers keep running, there's a defective limit switch in the motor; replace the motor.

6   If the wipers won't retract below the hoodline, check for mechanical obstructions in the wiper linkage or on the vehicle's body which would prevent the wipers from parking. If there are no obstructions,

19.9 Remove the wiper linkage access cover screws (arrows) and the wiper motor retaining bolts (arrows)

check the wiring between the switch and motor for continuity. If the wiring is OK, replace the wiper motor.

Wiper motor replacement

Refer to illustrations 19.9 and 19.10

7   Disconnect the electrical connector from the wiper motor.
8   On 1994 and later models, remove the EGR solenoid bracket from the wiper linkage access panel.
9   Remove the wiper linkage access panel screws and the wiper motor retaining bolts **(see illustration)**.
10   Pull the wiper motor outward slightly and detach the nut securing the wiper arm linkage to the backside of the motor **(see illustration)**, then remove the motor from the vehicle.
11   Installation is the reverse of removal.

20   Horn - check and replacement

Check

Refer to illustration 20.1

Note: *Check the fuses before beginning electrical diagnosis.*

1   Disconnect the electrical connector from the horn **(see illustration)**.
2   To test the horn(s), connect battery voltage to the horn terminal with a pair of jumper wires. If the horn doesn't sound, replace it.

Chapter 12 Chassis electrical system

19.10 Pull the wiper linkage access cover outward slightly to access the wiper motor spindle nut (arrow)

20.1 The horn (arrow) is located behind the bumper on the right side of the vehicle

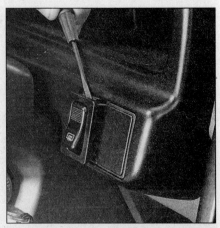

21.2 Use a small screwdriver to pry out the defogger switch from the instrument panel (1993 and earlier shown, 1994 and later similar)

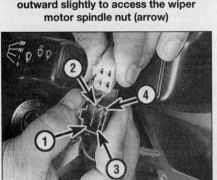

21.3a 1993 and earlier rear defogger switch terminal identification guide

Switch positions	Test terminals	Ohmmeter readings
OFF	1 and 3	open circuit
	2 and 4	closed circuit
ON	1 and 3	closed circuit
	2 and 4	closed circuit

21.3b 1993 and earlier rear defogger switch continuity chart

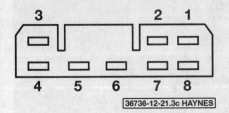

21.3c 1994 and later rear defogger switch terminal identification guide

Switch positions	Test terminals	Ohmmeter readings
OFF	5 and 6	open circuit
	7 and 8	open circuit
ON	5 and 6	closed circuit
	7 and 8	closed circuit

21.3d 1994 and later rear defogger switch continuity chart

3 If the horn does sound, check for voltage at the terminal when the horn button is depressed. If there's voltage at the terminal, check for a bad ground at the horn.
4 If there's no voltage at the horn, check the relay (see Section 5). Note that most horn relays are either the four-terminal or externally grounded three-terminal type.
5 If the relay is OK, check for voltage to the relay power and control circuits. If either of the circuits is not receiving voltage, inspect the wiring between the relay and the fuse panel.
6 If both relay circuits are receiving voltage, depress the horn button and check the circuit from the relay to the horn button for continuity to ground. If there's no continuity, check the circuit for an open. If there's no open circuit, replace the horn button.
7 If there's continuity to ground through the horn button, check for an open or short in the circuit from the relay to the horn.

Replacement

8 Disconnect the electrical connector and remove the bracket bolt.
9 Installation is the reverse of removal.

21 Rear window defogger switch - check and replacement

Refer to illustrations 21.2, 21.3a, 21.3b, 21.3c and 21.3d

Warning: *Some models covered by this manual are equipped with Supplemental Restraint systems (SRS), more commonly known as airbags. Always disconnect the negative battery cable, then the positive battery cable and wait two minutes before working in the vicinity of the impact sensors, steering column or instrument panel to avoid the possibility of accidental deployment of the airbag, which could cause personal injury (see Section 24). Do not use electrical test equipment on any of the airbag system wiring or tamper with them in any way.*

1 Disconnect the negative battery cable, then the positive battery cable and wait two minutes before proceeding any further.
2 Using a small screwdriver carefully pry out the rear defogger switch from the instrument panel **(see illustration)**.
3 Use an ohmmeter to check for continuity at the indicated terminals with the switch in the indicated positions **(see illustrations)**.
4 Replace the switch if the continuity is not as specified.

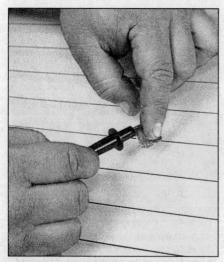

22.4 When measuring the voltage at the rear window defogger grid, wrap a piece of aluminum foil around the positive probe of the voltmeter and press the foil against the element with your finger

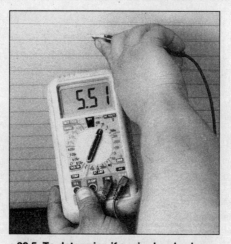

22.5 To determine if a wire has broken, check the voltage at the center of each wire. If the voltage is 5-volts, the wire is unbroken; if the voltage is 10-volts, the wire is broken between the center of the wire and the ground side; if the voltage is 0-volts, the wire is broken between the center of the wire and the power side

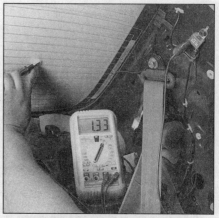

22.7 To find the break, place the voltmeter negative lead against the defogger ground terminal, place the voltmeter positive lead with the foil strip against the heat wire at the positive terminal end and slide it toward the negative terminal end - the point at which the voltmeter deflects from 10-volts to zero volts is the point at which the wire is broken

22 Rear window defogger - check and repair

1   The rear window defogger consists of a number of horizontal elements baked onto the glass surface.
2   Small breaks in the element can be repaired without removing the rear window.

Check

Refer to illustrations 22.4, 22.5 and 22.7

3   Turn the ignition switch and defogger system switches to the ON position. Using a voltmeter, place the positive probe against the defogger grid positive terminal and the negative lead against the ground terminal. If battery voltage is not indicated, check the fuse, defogger switch and related wiring.
4   When measuring voltage during the next two tests, wrap a piece of aluminum foil around the tip of the voltmeter positive probe and press the foil against the heating element with your finger **(see illustration)**.
5   Check the voltage at the center of each heating element **(see illustration)**. If the voltage is 5-volts, the element is okay (there is no break). If the voltage is 10-volts, the element is broken between the center of the element and the ground side. If the voltage is 0-volts the element is broken between the center of the element and the positive side.
6   If none of the elements are broken, connect the negative lead to a good body ground. The voltage reading should stay the same, if it doesn't the ground connection is bad.
7   To find the break, place the voltmeter negative lead against the defogger ground terminal. Place the voltmeter positive lead with the foil strip against the heating element at the positive terminal end and slide it toward the negative terminal end. The point at which the voltmeter deflects from several volts to zero is the point at which the heating element is broken **(see illustration)**.

Repair

Refer to illustration 22.13

8   Repair the break in the element using a repair kit specifically recommended for this purpose, such as Dupont paste No. 4817 (or equivalent). Included in this kit is plastic conductive epoxy.
9   Prior to repairing a break, turn off the system and allow it to cool

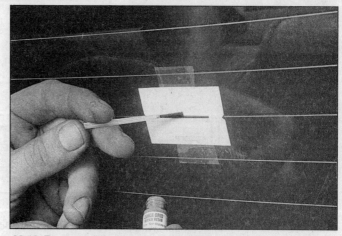

22.13 To use a defogger repair kit, apply masking to the inside of the window at the damaged area, then brush on the special conductive coating

off for a few minutes.
10   Lightly buff the element area with fine steel wool, then clean it thoroughly with rubbing alcohol.
11   Use masking tape to mask off the area being repaired.
12   Thoroughly mix the epoxy, following the instructions provided with the repair kit.
13   Apply the epoxy material to the slit in the masking tape, overlapping the undamaged area about 3/4-inch on either end **(see illustration)**.
14   Allow the repair to cure for 24 hours before removing the tape and using the system.

23 Electric side view mirrors - description and check

1   Most electric rear view mirrors use two motors to move the glass; one for up and down adjustments and one for left-right adjustments.
2   The control switch has a selector portion which sends voltage to the left or right side mirror. With the ignition ON but the engine OFF, roll

Chapter 12 Chassis electrical system

24.1a Crash sensors (arrow) are mounted behind the radiator grille and behind the bumper near the bumper cover support brackets

24.1b The airbag diagnostic module (arrow) is mounted behind the center console

down the windows and operate the mirror control switch through all functions (left-right and up-down) for both the left and right side mirrors.
3 Listen carefully for the sound of the electric motors running in the mirrors.
4 If the motors can be heard but the mirror glass doesn't move, there's probably a problem with the drive mechanism inside the mirror. Remove and disassemble the mirror to locate the problem.
5 If the mirrors don't operate and no sound comes from the mirrors, check the fuse (see Section 3).
6 If the fuse is OK, remove the mirror control switch from its mounting without disconnecting the wires attached to it. Turn the ignition ON and check for voltage at the switch. There should be voltage at one terminal. If there's no voltage at the switch, check for an open or short in the wiring between the fuse panel and the switch.
7 If there's voltage at the switch, disconnect it. Check the switch for continuity in all its operating positions. If the switch does not have continuity, replace it.
8 Re-connect the switch. Locate the wire going from the switch to ground. Leaving the switch connected, connect a jumper wire between this wire and ground. If the mirror works normally with this wire in place, repair the faulty ground connection.
9 If the mirror still doesn't work, remove the mirror and check the wires at the mirror for voltage. Check with ignition ON and the mirror selector switch on the appropriate side. Operate the mirror switch in all its positions. There should be voltage at one of the switch-to-mirror wires in each switch position (except the neutral "off" position).
10 If there's not voltage in each switch position, check the wiring between the mirror and control switch for opens and shorts.
11 If there's voltage, remove the mirror and test it off the vehicle with jumper wires. Replace the mirror if it fails this test.

24 Airbag system - general information

Refer to illustrations 24.1a and 24.1b

1994 and later models are equipped with a Supplemental Restraint System (SRS), more commonly known as an airbag. This system is designed to protect the driver, and the front seat passenger, from serious injury in the event of a head-on or frontal collision. It consists of an airbag module in the center of the steering wheel and the right side of the instrument panel, three crash sensors mounted at the front of the vehicle and a diagnostic module which also contains a safing sensor is located inside the passenger compartment **(see illustrations)**.

Airbag module

Steering wheel-mounted

The airbag inflator module contains a housing incorporating the cushion (airbag) and inflator unit, mounted in the center of the steering wheel The inflator assembly is mounted on the back of the housing over a hole through which gas is expelled, inflating the bag almost instantaneously when an electrical signal is sent from the system. A coil assembly on the steering column under the module carries this signal to the module.

This coil assembly can transmit an electrical signal regardless of steering wheel position. The igniter in the air bag converts the electrical signal to heat and ignites the sodium azide/copper oxide powder, producing nitrogen gas, which inflates the bag.

Instrument panel-mounted

The airbag is mounted above the glove compartment and designated by the letters SRS (Supplemental Restraint System). It consists of an inflator containing an igniter, a bag assembly, a reaction housing and a trim cover.

The air bag is considerably larger that the steering wheel-mounted unit (approximately 8 cu ft vs. 2.3 cu ft) and is supported by the steel reaction housing. The trim cover is textured and painted to match the instrument panel and has a molded seam which splits when the bag inflates. As with the steering-wheel-mounted air bag, the igniter electrical signal converts to heat, converting sodium azide/iron oxide powder to nitrogen gas, inflating the bag.

Sensors

The system has three sensors: two forward crash sensors at the front of the vehicle and a safing sensor mounted inside the airbag diagnostic module which is located behind the right kick panel.

The forward and passenger compartment sensors are basically pressure sensitive switches that complete an electrical circuit during an impact of sufficient G force. The electrical signal from these sensors is sent to the electronic diagnostic monitor which then completes the circuit and inflates the airbag(s).

Electronic diagnostic monitor

The electronic diagnostic monitor supplies the current to the airbag system in the event of the collision, even if battery power is cut off. It checks this system every time the vehicle is started, causing the "AIR BAG" light to go on then off, if the system is operating properly. If there is a fault in the system, the light will go on and stay on, flash, or the dash will make a beeping sound. If this happens, the vehicle should be taken to your dealer immediately for service.

Disabling the system

Whenever working in the vicinity of the steering wheel, steering column or near other components of the airbag system, the system should be disarmed. To do this, perform the following steps:
a) Turn the ignition switch to Off.
b) Detach the cable from the negative battery terminal then detach the positive cable. Wait 2 minutes for the electronic module backup power supply to be depleted.

Enabling the system

a) Turn the ignition switch to the Off position.
b) Connect the positive battery cable first, then connect the negative cable.
c) Turn the ignition key to On and verify that the "AIRBAG" warning light comes on for approximately six seconds, then goes off.

25 Wiring diagrams

Since it isn't possible to include all wiring diagrams for every year covered by this manual, the following diagrams are those that are typical and most commonly needed.

Prior to troubleshooting any circuits, check the fuse and circuit breakers (if equipped) to make sure they're in good condition. Make sure the battery is properly charged and check the cable connections (see Chapter 1).

When checking a circuit, make sure that all connectors are clean, with no broken or loose terminals. When unplugging a connector, do not pull on the wires. Pull only on the connector housings themselves.

Chapter 12 Chassis electrical system

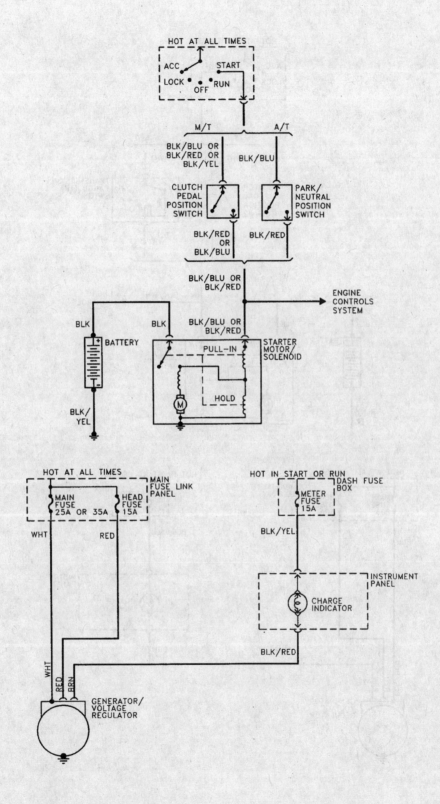

Typical 1993 and earlier starting and charging system

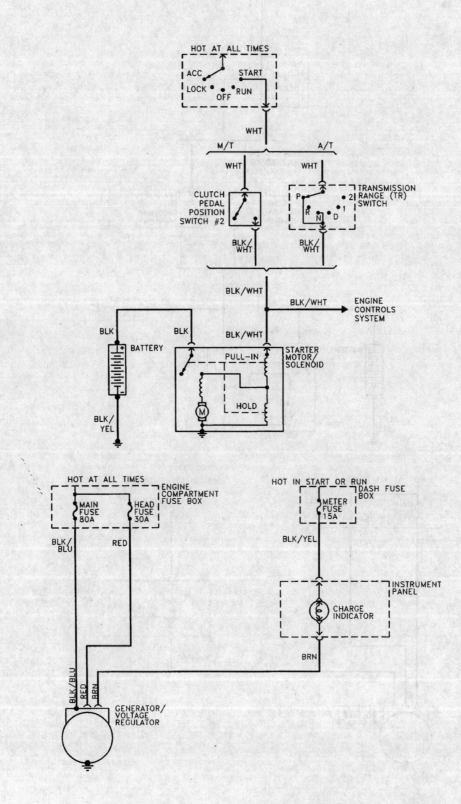

Typical 1994 and later starting and charging system

Chapter 12 Chassis electrical system

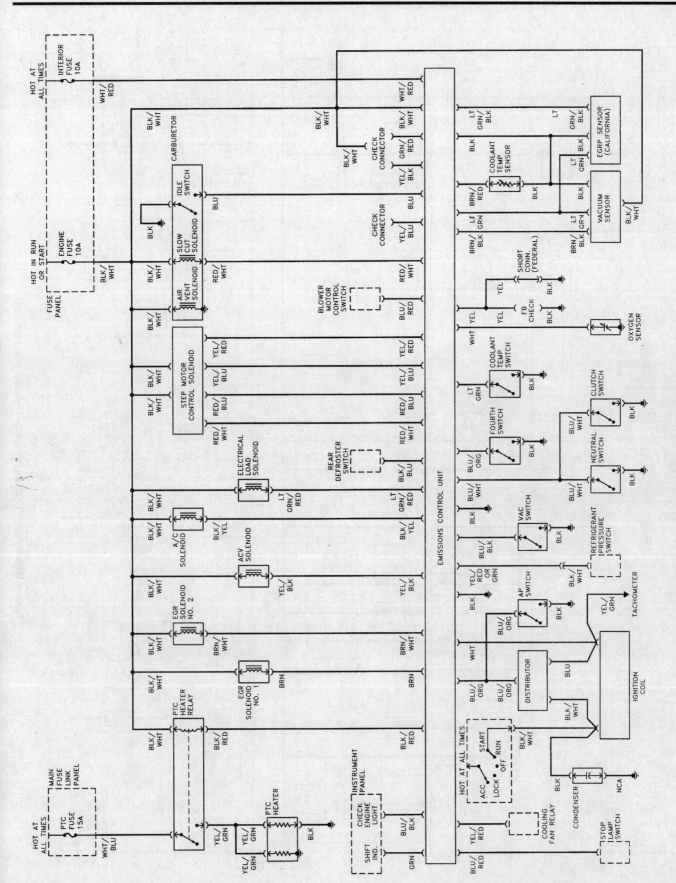

Typical 1988 and 1989 engine control system (carbureted)

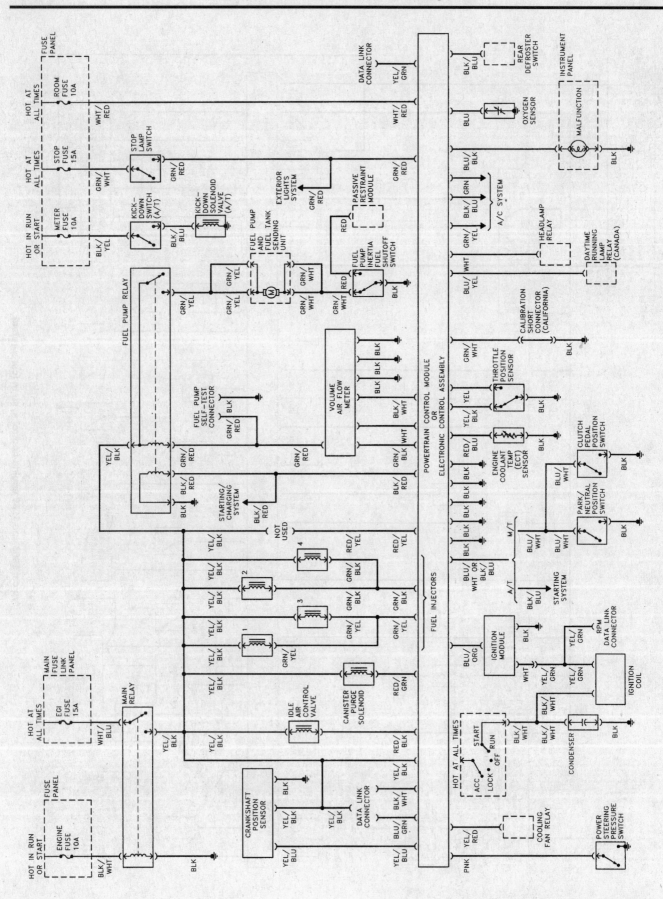

Typical 1989 through 1993 engine control system (fuel injection)

Chapter 12 Chassis electrical system

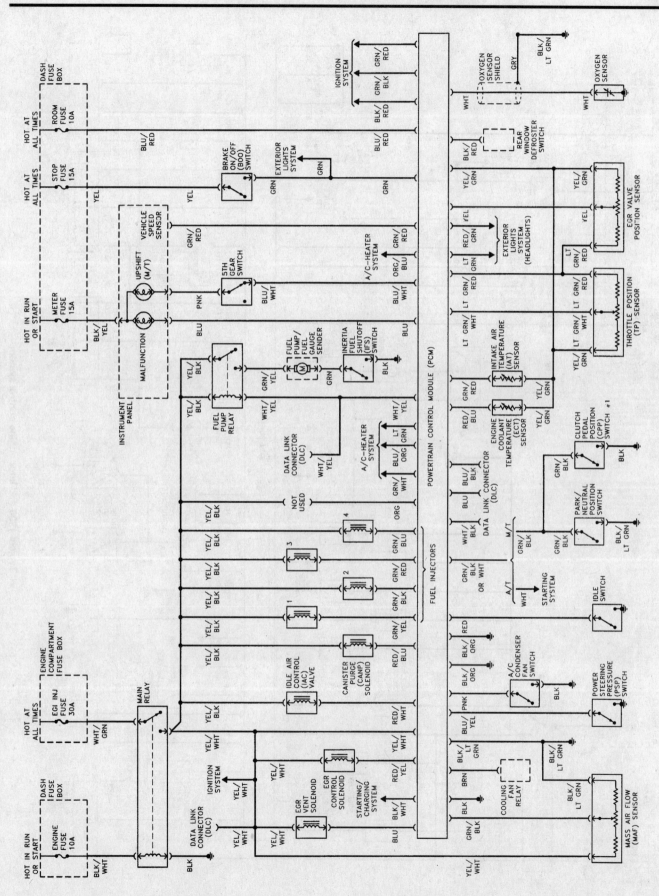

Typical 1994 and 1995 engine control system

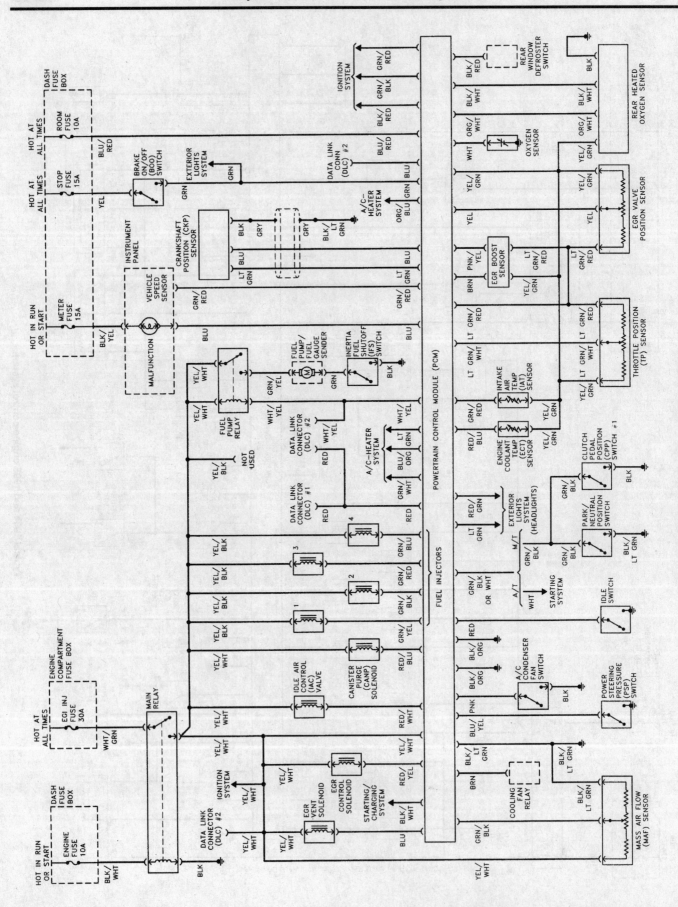

Typical 1996 and later engine control system

Chapter 12 Chassis electrical system

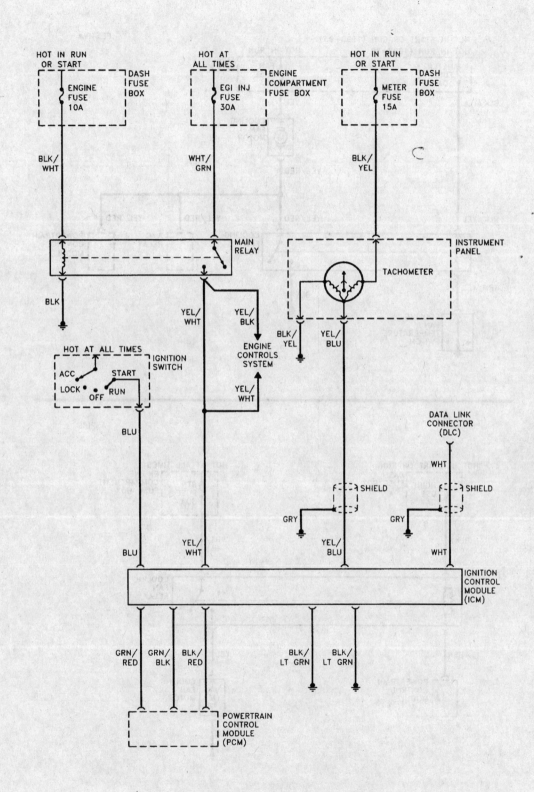

Typical 1994 and later ignition system

Chapter 12 Chassis electrical system

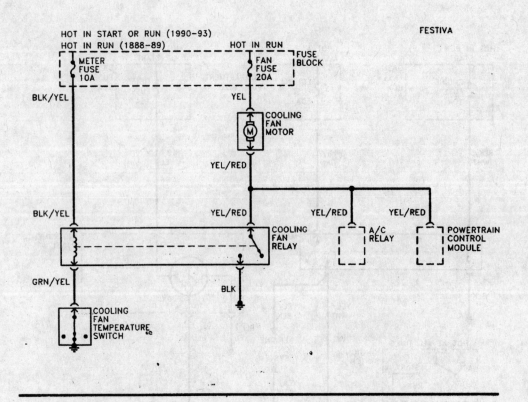

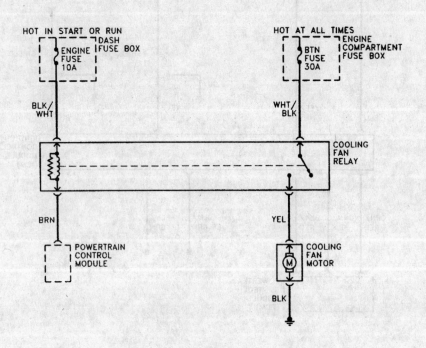

Typical engine cooling fan system

Chapter 12 Chassis electrical system

12-27

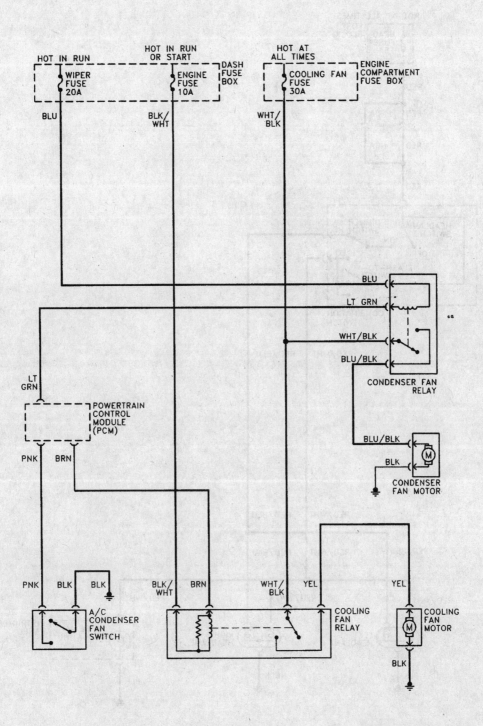

Typical 1994 and later air conditioning condenser fan system

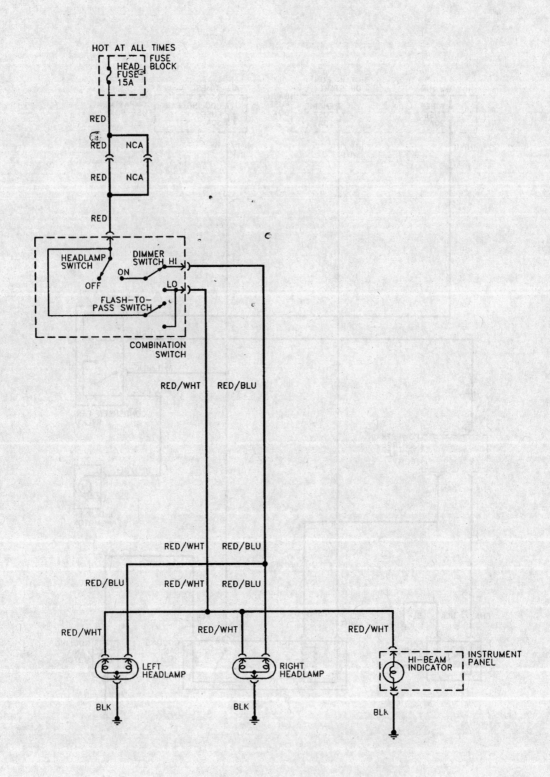

Typical 1988 and 1989 headlight system

Chapter 12 Chassis electrical system

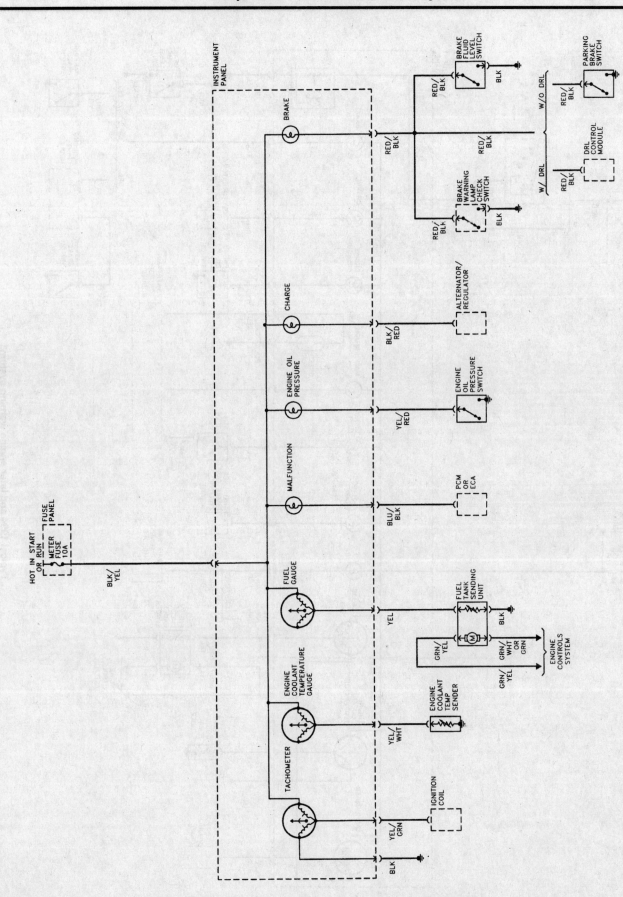

Typical 1993 and earlier engine warning system

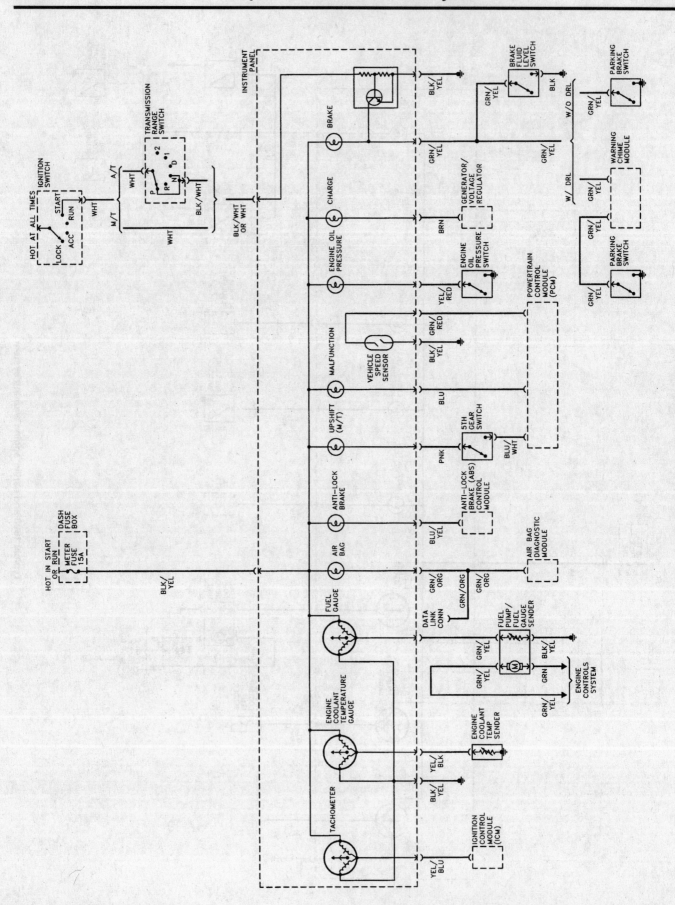

Typical 1994 and later engine warning system

Chapter 12 Chassis electrical system

12-31

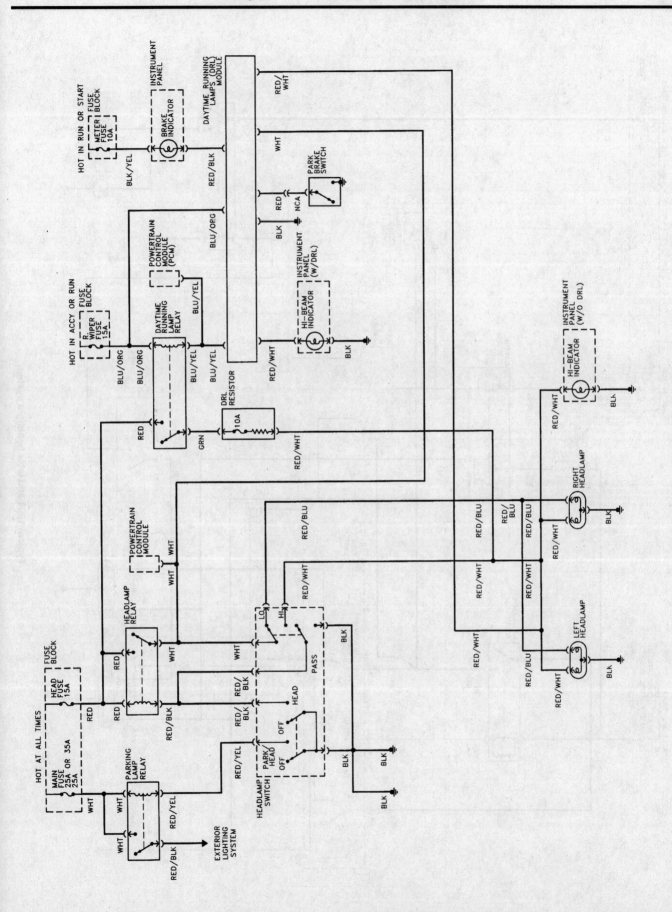

Typical 1990 through 1993 headlight system

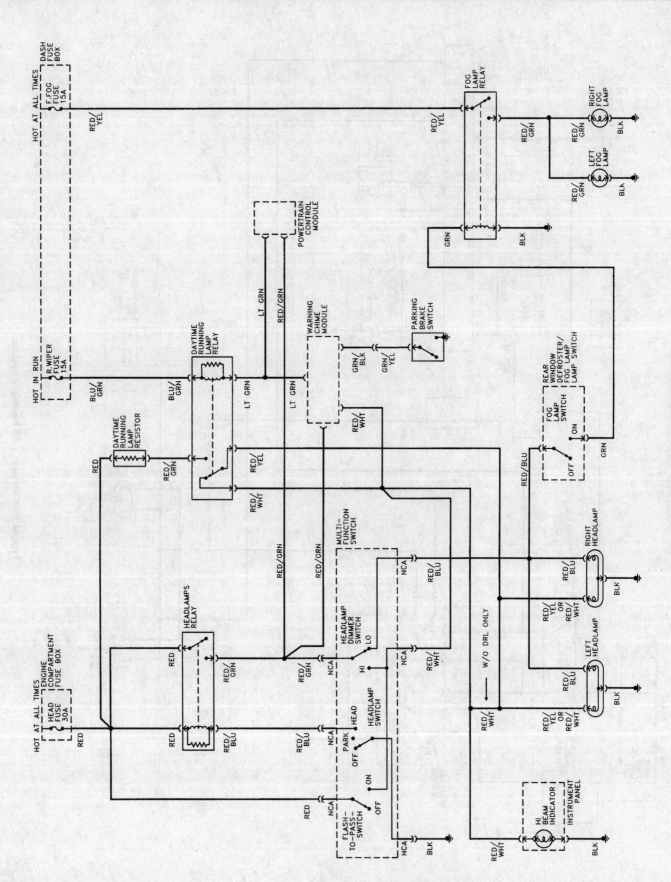

Typical 1994 and later headlight system

Chapter 12 Chassis electrical system

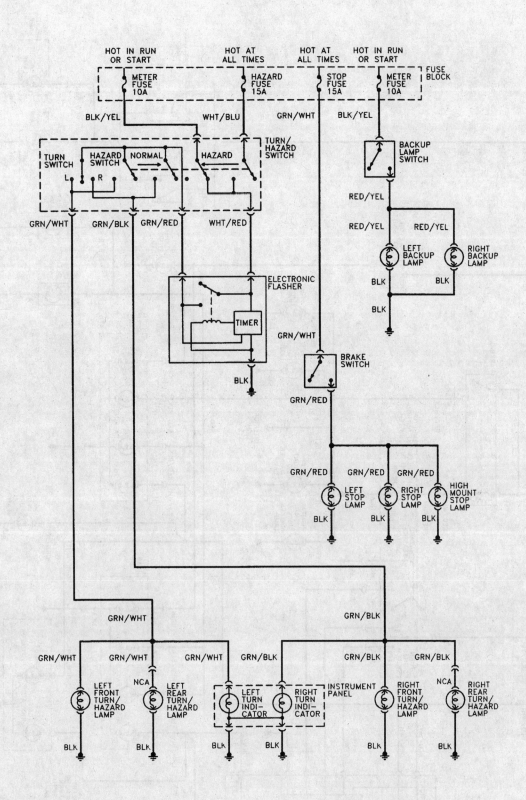

Typical 1993 and earlier exterior lighting system (except headlights)

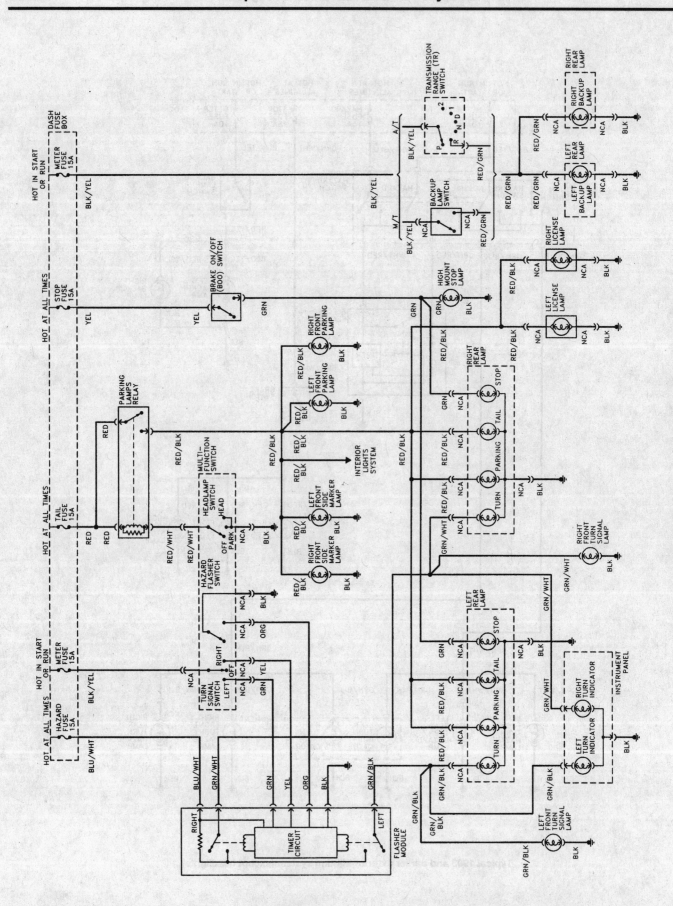

12-34 Chapter 12 Chassis electrical system

Typical 1994 and later exterior lighting system (except headlights)

Chapter 12 Chassis electrical system

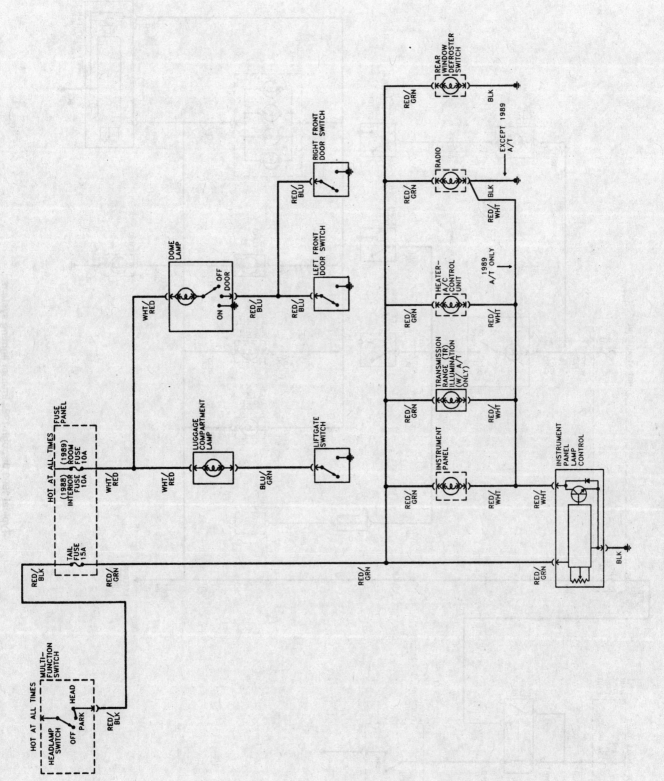

Typical 1988 and 1989 interior lighting system

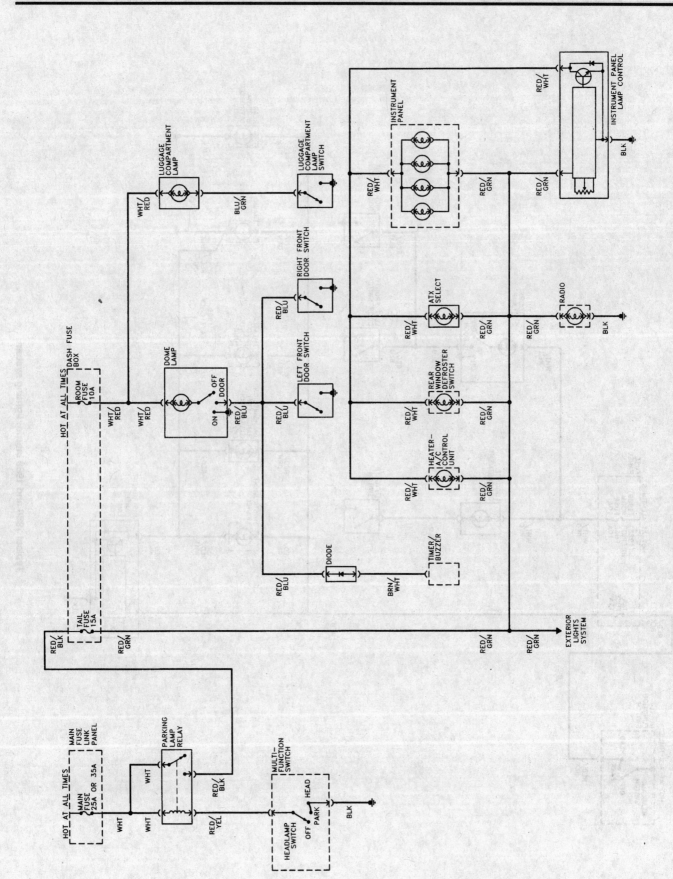

Typical 1990 through 1993 interior lighting system

Chapter 12 Chassis electrical system

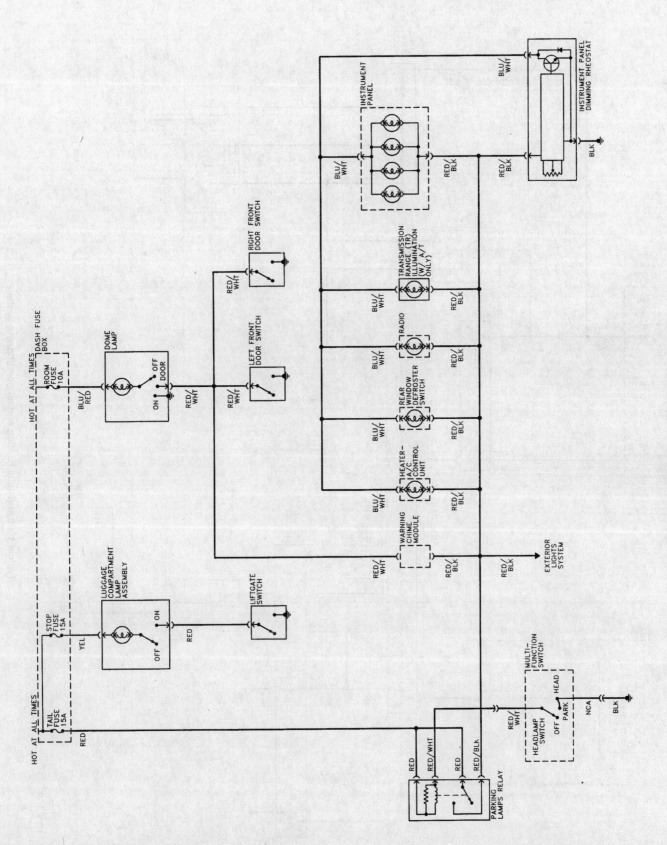

Typical 1994 and later interior lighting system

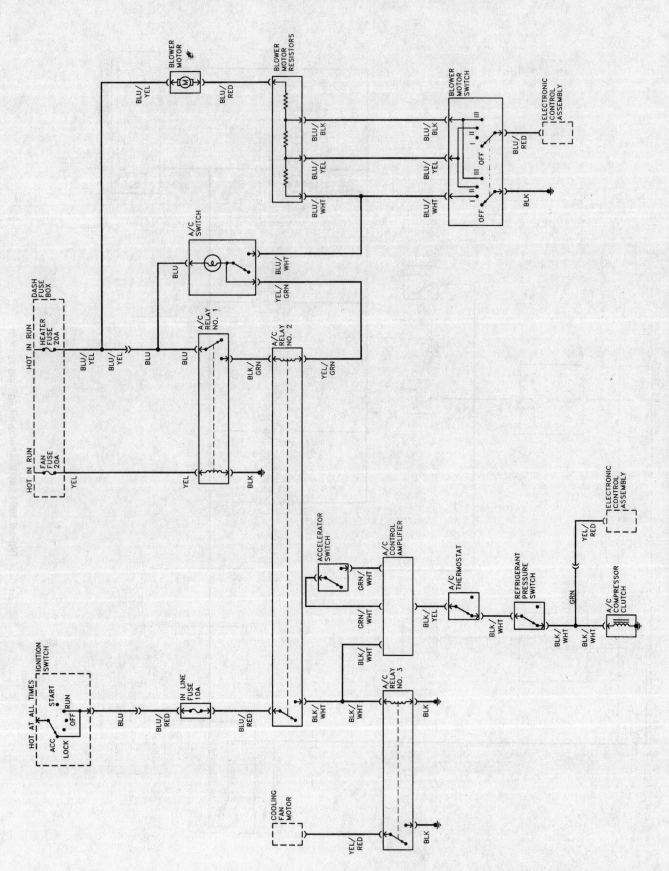

Typical 1988 and 1989 heating and air conditioning system

Chapter 12 Chassis electrical system

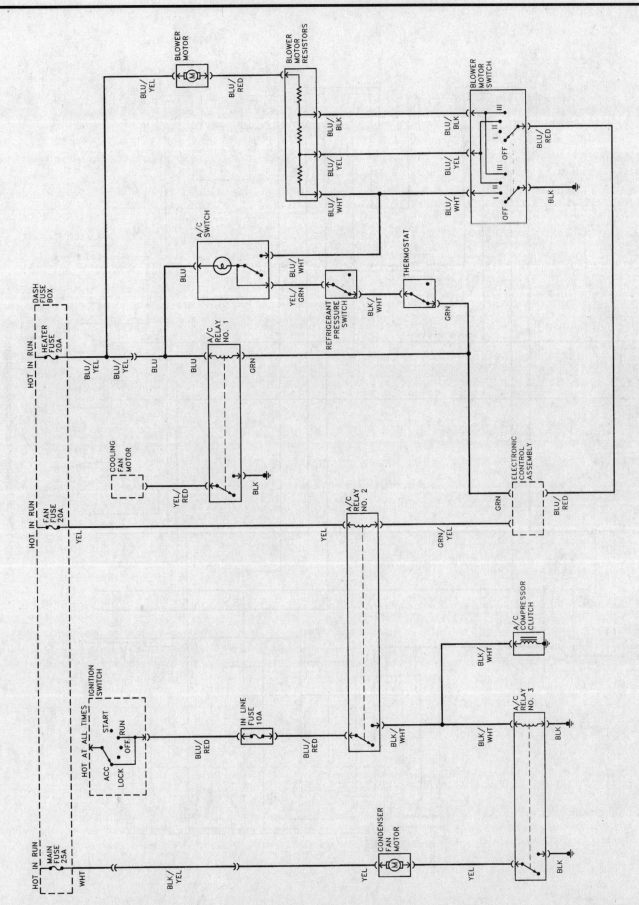

Typical 1990 through 1993 heating and air conditioning system

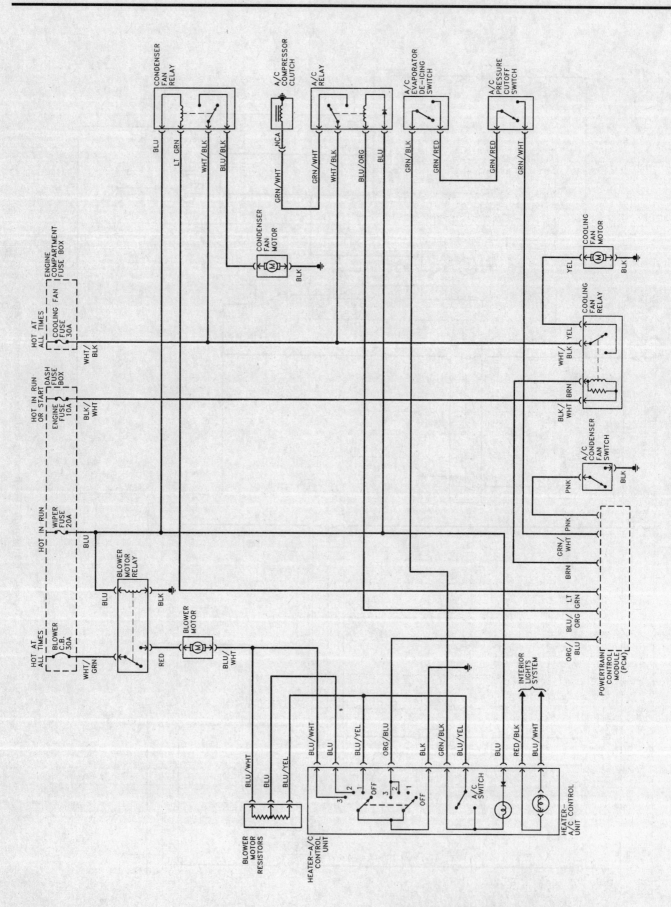

Typical 1994 and later air conditioning system

Chapter 12 Chassis electrical system

12-41

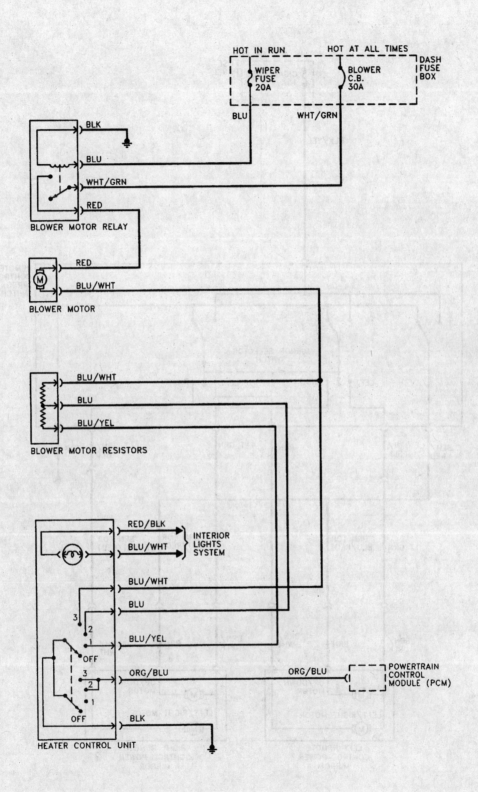

Typical 1994 and later heating system

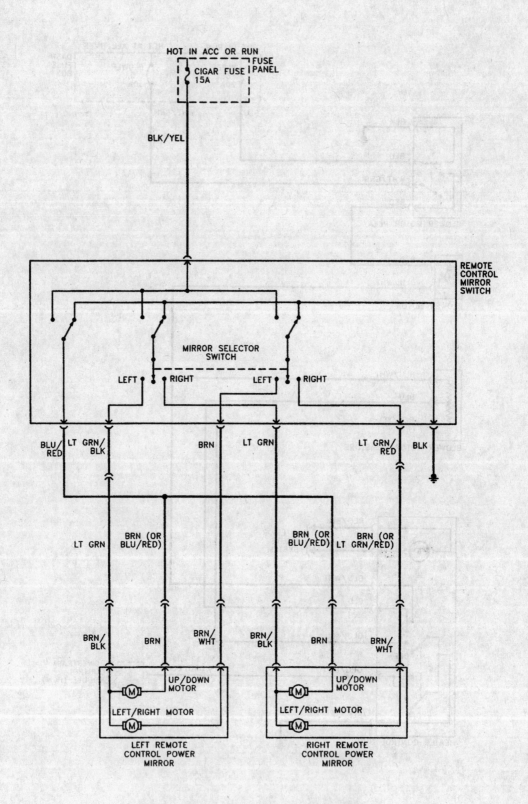

Typical 1993 and earlier power mirror system

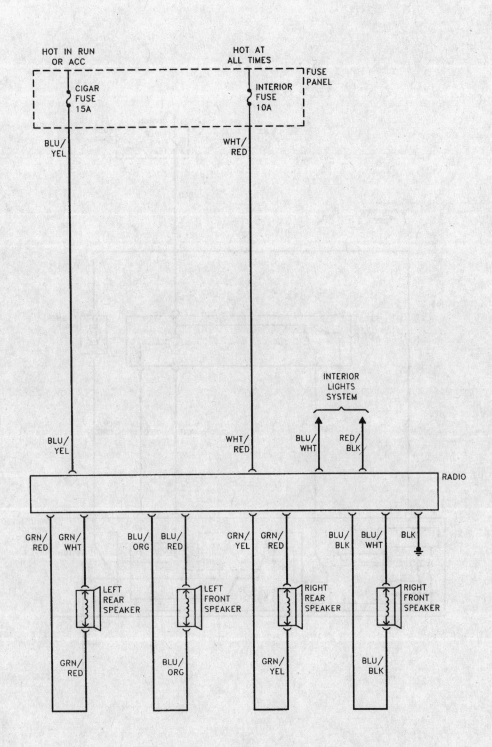

Typical audio system

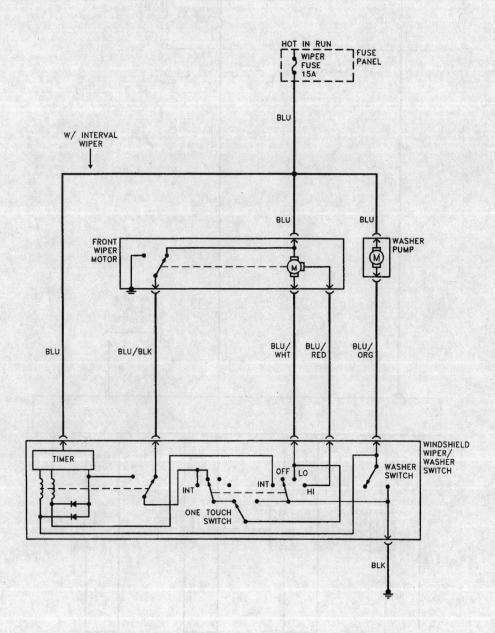

Typical 1993 and earlier windshield wiper and washer system

Chapter 12 Chassis electrical system

12-45

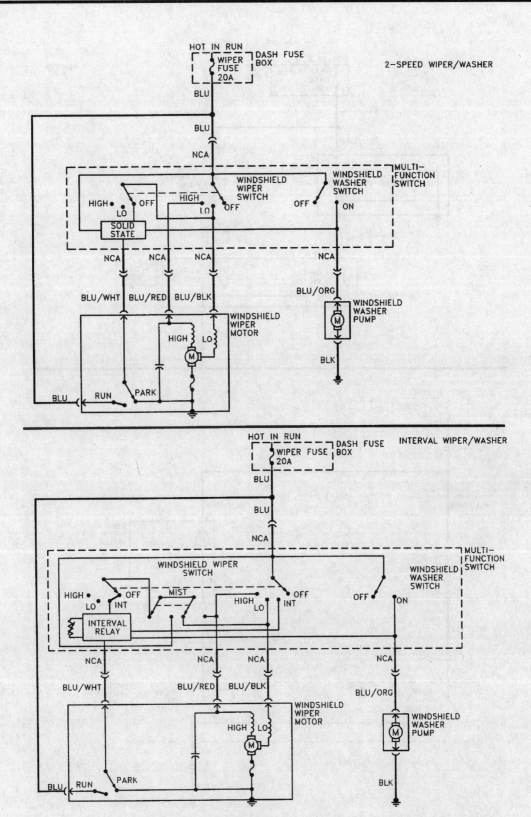

Typical 1994 and later windshield wiper and washer system

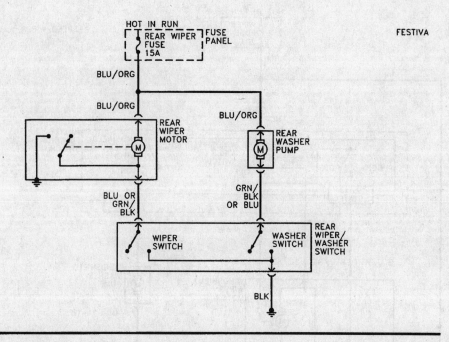
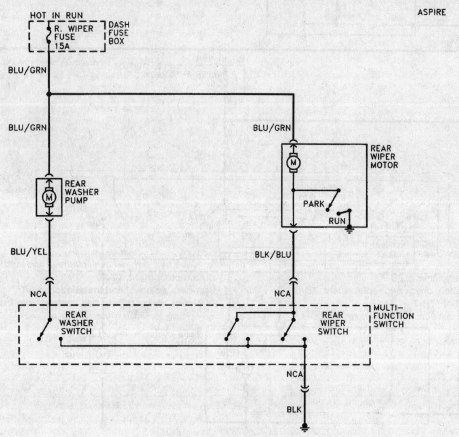

Typical rear windshield wiper and washer system

Index

A

About this manual, 0-5
Accelerator cable, removal, installation and adjustment, 4-8
Antenna, removal and installation, 12-13
Air cleaner assembly, removal and installation, 4-8
Air conditioning and heating system, check and maintenance, 3-12
Air conditioning system
  compressor, removal and installation, 3-15
  condenser removal and installation, 3-16
  evaporator and expansion valve, removal and installation, 3-17
  receiver/drier, removal and installation, 3-14
Air filter replacement, 1-25
Air inlet plenum, 4-13
Airbag system, general information, 12-17
Airflow sensor/intake air temperature sensor, 6-11
Alternator, removal and installation, 5-7
Antifreeze, general information, 3-2
Anti-lock Brake System (ABS), general information, 9-2
Automatic transaxle, 7B-1 through 7B-8
  diagnosis, general, 7B-2
  fluid and filter change, 1-29
  fluid level check, 1-12
  intermediate band, adjustment, 7B-7
  park/neutral start switch, check and replacement, 7B-5
  removal and installation, 7B-7
  shift cable, removal, installation and adjustment, 7B-4
  shift lever, removal and installation, 7B-2
  shift-lock system, description, check and cable replacement, 7B-6
Automotive chemicals and lubricants, 0-16
Axle beam (rear), removal, inspection and installation, 10-7

B

Back-up light switch, check and replacement, 7A-2
Balljoints, check and replacement, 10-6
Battery
  cables, check and replacement, 5-2
  electrolyte, check, 1-8
  jump starting, 0-14
  removal and installation, 5-2
Blower motor and circuit, check and component replacement, 3-7

Body, 11-1 through 11-18
  damage repair
    major, 11-5
    minor, 11-4
  maintenance, 11-1
Booster battery (jump) starting, 0-14
Brakes, 9-1 through 9-20
  booster, check, removal and installation, 9-16
  caliper, removal, overhaul and installation, 9-4
  check, 1-19
  disc, inspection, removal and installation, 9-6
  fluid, check, 1-8
  general information, 9-2
  hoses and lines, inspection and replacement, 9-15
  hydraulic system, bleeding, 9-16
  light switch, check, adjustment and replacement, 9-20
  master cylinder, removal, overhaul and installation, 9-13
  pads (front), replacement, 9-3
  parking
    adjustment, 9-18
    cables, replacement, 9-18
  pedal, adjustment, 9-19
  shoes (rear), replacement, 9-7
  wheel cylinder, removal, overhaul and installation, 9-12
Bulb replacement, 12-10
Bumpers, removal and installation, 11-7
Bushings and stabilizer bar, removal and installation, 10-5
Buying parts, 0-7

C

Camshaft oil seal, replacement, 2A-8
Camshaft Position Sensor (CMP), 6-13
Camshaft, removal, inspection, and installation, 2A-10
Capacities, 1-2
Carburetor
  diagnosis, overhaul and adjustment, 4-9
  feedback system, 6-16
  removal and installation, 4-10
Carpets and upholstery, maintenance, 11-1
Catalytic converter, 6-17
Center console, removal and installation, 11-14

IND-2   Index

Charging system
  check, 5-7
  general information and precautions, 5-6
Chassis electrical system, 12-1 through 12-46
Clearing codes, 6-8
Clutch and driveaxles, 8-1 through 8-12
Clutch
  cable, replacement, 8-2
  components, removal, inspection and installation, 8-2
  description and check, 8-2
  pedal height and freeplay, check and adjustment, 1-15
  release bearing and lever, removal, inspection and installation, 8-4
  start switch, check and replacement, 8-6
Clutch Pedal Position switch (manual transaxle), 6-13
Coil spring or strut/shock absorber, replacement, 10-4
Coil spring/shock absorber (rear), removal, inspection and installation, 10-7
Combination switch
  check, 12-4
  removal and installation, 12-4
Compression check, 2B-4
Control arm, removal, inspection and installation, 10-5
Conversion factors, 0-17
Cooling system
  check, 1-16
  coolant temperature
    gauge sending unit, check and replacement, 3-6
    sensor, 6-9
  servicing (draining, flushing and refilling), 1-26
Cooling, heating and air conditioning system, 3-1 through 3-18
Crankshaft
  front oil seal, replacement, 2A-16
  inspection, 2B-14
  installation and main bearing oil clearance check, 2B-16
  removal, 2B-10
Crankshaft Position Sensor (CKP), 6-13
Cylinder head
  cleaning and inspection, 2B-7
  disassembly, 2B-7
  reassembly, 2B-9
  removal and installation, 2A-13
Cylinder honing, 2B-12

D

Dashboard trim panels, removal and installation, 11-15
Daytime Running Lights (DRL), general information, 12-12
Deceleration Control system, 6-3
Diagnosis, 0-19
Diagnostic tool information, 6-4
Disc brake
  caliper, removal, overhaul and installation, 9-4
  pads (front), replacement, 9-3
Distributor, removal and installation, 5-5
Door
  latch, lock cylinder and handle, removal and installation, 11-11
  trim panel, removal and installation, 11-9
  window glass, removal and installation, 11-12
  removal, installation and adjustment, 11-10
Driveaxle
  boot
    check, 1-28
    replacement, 8-8
  removal and installation, 8-7
  general information and inspection, 8-6
Drivebelt check, adjustment and replacement, 1-20
Drum brake shoes, replacement, 9-7
Early Fuel Evaporation (EFE) system, 6-2

EGR gas temperature sensor, 6-13
Electric pump (fuel-injected engines), 4-3
Electric side view mirrors, description and check, 12-16
Electrical troubleshooting, general information, 12-1
Electronic Control Module (ECM), removal and installation, 6-14
Electronic Fuel Injection (EFI) system,
  check, 4-11
  component check and replacement, 4-12
  general information, 4-11
Emissions and engine control systems, 6-1 through 6-18
Engines, 2A-1 through 2A-18
  block
    cleaning, 2B-11
    inspection, 2B-12
  camshaft
    oil seal, replacement, 2A-8
    removal, inspection, and installation, 2A-10
  compression check, 2B-4
  coolant, check, 1-7
  cooling fan and circuit, check and component replacement, 3-3
  crankshaft
    front oil seal, replacement, 2A-16
    inspection, 2B-14
    installation and main bearing oil clearance check, 2B-16
    removal, 2B-10
  cylinder head
    cleaning and inspection, 2B-7
    disassembly, 2B-7
    reassembly, 2B-9
    removal and installation, 2A-13
  cylinder honing, 2B-12
  electrical systems, 5-1 through 5-8
  exhaust manifold, removal and installation, 2A-5
  flywheel/driveplate, removal and installation, 2A-16
  initial start-up and break-in after overhaul, 2B-20
  intake manifold, removal and installation, 2A-4
  main and connecting rod bearings, inspection and selection, 2B-14
  mounts, check and replacement, 2A-17
  oil and oil filter change, 1-10
  oil
    check, 1-6
    pan, removal and installation, 2A-14
    pressure check, 2B-3
    pump, removal, inspection and installation, 2A-14
    type, 1-1
  overhaul
    disassembly sequence, 2B-6
    reassembly sequence, 2B-15
  piston rings, installation, 2B-16
  pistons/connecting rods
    inspection, 2B-13
    installation and rod bearing oil clearance check, 2B-19
    removal, 2B-9
  rear main oil seal
    installation, 2B-18
    replacement, 2A-17
  rebuilding alternatives, 2B-6
  removal and installation, 2B-5
  removal, methods and precautions, 2B-4
  repair operations possible with the engine in the vehicle, 2A-2
  rocker arm assembly and lash adjusters, removal, inspection and installation, 2A-9
  timing belt and sprockets, removal, inspection and installation, 2A-6
  Top Dead Center (TDC), locating, 2A-3
  vacuum gauge diagnostic check, 2B-3
  valve
    cover, removal and installation, 2A-4
    servicing, 2B-9
    springs, retainers and seals, replacement, 2A-11

Index

Evaporative Emission Control (EVAP) system, 1-26, 6-14
Exhaust Gas Recirculation (EGR) system, 6-15
Exhaust manifold, removal and installation, 2A-5
Exhaust system
    check, 1-28
    general information, 4-15

F

Fault finding, 0-19
Feedback carburetor system, 6-16
Fluid level checks, 1-6
Flywheel/driveplate, removal and installation, 2A-16
Front fender, removal and installation, 11-7
Fuel and exhaust systems, 4-1 through 4-16
Fuel system
    check, 1-28
    filter replacement, 1-31
    level sending unit, check and replacement, 4-7
    lines and fittings, inspection and replacement, 4-5
    pressure regulator, 4-13
    pressure relief procedure, fuel-injected models, 4-2
    pump
        electric (fuel-injected engines), 4-3
        mechanical (carbureted engines), 4-3
        pump, removal and installation, 4-6
    pump/fuel pressure, check, 4-3
    tank
        cleaning and repair, general information, 4-6
        removal and installation, 4-5
Fuses, general information, 12-2
Fusible links, general information, 12-3

G

General engine overhaul procedures, 2B-1 through 2B-20

H

Handle, door latch and lock cylinder, removal and installation, 11-11
Headlight
    adjustment, 12-9
    bulb, replacement, 12-8
    housing, removal and installation, 12-9
Heated Air Inlet (HAI) system, 6-3
Heater and air conditioning control assembly, removal, installation, check and adjustment, 3-10
Heater core, replacement, 3-9
High-Altitude Compensation (HAC) system, 6-4
Hinges and locks, maintenance, 11-5
Hood
    release latch and cable, removal and installation, 11-6
    removal, installation and adjustment, 11-5
Horn, check and replacement, 12-14
Hub and bearing assembly (front), removal and installation, 10-6

I

Idle speed check and adjustment, 1-17
Idle speed control-bypass air (ISC-BPA), 4-13
Ignition system, 1-2
    check, 5-3
    coil, check and replacement, 5-4
    control module, replacement, 5-5
    general information and precautions, 5-2
    key lock cylinder, removal and installation, 12-8
    switch, check and replacement, 12-6
    timing, check and adjustment, 5-6
Information sensors, 6-7, 6-9
    Airflow sensor/intake air temperature sensor, 6-11
    Camshaft Position Sensor (CMP), 1996 and later, 6-13
    Clutch Pedal Position switch (manual transaxle), 6-13
    Coolant temperature sensor, 6-9
    Crankshaft Position Sensor (CKP), external module ignition systems, 6-13
    EGR gas temperature sensor, 6-13
    Intake Air Temperature (IAT) sensor, 6-12
    Mass Airflow (MAF) sensor, 6-12
    Neutral Start switch (automatic transmission), 6-14
    Oxygen sensor, 6-10
    Power steering pressure switch, 6-13
    Throttle Position Sensor (TPS), 6-11
    Vehicle Speed Sensor (VSS), 6-13
Initial start-up and break-in after overhaul, 2B-20
Instrument cluster, removal and installation, 12-13
Instrument panel, removal and installation, 11-15
Intake Air Temperature (IAT) sensor, 6-12
Intake manifold, removal and installation, 2A-4
Intermediate band, adjustment, 7B-7
Introduction, 0-5

J

Jacking and towing, 0-15

L

Liftgate
    support strut, replacement, 11-9
    removal, installation and adjustment, 11-8

M

Main and connecting rod bearings, inspection and selection, 2B-14
Maintenance
    schedule, 1-5
    techniques, tools and working facilities, 0-7
Manual transaxle, 7A-1 through 7A-6
    back-up light switch, check and replacement, 7A-2
    lubricant
        change, 1-32
        level check, 1-20
    mount, check and replacement, 7A-4
    oil seal, replacement, 7A-3
    overhaul, general information, 7A-6
    removal and installation, 7A-4
    shift lever, removal and installation, 7A-2
Mass Airflow (MAF) sensor, 6-12
Master cylinder, removal, overhaul and installation, 9-13
Mechanical pump (carbureted engines), 4-3
Mirror, outside (manual), removal and installation, 11-13

N

Neutral Start switch (automatic transmission), 6-14
Neutral/park start switch, check and replacement, 7B-5

O

OBD general information, 6-5
Oil
    pan, removal and installation, 2A-14
    pressure check, 2B-3
    pump, removal, inspection and installation, 2A-14
    seal, replacement, 7A-3
On-Board Diagnostic (OBD) System and trouble codes, 6-4
Output actuators, 6-8
Outside mirror (manual), removal and installation, 11-13
Oxygen sensor, 6-10

P

Park/neutral start switch, check and replacement, 7B-5
Parking brake
    adjustment, 9-18
    cables, replacement, 9-18
Pilot bearing, inspection and replacement, 8-5
Piston rings, installation, 2B-16
Pistons/connecting rods
    inspection, 2B-13
    installation and rod bearing oil clearance check, 2B-19
    removal, 2B-9
Positive Crankcase Ventilation (PCV)
    system, 6-15
    valve and hose check and replacement, 1-26
Power brake booster, check, removal and installation, 9-16
Power steering system
    bleeding, 10-11
    fluid level check, 1-11
    pressure switch, 6-13
    pump, removal and installation, 10-11
Pulse Air system, 6-2

R

Radiator and coolant reservoir, removal and installation, 3-5
Radiator grille, removal and installation, 11-6
Radio and speakers, removal and installation, 12-12
Rear main oil seal
    installation, 2B-18
    replacement, 2A-17
Rear quarter window, replacement (1993 and earlier), 11-12
Rear wheel bearing check, repack and adjustment, 1-30
Rear window defogger
    switch, check and replacement, 12-15
    check and repair, 12-16
Recommended lubricants and fluids, 1-1
Relays, general information and testing, 12-3
Release bearing and lever, clutch, removal, inspection and installation, 8-4
Repair operations possible with the engine in the vehicle, 2A-2
Rocker arm assembly and lash adjusters, removal, inspection and installation, 2A-9

S

Safety first, 0-18
Seat
    belts, check, 11-18
    removal and installation, 11-17

Shift cable, removal, installation and adjustment, 7B-4
Shift lever, removal and installation, 7A-2, 7B-2
Shift-lock system, description, check and cable replacement, 7B-6
Shock absorber/coil spring (rear), removal, inspection and installation, 10-7
Spark plug
    check and replacement, 1-22
    wire, distributor cap and rotor check and replacement, 1-23
Stabilizer bar and bushings, removal and installation, 10-5
Starter motor
    in-vehicle testing, 5-8
    removal and installation, 5-8
Starting system, general information and precautions, 5-8
Steering and suspension check, 1-27
Steering system
    column covers, removal and installation, 11-14
    gear boots, replacement, 10-9
    gear, removal and installation, 10-10
    knuckle and hub, removal and installation, 10-6
    wheel, removal and installation, 10-8
Strut assembly (front), removal, inspection and installation, 10-3
Strut/shock absorber or spring assembly, replacement, 10-4
Suspension and steering systems, 10-1 through 10-12

T

Thermostat, check and replacement, 3-2
Throttle body, 4-12
Throttle Position Sensor (TPS), 4-12, 6-11
Tie-rod ends, removal and installation, 10-9
Timing belt and sprockets, removal, inspection and installation, 2A-6
Tire and tire pressure checks, 1-8
Tire rotation, 1-15
Tires and wheels, general information, 10-11
Tools, 0-10
Top Dead Center (TDC), locating, 2A-3
Towing the vehicle, 0-15
Transaxle
    automatic, 7B-1 through 7B-8
        diagnosis, general, 7B-2
        fluid and filter change, 1-29
        fluid level check, 1-12
        intermediate band, adjustment, 7B-7
        park/neutral start switch, check and replacement, 7B-5
        removal and installation, 7B-7
        shift cable, removal, installation and adjustment, 7B-4
        shift lever, removal and installation, 7B-2
        shift-lock system, description, check and cable replacement, 7B-6
    manual
        back-up light switch, check and replacement, 7A-2
        lubricant change, 1-32
        lubricant level check, 1-20
        mount, check and replacement, 7A-4
        oil seal, replacement, 7A-3
        overhaul, general information, 7A-6
        removal and installation, 7A-4
        shift lever, removal and installation, 7A-2
Trouble codes, 6-4
Troubleshooting, 0-19
Tune-up and routine maintenance, 1-1 through 1-32
Tune-up, general information, 1-6
Turn signal/hazard flashers, check and replacement, 12-3

Index

U
Underhood hose check and replacement, 1-16
Upholstery and carpets, maintenance, 11-1

V
Vacuum gauge diagnostic check, 2B-3
Valve
    clearance check and adjustment (carbureted engines), 1-18
    cover, removal and installation, 2A-4
    servicing, 2B-9
    springs, retainers and seals, replacement, 2A-11
Vehicle identification numbers, 0-6
Vehicle Speed Sensor (VSS), 6-13
Vinyl trim, maintenance, 11-1

W
Water pump
    check, 3-6
    removal and installation, 3-6
Wheel
    alignment, general information, 10-12
    cylinder, removal, overhaul and installation, 9-12
    rear, bearing check, repack and adjustment, 1-30
Wheels and tires, general information, 10-11
Window regulator, removal and installation, 11-12
Windshield and fixed glass, replacement, 11-5
Windshield wiper blade inspection and replacement, 1-13
Windshield/window washer fluid, check, 1-7
Wiper motor, check and replacement, 12-14
Wiring diagrams, 12-18

Haynes Automotive Manuals

NOTE: If you do not see a listing for your vehicle, consult your local Haynes dealer for the latest product information.

ACURA
- 12020 Integra '86 thru '89 & Legend '86 thru '90
- 12021 Integra '90 thru '93 & Legend '91 thru '95
- Integra '94 thru '00 - see HONDA Civic (42025)
- MDX '01 thru '07 - see HONDA Pilot (42037)
- 12050 Acura TL all models '99 thru '08

AMC
- Jeep CJ - see JEEP (50020)
- 14020 Mid-size models '70 thru '83
- 14025 (Renault) Alliance & Encore '83 thru '87

AUDI
- 15020 4000 all models '80 thru '87
- 15025 5000 all models '77 thru '83
- 15026 5000 all models '84 thru '88
- Audi A4 '96 thru '01 - see VW Passat (96023)
- 15030 Audi A4 '02 thru '08

AUSTIN-HEALEY
- Sprite - see MG Midget (66015)

BMW
- 18020 3/5 Series '82 thru '92
- 18021 3-Series incl. Z3 models '92 thru '98
- 18022 3-Series incl. Z4 models '99 thru '05
- 18023 3-Series '06 thru '10
- 18025 320i all 4 cyl models '75 thru '83
- 18050 1500 thru 2002 except Turbo '59 thru '77

BUICK
- 19010 Buick Century '97 thru '05
- Century (front-wheel drive) - see GM (38005)
- 19020 Buick, Oldsmobile & Pontiac Full-size (Front-wheel drive) '85 thru '05
- Buick Electra, LeSabre and Park Avenue; Oldsmobile Delta 88 Royale, Ninety Eight and Regency; Pontiac Bonneville
- 19025 Buick, Oldsmobile & Pontiac Full-size (Rear wheel drive) '70 thru '90
- Buick Estate, Electra, LeSabre, Limited, Oldsmobile Custom Cruiser, Delta 88, Ninety-eight, Pontiac Bonneville, Catalina, Grandville, Parisienne
- 19030 Mid-size Regal & Century all rear-drive models with V6, V8 and Turbo '74 thru '87
- Regal - see GENERAL MOTORS (38010)
- Riviera - see GENERAL MOTORS (38030)
- Roadmaster - see CHEVROLET (24046)
- Skyhawk - see GENERAL MOTORS (38015)
- Skylark - see GM (38020, 38025)
- Somerset - see GENERAL MOTORS (38025)

CADILLAC
- 21015 CTS & CTS-V '03 thru '12
- 21030 Cadillac Rear Wheel Drive '70 thru '93
- Cimarron - see GENERAL MOTORS (38015)
- DeVille - see GM (38031 & 38032)
- Eldorado - see GM (38030 & 38031)
- Fleetwood - see GM (38031)
- Seville - see GM (38030, 38031 & 38032)

CHEVROLET
- 10305 Chevrolet Engine Overhaul Manual
- 24010 Astro & GMC Safari Mini-vans '85 thru '05
- 24015 Camaro V8 all models '70 thru '81
- 24016 Camaro all models '82 thru '92
- 24017 Camaro & Firebird '93 thru '02
- Cavalier - see GENERAL MOTORS (38016)
- Celebrity - see GENERAL MOTORS (38005)
- 24020 Chevelle, Malibu & El Camino '69 thru '87
- 24024 Chevette & Pontiac T1000 '76 thru '87
- Citation - see GENERAL MOTORS (38020)
- 24027 Colorado & GMC Canyon '04 thru '10
- 24032 Corsica/Beretta all models '87 thru '96
- 24040 Corvette all V8 models '68 thru '82
- 24041 Corvette all models '84 thru '96
- 24045 Full-size Sedans Caprice, Impala, Biscayne, Bel Air & Wagons '69 thru '90
- 24046 Impala SS & Caprice and Buick Roadmaster '91 thru '96
- Impala '00 thru '05 - see LUMINA (24048)
- 24047 Impala & Monte Carlo all models '06 thru '11
- Lumina '90 thru '94 - see GM (38010)
- 24048 Lumina & Monte Carlo '95 thru '05
- Lumina APV - see GM (38035)
- 24050 Luv Pick-up all 2WD & 4WD '72 thru '82
- Malibu '97 thru '00 - see GM (38026)
- 24055 Monte Carlo all models '70 thru '88
- Monte Carlo '95 thru '01 - see LUMINA (24048)
- 24059 Nova all V8 models '69 thru '79
- 24060 Nova and Geo Prizm '85 thru '92
- 24064 Pick-ups '67 thru '87 - Chevrolet & GMC
- 24065 Pick-ups '88 thru '98 - Chevrolet & GMC
- 24066 Pick-ups '99 thru '06 - Chevrolet & GMC
- 24067 Chevrolet Silverado & GMC Sierra '07 thru '12
- 24070 S-10 & S-15 Pick-ups '82 thru '93, Blazer & Jimmy '83 thru '94,
- 24071 S-10 & Sonoma Pick-ups '94 thru '04, including Blazer, Jimmy & Hombre
- 24072 Chevrolet TrailBlazer, GMC Envoy & Oldsmobile Bravada '02 thru '09
- 24075 Sprint '85 thru '88 & Geo Metro '89 thru '01
- 24080 Vans - Chevrolet & GMC '68 thru '96
- 24081 Chevrolet Express & GMC Savana Full-size Vans '96 thru '10

CHRYSLER
- 10310 Chrysler Engine Overhaul Manual
- 25015 Chrysler Cirrus, Dodge Stratus, Plymouth Breeze '95 thru '00
- 25020 Full-size Front-Wheel Drive '88 thru '93
- K-Cars - see DODGE Aries (30008)
- Laser - see DODGE Daytona (30030)
- 25025 Chrysler LHS, Concorde, New Yorker, Dodge Intrepid, Eagle Vision, '93 thru '97
- 25026 Chrysler LHS, Concorde, 300M, Dodge Intrepid, '98 thru '04
- 25027 Chrysler 300, Dodge Charger & Magnum '05 thru '09
- 25030 Chrysler & Plymouth Mid-size front wheel drive '82 thru '95
- Rear-wheel Drive - see Dodge (30050)
- 25035 PT Cruiser all models '01 thru '10
- 25040 Chrysler Sebring '95 thru '06, Dodge Stratus '01 thru '06, Dodge Avenger '95 thru '00

DATSUN
- 28005 200SX all models '80 thru '83
- 28007 B-210 all models '73 thru '78
- 28009 210 all models '79 thru '82
- 28012 240Z, 260Z & 280Z Coupe '70 thru '78
- 28014 280ZX Coupe & 2+2 '79 thru '83
- 300ZX - see NISSAN (72010)
- 28018 510 & PL521 Pick-up '68 thru '73
- 28020 510 all models '78 thru '81
- 28022 620 Series Pick-up all models '73 thru '79
- 720 Series Pick-up - see NISSAN (72030)
- 28025 810/Maxima all gasoline models '77 thru '84

DODGE
- 400 & 600 - see CHRYSLER (25030)
- 30008 Aries & Plymouth Reliant '81 thru '89
- 30010 Caravan & Plymouth Voyager '84 thru '95
- 30011 Caravan & Plymouth Voyager '96 thru '02
- 30012 Challenger/Plymouth Saporro '78 thru '83
- 30013 Caravan, Chrysler Voyager, Town & Country '03 thru '07
- 30016 Colt & Plymouth Champ '78 thru '87
- 30020 Dakota Pick-ups all models '87 thru '96
- 30021 Durango '98 & '99, Dakota '97 thru '99
- 30022 Durango '00 thru '03 Dakota '00 thru '04
- 30023 Durango '04 thru '09, Dakota '05 thru '11
- 30025 Dart, Demon, Plymouth Barracuda, Duster & Valiant 6 cyl models '67 thru '76
- 30030 Daytona & Chrysler Laser '84 thru '89
- Intrepid - see CHRYSLER (25025, 25026)
- 30034 Neon all models '95 thru '99
- 30035 Omni & Plymouth Horizon '78 thru '90
- 30036 Dodge and Plymouth Neon '00 thru '05
- 30040 Pick-ups all full-size models '74 thru '93
- 30041 Pick-ups all full-size models '94 thru '01
- 30042 Pick-ups full-size models '02 thru '08
- 30045 Ram 50/D50 Pick-ups & Raider and Plymouth Arrow Pick-ups '79 thru '93
- 30050 Dodge/Plymouth/Chrysler RWD '71 thru '89
- 30055 Shadow & Plymouth Sundance '87 thru '94
- 30060 Spirit & Plymouth Acclaim '89 thru '95
- 30065 Vans - Dodge & Plymouth '71 thru '03

EAGLE
- Talon - see MITSUBISHI (68030, 68031)
- Vision - see CHRYSLER (25025)

FIAT
- 34010 124 Sport Coupe & Spider '68 thru '78
- 34025 X1/9 all models '74 thru '80

FORD
- 10320 Ford Engine Overhaul Manual
- 10355 Ford Automatic Transmission Overhaul
- 11500 Mustang '64-1/2 thru '70 Restoration Guide
- 36004 Aerostar Mini-vans all models '86 thru '97
- 36006 Contour & Mercury Mystique '95 thru '00
- 36008 Courier Pick-up all models '72 thru '82
- 36012 Crown Victoria & Mercury Grand Marquis '88 thru '10
- 36016 Escort/Mercury Lynx all models '81 thru '90
- 36020 Escort/Mercury Tracer '91 thru '02
- 36022 Escape & Mazda Tribute '01 thru '11
- 36024 Explorer & Mazda Navajo '91 thru '01
- 36025 Explorer/Mercury Mountaineer '02 thru '10
- 36028 Fairmont & Mercury Zephyr '78 thru '83
- 36030 Festiva & Aspire '88 thru '97
- 36032 Fiesta all models '77 thru '80
- 36034 Focus all models '00 thru '11
- 36036 Ford & Mercury Full-size '75 thru '87
- 36044 Ford & Mercury Mid-size '75 thru '86
- 36045 Fusion & Mercury Milan '06 thru '10
- 36048 Mustang V8 all models '64-1/2 thru '73
- 36049 Mustang II 4 cyl, V6 & V8 models '74 thru '78
- 36050 Mustang & Mercury Capri '79 thru '93
- 36051 Mustang all models '94 thru '04
- 36052 Mustang '05 thru '10
- 36054 Pick-ups & Bronco '73 thru '79
- 36058 Pick-ups & Bronco '80 thru '96
- 36059 F-150 & Expedition '97 thru '09, F-250 '97 thru '99 & Lincoln Navigator '90 thru '09
- 36060 Super Duty Pick-ups, Excursion '99 thru '10
- 36061 F-150 full-size '04 thru '10
- 36062 Pinto & Mercury Bobcat '75 thru '80
- 36066 Probe all models '89 thru '92
- Probe '93 thru '97 - see MAZDA 626 (61042)
- 36070 Ranger/Bronco II gasoline models '83 thru '92
- 36071 Ranger '93 thru '10 & Mazda Pick-ups '94 thru '09
- 36074 Taurus & Mercury Sable '86 thru '95
- 36075 Taurus & Mercury Sable '96 thru '05
- 36078 Tempo & Mercury Topaz '84 thru '94
- 36082 Thunderbird/Mercury Cougar '83 thru '88
- 36086 Thunderbird/Mercury Cougar '89 thru '97
- 36090 Vans all V8 Econoline models '69 thru '91
- 36094 Vans full size '92 thru '10
- 36097 Windstar Mini-van '95 thru '07

GENERAL MOTORS
- 10360 GM Automatic Transmission Overhaul
- 38005 Buick Century, Chevrolet Celebrity, Oldsmobile Cutlass Ciera & Pontiac 6000 all models '82 thru '96
- 38010 Buick Regal, Chevrolet Lumina, Oldsmobile Cutlass Supreme & Pontiac Grand Prix (FWD) '88 thru '07
- 38015 Buick Skyhawk, Cadillac Cimarron, Chevrolet Cavalier, Oldsmobile Firenza & Pontiac J-2000 & Sunbird '82 thru '94
- 38016 Chevrolet Cavalier & Pontiac Sunfire '95 thru '05
- 38017 Chevrolet Cobalt & Pontiac G5 '05 thru '11
- 38020 Buick Skylark, Chevrolet Citation, Olds Omega, Pontiac Phoenix '80 thru '85
- 38025 Buick Skylark & Somerset, Oldsmobile Achieva & Calais and Pontiac Grand Am all models '85 thru '98
- 38026 Chevrolet Malibu, Olds Alero & Cutlass, Pontiac Grand Am '97 thru '03
- 38027 Chevrolet Malibu '04 thru '10
- 38030 Cadillac Eldorado, Seville, Oldsmobile Toronado, Buick Riviera '71 thru '85
- 38031 Cadillac Eldorado & Seville, DeVille, Fleetwood & Olds Toronado, Buick Riviera '86 thru '93
- 38032 Cadillac DeVille '94 thru '05 & Seville '92 thru '04 Cadillac DTS '06 thru '10
- 38035 Chevrolet Lumina APV, Olds Silhouette & Pontiac Trans Sport all models '90 thru '96
- 38036 Chevrolet Venture, Olds Silhouette, Pontiac Trans Sport & Montana '97 thru '05
- General Motors Full-size Rear-wheel Drive - see BUICK (19025)
- 38040 Chevrolet Equinox '05 thru '09 Pontiac Torrent '06 thru '09
- 38070 Chevrolet HHR '06 thru '11

GEO
- Metro - see CHEVROLET Sprint (24075)
- Prizm - '85 thru '92 see CHEVY (24060), '93 thru '02 see TOYOTA Corolla (92036)
- 40030 Storm all models '90 thru '93
- Tracker - see SUZUKI Samurai (90010)

GMC
- Vans & Pick-ups - see CHEVROLET

HONDA
- 42010 Accord CVCC all models '76 thru '83
- 42011 Accord all models '84 thru '89
- 42012 Accord all models '90 thru '93
- 42013 Accord all models '94 thru '97
- 42014 Accord all models '98 thru '02
- 42015 Accord '03 thru '07
- 42020 Civic 1200 all models '73 thru '79
- 42021 Civic 1300 & 1500 CVCC '80 thru '83
- 42022 Civic 1500 CVCC all models '75 thru '79

(Continued on other side)

Haynes North America, Inc., 861 Lawrence Drive, Newbury Park, CA 91320-1514 • (805) 498-6703 • http://www.haynes.com

Haynes Automotive Manuals (continued)

NOTE: If you do not see a listing for your vehicle, consult your local Haynes dealer for the latest product information.

- 42023 Civic all models '84 thru '91
- 42024 Civic & del Sol '92 thru '95
- 42025 Civic '96 thru '00, CR-V '97 thru '01, Acura Integra '94 thru '00
- 42026 Civic '01 thru '10, CR-V '02 thru '09
- 42035 Odyssey all models '99 thru '10
- Passport - see ISUZU Rodeo (47017)
- 42037 Honda Pilot '03 thru '07, Acura MDX '01 thru '07
- 42040 Prelude CVCC all models '79 thru '89

HYUNDAI
- 43010 Elantra all models '96 thru '10
- 43015 Excel & Accent all models '86 thru '09
- 43050 Santa Fe all models '01 thru '06
- 43055 Sonata all models '99 thru '08

INFINITI
- G35 '03 thru '08 - see NISSAN 350Z (72011)

ISUZU
- Hombre - see CHEVROLET S-10 (24071)
- 47017 Rodeo, Amigo & Honda Passport '89 thru '02
- 47020 Trooper & Pick-up '81 thru '93

JAGUAR
- 49010 XJ6 all 6 cyl models '68 thru '86
- 49011 XJ6 all models '88 thru '94
- 49015 XJ12 & XJS all 12 cyl models '72 thru '85

JEEP
- 50010 Cherokee, Comanche & Wagoneer Limited all models '84 thru '01
- 50020 CJ all models '49 thru '86
- 50025 Grand Cherokee all models '93 thru '04
- 50026 Grand Cherokee '05 thru '09
- 50029 Grand Wagoneer & Pick-up '72 thru '91 Grand Wagoneer '84 thru '91, Cherokee & Wagoneer '72 thru '83, Pick-up '72 thru '88
- 50030 Wrangler all models '87 thru '11
- 50035 Liberty '02 thru '07

KIA
- 54050 Optima '01 thru '10
- 54070 Sephia '94 thru '01, Spectra '00 thru '09, Sportage '05 thru '10

LEXUS
- ES 300/330 - see TOYOTA Camry (92007) (92008)
- RX 330 - see TOYOTA Highlander (92095)

LINCOLN
- Navigator - see FORD Pick-up (36059)
- 59010 Rear-Wheel Drive all models '70 thru '10

MAZDA
- 61010 GLC Hatchback (rear-wheel drive) '77 thru '83
- 61011 GLC (front-wheel drive) '81 thru '85
- 61012 Mazda3 '04 thru '11
- 61015 323 & Protegé '90 thru '03
- 61016 MX-5 Miata '90 thru '09
- 61020 MPV all models '89 thru '98
- Navajo - see Ford Explorer (36024)
- 61030 Pick-ups '72 thru '93
- Pick-ups '94 thru '00 - see Ford Ranger (36071)
- 61035 RX-7 all models '79 thru '85
- 61036 RX-7 all models '86 thru '91
- 61040 626 (rear-wheel drive) all models '79 thru '82
- 61041 626/MX-6 (front-wheel drive) '83 thru '92
- 61042 626, MX-6/Ford Probe '93 thru '02
- 61043 Mazda6 '03 thru '11

MERCEDES-BENZ
- 63012 123 Series Diesel '76 thru '85
- 63015 190 Series four-cyl gas models, '84 thru '88
- 63020 230/250/280 6 cyl models '68 thru '72
- 63025 280 123 Series gasoline models '77 thru '81
- 63030 350 & 450 all models '71 thru '80
- 63040 C-Class: C230/C240/C280/C320/C350 '01 thru '07

MERCURY
- 64200 Villager & Nissan Quest '93 thru '01
- All other titles, see FORD Listing.

MG
- 66010 MGB Roadster & GT Coupe '62 thru '80
- 66015 MG Midget, Austin Healey Sprite '58 thru '80

MINI
- 67020 Mini '02 thru '11

MITSUBISHI
- 68020 Cordia, Tredia, Galant, Precis & Mirage '83 thru '93
- 68030 Eclipse, Eagle Talon & Ply. Laser '90 thru '94
- 68031 Eclipse '95 thru '05, Eagle Talon '95 thru '98
- 68035 Galant '94 thru '10
- 68040 Pick-up '83 thru '96 & Montero '83 thru '93

NISSAN
- 72010 300ZX all models including Turbo '84 thru '89
- 72011 350Z & Infiniti G35 all models '03 thru '08
- 72015 Altima all models '93 thru '06
- 72016 Altima '07 thru '10
- 72020 Maxima all models '85 thru '92
- 72021 Maxima all models '93 thru '04
- 72025 Murano '03 thru '10
- 72030 Pick-ups '80 thru '97 Pathfinder '87 thru '95
- 72031 Frontier Pick-up, Xterra, Pathfinder '96 thru '04
- 72032 Frontier & Xterra '05 thru '11
- 72040 Pulsar all models '83 thru '86
- Quest - see MERCURY Villager (64200)
- 72050 Sentra all models '82 thru '94
- 72051 Sentra & 200SX all models '95 thru '06
- 72060 Stanza all models '82 thru '90
- 72070 Titan pick-ups '04 thru '10 Armada '05 thru '10

OLDSMOBILE
- 73015 Cutlass V6 & V8 gas models '74 thru '88
- For other OLDSMOBILE titles, see BUICK, CHEVROLET or GENERAL MOTORS listing.

PLYMOUTH
- For PLYMOUTH titles, see DODGE listing.

PONTIAC
- 79008 Fiero all models '84 thru '88
- 79018 Firebird V8 models except Turbo '70 thru '81
- 79019 Firebird all models '82 thru '92
- 79025 G6 all models '05 thru '09
- 79040 Mid-size Rear-wheel Drive '70 thru '87
- Vibe '03 thru '11 - see TOYOTA Matrix (92060)
- For other PONTIAC titles, see BUICK, CHEVROLET or GENERAL MOTORS listing.

PORSCHE
- 80020 911 except Turbo & Carrera 4 '65 thru '89
- 80025 914 all 4 cyl models '69 thru '76
- 80030 924 all models including Turbo '76 thru '82
- 80035 944 all models including Turbo '83 thru '89

RENAULT
- Alliance & Encore - see AMC (14020)

SAAB
- 84010 900 all models including Turbo '79 thru '88

SATURN
- 87010 Saturn all S-series models '91 thru '02
- 87011 Saturn Ion '03 thru '07
- 87020 Saturn all L-series models '00 thru '04
- 87040 Saturn VUE '02 thru '07

SUBARU
- 89002 1100, 1300, 1400 & 1600 '71 thru '79
- 89003 1600 & 1800 2WD & 4WD '80 thru '94
- 89100 Legacy all models '90 thru '99
- 89101 Legacy & Forester '00 thru '09

SUZUKI
- 90010 Samurai/Sidekick & Geo Tracker '86 thru '01

TOYOTA
- 92005 Camry all models '83 thru '91
- 92006 Camry all models '92 thru '96
- 92007 Camry, Avalon, Solara, Lexus ES 300 '97 thru '01
- 92008 Toyota Camry, Avalon and Solara and Lexus ES 300/330 all models '02 thru '06
- 92009 Camry '07 thru '11
- 92015 Celica Rear Wheel Drive '71 thru '85
- 92020 Celica Front Wheel Drive '86 thru '99
- 92025 Celica Supra all models '79 thru '92
- 92030 Corolla all models '75 thru '79
- 92032 Corolla all rear wheel drive models '80 thru '87
- 92035 Corolla all front wheel drive models '84 thru '92
- 92036 Corolla & Geo Prizm '93 thru '02
- 92037 Corolla models '03 thru '11
- 92040 Corolla Tercel all models '80 thru '82
- 92045 Corona all models '74 thru '82
- 92050 Cressida all models '78 thru '82
- 92055 Land Cruiser FJ40, 43, 45, 55 '68 thru '82
- 92056 Land Cruiser FJ60, 62, 80, FZJ80 '80 thru '96
- 92060 Matrix & Pontiac Vibe '03 thru '11
- 92065 MR2 all models '85 thru '87
- 92070 Pick-up all models '69 thru '78
- 92075 Pick-up all models '79 thru '95
- 92076 Tacoma, 4Runner, & T100 '93 thru '04
- 92077 Tacoma all models '05 thru '09
- 92078 Tundra '00 thru '06 & Sequoia '01 thru '07
- 92079 4Runner all models '03 thru '09
- 92080 Previa all models '91 thru '95
- 92081 Prius all models '01 thru '08
- 92082 RAV4 all models '96 thru '10
- 92085 Tercel all models '87 thru '94
- 92090 Sienna all models '98 thru '09
- 92095 Highlander & Lexus RX-330 '99 thru '07

TRIUMPH
- 94007 Spitfire all models '62 thru '81
- 94010 TR7 all models '75 thru '81

VW
- 96008 Beetle & Karmann Ghia '54 thru '79
- 96009 New Beetle '98 thru '11
- 96016 Rabbit, Jetta, Scirocco & Pick-up gas models '75 thru '92 & Convertible '80 thru '92
- 96017 Golf, GTI & Jetta '93 thru '98, Cabrio '95 thru '02
- 96018 Golf, GTI, Jetta '99 thru '05
- 96019 Jetta, Rabbit, GTI & Golf '05 thru '11
- 96020 Rabbit, Jetta & Pick-up diesel '77 thru '84
- 96023 Passat '98 thru '05, Audi A4 '96 thru '01
- 96030 Transporter 1600 all models '68 thru '79
- 96035 Transporter 1700, 1800 & 2000 '72 thru '79
- 96040 Type 3 1500 & 1600 all models '63 thru '73
- 96045 Vanagon all air-cooled models '80 thru '83

VOLVO
- 97010 120, 130 Series & 1800 Sports '61 thru '73
- 97015 140 Series all models '66 thru '74
- 97020 240 Series all models '76 thru '93
- 97040 740 & 760 Series all models '82 thru '88
- 97050 850 Series all models '93 thru '97

TECHBOOK MANUALS
- 10205 Automotive Computer Codes
- 10206 OBD-II & Electronic Engine Management
- 10210 Automotive Emissions Control Manual
- 10215 Fuel Injection Manual '78 thru '85
- 10220 Fuel Injection Manual '86 thru '99
- 10225 Holley Carburetor Manual
- 10230 Rochester Carburetor Manual
- 10240 Weber/Zenith/Stromberg/SU Carburetors
- 10305 Chevrolet Engine Overhaul Manual
- 10310 Chrysler Engine Overhaul Manual
- 10320 Ford Engine Overhaul Manual
- 10330 GM and Ford Diesel Engine Repair Manual
- 10333 Engine Performance Manual
- 10340 Small Engine Repair Manual, 5 HP & Less
- 10341 Small Engine Repair Manual, 5.5 - 20 HP
- 10345 Suspension, Steering & Driveline Manual
- 10355 Ford Automatic Transmission Overhaul
- 10360 GM Automatic Transmission Overhaul
- 10405 Automotive Body Repair & Painting
- 10410 Automotive Brake Manual
- 10411 Automotive Anti-lock Brake (ABS) Systems
- 10415 Automotive Detailing Manual
- 10420 Automotive Electrical Manual
- 10425 Automotive Heating & Air Conditioning
- 10430 Automotive Reference Manual & Dictionary
- 10435 Automotive Tools Manual
- 10440 Used Car Buying Guide
- 10445 Welding Manual
- 10450 ATV Basics
- 10452 Scooters 50cc to 250cc

SPANISH MANUALS
- 98903 Reparación de Carrocería & Pintura
- 98904 Manual de Carburador Modelos Holley & Rochester
- 98905 Códigos Automotrices de la Computadora
- 98906 OBD-II & Sistemas de Control Electrónico del Motor
- 98910 Frenos Automotriz
- 98913 Electricidad Automotriz
- 98915 Inyección de Combustible '86 al '99
- 99040 Chevrolet & GMC Camionetas '67 al '87
- 99041 Chevrolet & GMC Camionetas '88 al '98
- 99042 Chevrolet & GMC Camionetas Cerradas '68 al '95
- 99043 Chevrolet/GMC Camionetas '94 al '04
- 99048 Chevrolet/GMC Camionetas '99 al '06
- 99055 Dodge Caravan & Plymouth Voyager '84 al '95
- 99075 Ford Camionetas y Bronco '80 al '94
- 99076 Ford F-150 '97 al '09
- 99077 Ford Camionetas Cerradas '69 al '91
- 99088 Ford Modelos de Tamaño Mediano '75 al '86
- 99089 Ford Camionetas Ranger '93 al '10
- 99091 Ford Taurus & Mercury Sable '86 al '95
- 99095 GM Modelos de Tamaño Grande '70 al '90
- 99100 GM Modelos de Tamaño Mediano '70 al '88
- 99106 Jeep Cherokee, Wagoneer & Comanche '84 al '00
- 99110 Nissan Camioneta '80 al '96, Pathfinder '87 al '95
- 99118 Nissan Sentra '82 al '94
- 99125 Toyota Camionetas y 4Runner '79 al '95

Over 100 Haynes motorcycle manuals also available

Haynes North America, Inc., 861 Lawrence Drive, Newbury Park, CA 91320-1514 • (805) 498-6703 • http://www.haynes.com